应用化学专业英语

ENGLISH IN APPLIED CHEMISTRY

（修订版）

主　编　唐冬雁　刘本才
副主编　黎　钢　潘　丽
　　　　李　欣　龙　军
主　审　李玉铭

哈尔滨工业大学出版社

内 容 提 要

本书由哈尔滨工业大学牵头,联合东北林业大学、哈尔滨理工大学、大庆石油学院、哈尔滨工程大学、黑龙江大学等院校,在总结多年专业英语教学经验的基础上编写而成。

全书共八个单元。其中:第一单元为化学化工一般知识;第二单元为化学化工基本理论;第三单元为化学化工技术;第四单元为仪器分析和化学分析方法;第五单元为化学命名法;第六单元为专业英语阅读材料;第七单元为化学化工文献介绍;第八单元为科技英语写作及翻译。另外还附有化学元素表、常用数学符号和数学式写法、SI 单位、实验室常用仪器名称及词汇索引。

图书在版编目(CIP)数据

应用化学专业英语/唐冬雁,刘本才主编.修订版.—哈尔滨:哈尔滨工业大学出版社,2004.12(2012.7 重印)
ISBN 978-7-5603-1395-5

Ⅰ.应… Ⅱ.①唐…②刘… Ⅲ.应用化学-英语-高等学校-教材 Ⅳ.H31

中国版本图书馆 CIP 数据核字(2007)第 014116 号

责任编辑	黄菊英
封面设计	卞秉利
出版发行	哈尔滨工业大学出版社
社　　址	哈尔滨市南岗区复华四道街 10 号　邮编 150006
传　　真	0451-86414749
网　　址	http://hitpress.hit.edu.cn
印　　刷	黑龙江省地质测绘印制中心印刷厂
开　　本	880mm×1230mm　1/32　印张 14.25　字数 412 千字
版　　次	2005 年 1 月第 3 版　2012 年 7 月第 8 次印刷
书　　号	ISBN 978-7-5603-1395-5
定　　价	21.00 元

(如因印装质量问题影响阅读,我社负责调换)

修订版前言

应用化学专业英语一书自1999年5月出版以来,受到了广大高校师生和英语爱好者的欢迎,目前国内已有20余所高等院校将此书作为教材,并在教学中收到良好效果。还有许多化学化工人员将此书作为科研工作中查阅英文资料的重要参考书。本书自出版以来已修订1次,重印2次。为进一步满足教学的要求,哈尔滨工业大学出版社建议我们根据目前使用的情况,作必要的充实和修改,再次修订再版。

为增强本书的科学性、新颖性和实用性,综合各校使用的情况和体会,在保证本书基本体系和主要内容的前提下,作了如下补充和修改。

(1) 为突出体系的完整性,在内容编排上进行了调整,删除了内容有交叠的课文,充实了能突出反映与化学各学科紧密结合的内容,并增添了化学实验的内容。

(2) 为增强本书的规范性,便于阅读和掌握,在课文后增添了惯用短语(Phrases)和难句注释(Notes)部分,并对各课生词进行了核准。

(3) 为使构词部分内容进一步完善,第五单元的无机物命名法中新增了配合物构词法介绍。

(4) 考虑到近年来利用Internet查阅化学化工文献已成为必不可少的获取文献的方法,补充了Internet上化学化工文献资源及其检索方法。

(5) 附录中补充了化学实验常见仪器、装置的中、英文对照内容和元素周期表中各元素的音标。

(6) 书后给出词汇索引(按字母顺序排列)，并给出其首次出现的课文序号，以方便查阅。

为真实地反映每位编者对本书的贡献，本次修订再版对个别副主编参编的顺序进行了调整，参加本次修订再版工作的有哈尔滨工业大学唐冬雁、李欣、刘玉文、潘丽等。

由于编者水平所限，即使进行了较为细致的修改，也还难免有疏漏和不妥之处，恳请读者提出宝贵意见，以便进一步完善。

作 者
2005 年 1 月

前　言

　　应用化学专业英语是继大学基础英语之后为提高应用化学各专业学生专业英文文献的读写能力而开设的必修课，是大学英语学习的重要环节。通过本课程的学习，可以进一步增加学生对专业英语词汇、语法和结构的了解，为将来从事化学化工方面英文资料的查阅、翻译和写作打下坚实的基础。

　　随着我国化学工业的发展，各类应用化学信息的交流日益广泛，全社会对化学化工人才素质的要求也越来越高，尤其是通过英文文献熟练获取和共享信息资源的能力。全国现有高等工科院校多数设有应用化学方面的专业，由于地域、历史发展等方面的原因，各校的专业方向又不尽相同，专业英语教学大都自成体系，且偏重于各自的专业方向，覆盖面较窄。急需一本能够兼顾各校实际情况、通用性较强的应用化学专业英语教材。为此，我们几所院校的教师总结多年的专业英语教学经验，联合编写了这本应用化学专业英语教材。本教材的主要特点是：

　　1.有较强的系统性和完整性，内容编排由浅入深，符合教学规律，且紧密结合应用化学学科各个专业，使学习化学知识和掌握专业英语融为一体。

　　2.信息涵盖面大，既包括化学化工基本知识、基本理论，也涉及化学化工文献检索方法、初步的写作技巧和翻译方法，还有结合各校专业实际难易适中的阅读资料。

　　3.突出了适用性和灵活性，各校可结合其专业实际，有针对性地选择教学内容，并在 30～120 学时的范围内组织教学。

4. 附有化学元素表、常用数学符号及数学式写法和 SI 单位，以方便查阅和对照。

本书内容广泛，知识介绍循序渐进，涵盖了化学化工的多个领域，既可作为高等院校应用化学各专业本、专科学生的专业英语教材，也可作为应用化学领域广大研究人员英语学习的参考书。

本书由哈尔滨工业大学唐冬雁、佳木斯大学张磊主编，大庆石油学院黎钢、哈尔滨建筑大学周胜绪、哈尔滨工业大学王鹏、龙军任副主编，参加编写的还有哈尔滨工程大学朱元凯、黑龙江大学闫鹏飞、付宏刚、哈尔滨师范大学附属中学唐海燕。全书由哈尔滨理工大学李玉铭主审。本书在编写过程中还有上述各校的一些同志做了工作，在此不一一列举，仅表谢意。

本书的编写得到了哈尔滨工业大学强亮生教授的关心和指导，在此表示衷心的感谢。哈尔滨工业大学夏保佳、王金玉、尹鸽平、姚杰、王铀等同志对本书的编写提出了许多很好的建议，在此一并表示感谢。

本书是为解决教学之急需编写的，加之参编者较多，水平有限，难免有疏漏和其他不妥之处，恳请读者提出宝贵意见，以便完善。

作 者
1999 年 5 月

CONTENTS

1 General Knowledge of Chemistry and Chemical Engineering

1.1 CHEMISTRY AND CHEMICAL ENGINEERING ·············· 1

1.2 ACIDS, BASES AND SALTS ·············· 8

1.3 THE PERIODIC LAW ·············· 13

1.4 ELEMENTS, COMPOUNDS AND MIXTURES ·············· 18

1.5 METALS, NONMETALS AND METALLOIDS ·············· 24

1.6 CHEMICAL EQUILIBRIUM ·············· 29

1.7 ENERGY AND CHEMICAL ENERGY ·············· 32

1.8 CHEMICAL BONDS ·············· 35

1.9 TYPES OF CHEMICAL EQUATIONS ·············· 40

1.10 ENTHALPY, ENTROPY AND GIBBS FREE ENERGY ·············· 44

1.11 FRIEDEL-CRAFTS ALKYLATION: PREPARATION OF 4,4′ − Di − t − BUTYLBIPHENYL ·············· 51

1.12 THE PREAPARATION OF FERROCENE〔BIS (CYCLOPENTADIENYL)IRON(II)〕 ·············· 54

2 Some Basic Chemical Theories

2.1 DALTON'S ATOMIC THEORY ·············· 59

2.2 VALENCE BOND THEORY ·············· 65

2.3 THE KINETIC THEORY OF GASES ·············· 68

2.4 COLLISION THEORY AND REACTION MECHANISMS ·············· 75

2.5 REACTIVE INTERMEDIATES OF ORGANIC REACTIONS ·············· 80

ii English in Applied Chemistry

    2.6 TYPES OF ORGANIC REACTION 84

    2.7 FREE RADICAL REACTIONS 92

    2.8 LAWS OF THERMODYNAMICS 96

3 Some Techniques of Chemistry and Chemical Engineering

    3.1 FLUID FLOW 102

    3.2 MOMENTUM, HEAT AND MASS TRANSFER 110

    3.3 PROCESS DESIGN 119

    3.4 REACTOR TYPES 126

    3.5 MEMBRANES AND MEMBRANE PROCESSES 138

    3.6 PROCESS SELECTION 141

4 Instrumental and Chemical Analysis Methods

    4.1 INTRODUCTION TO INSTRUMENTAL METHODS 145

    4.2 ELECTROMAGNETIC RADIATION 150

    4.3 MOLECULE SPECTROSCOPY 158

    4.4 IR AND RAMAN SPECTROSCOPY 161

    4.5 INTRODUCTION TO ELECTROANALYTICAL

           METHODS OF ANALYSIS 159

    4.6 THERMAL ANALYSIS 171

5 Nomenclature

    5.1 NOMENCLATURE OF INORGANIC COMPOUNDS 177

    5.2 NOMENCLATURE OF ORGANIC COMPOUNDS 194

6 Reading Materials

    6.1 THE GENESIS OF POLYMER 209

    6.2 MECHANISM OF DEVELOPMENT OF LARGE STRAINS 218

6.3	ELASTOMER	227
6.4	PLASTIC MATRIX MATERIALS	233
6.5	FABRICATION METHODS OF RESIN MATRIX COMPOSITES	240
6.6	METAL EXTRACTION AND REFINING	247
6.7	ELECTROPLATING	249
6.8	CORROSION OF IRON	252
6.9	FUEL CELLS	255
6.10	SECONDARY BATTERIES—NICKEL-CADMIUM BATTERY	259
6.11	FUNDAMENTAL PRINCIPLES OF ELECTROCHEMISTRY	265
6.12	ATMOSPHERIC CLEANSING PROCESSES	270
6.13	DEGREES OF WASTEWATER TREATMENT AND WATER QUALITY STANDARDS	273
6.14	SOLID WASTE	276
6.15	DIRECT AND INDIRECT REUSE OF WASTEWATER	281
6.16	WASTEWATER TREATMENT PROCESSES	285
6.17	TYPES OF WATER SUPPLY AND CLASSIFICATION OF WATER CONTAMINANTS	292
6.18	BIOLOGICAL EFFECTS OF RADIATION	295
6.19	WHAT CAN CHEMISTS DO FOR NANOSTRUCTURED MATERIALS	299
6.20	MODERN SYNTHETIC SURFACTANT SHAMPOOS	305
6.21	COSMETIC	311
6.22	INORGANIC PIGMENTS	315
6.23	SYNTHETIC DYES	323
6.24	ADHESION	328
6.25	COATINGS	332

7 化学化工文献介绍

- 7.1 印刷版期刊 ·· 359
- 7.2 印刷版文摘刊和索引刊 ·· 347
- 7.3 Internet 上的电子期刊和图书 ··· 371
- 7.4 Internet 上的化学数据库 ·· 376

8 科技英语写作及翻译

- 8.1 科技英语特点 ·· 382
- 8.2 科技英语写作素材 ··· 384
- 8.3 科技英语写作题材 ··· 384
- 8.4 科技英语翻译 ·· 391
- 8.5 科技论文范例 ·· 399

Appendixes

- 附录Ⅰ 化学元素表 ·· 416
- 附录Ⅱ 常用数学符号及数学式写法 ·· 420
- 附录Ⅲ SI 单位 ··· 422
- 附录Ⅳ 实验室常用化学仪器名称 ·· 424
- 附录Ⅴ 词汇索引 ··· 427

参考文献 ·· 443

1

General Knowledge of Chemistry and Chemical Engineering

1.1 CHEMISTRY AND CHEMICAL ENGINEERING

1.1.1 What is Chemistry about

The different kinds of matter that compose the universe are termed materials. Each material has its own distinguishing characteristics, which is termed its properties. These properties enable the material to be recognized or separated from other materials.

The study of materials is the joint concern of chemistry and physics. These two sciences are so closely related that no one can learn very much about either without considerable training in the other[1]. In many of their applications it is hard to tell where the one science leaves off and the other begins.

Roughly stated, physics is concerned with the general properties and energy and with events which result in what are termed physical changes[2]. Physical changes are those in which materials are not so thoroughly altered as to be converted into other materials distinct from those present at the beginning[3].

Chemistry, by contrast, is chiefly concerned with properties that distinguish materials from one another and with events which result in chemical changes. Chemical changes are those in which materials are transformed into completely different materials. Who but a chemist would ever guess that common salt can be resolved into a greenish gas and silvery metal[4]? Or that two odorless gases, nitrogen and hydrogen, can be combined to form ammonia? Or that ordinary air and water can be converted into nitric acid ? Or that coal tar contains ingredients that can be transformed into dyestuffs and perfumes?

Such thoroughgoing transformations, in which all the properties of a material are altered, so that a completely different material is obtained, are called chemical transformations, chemical changes or chemical reactions.

Chemistry as an art is concerned with identifying, separating and transforming materials, in applying them to definite uses[5].

Chemistry as a science is a manner of thinking about transformations of materials which helps us to understand, predict and control them. It furnishes directing intelligence in the use of materials.

1.1.2 The Scope of Chemistry

Chemistry is sometimes called the "central science" because it relates to so many areas of human endeavor and curiosity. Chemists who develop new materials to improve electronic devices such as solar cells, transistors, and fiber optic cables work at the interfaces of chemistry with physics and engineering. Those who develop new pharmaceuticals for use against cancer or AIDS work at the interfaces of chemistry with pharmacology and medicine.

Many chemists work in more traditional fields of chemistry. Biochemists are interested in chemical processes that occur in living organisms. Physical chemists work with fundamental principles of physics and chemistry in an attempt to answer the basic questions that apply to all of chemistry: Why do some substances react with one another while others do not? How fast will a particular chemical reaction occur? How much useful energy can be extracted from a chemical reaction? Analytical chemists are investigators; they study ways to separate and identify chemical substances. Many of the techniques developed by analytical chemists are used extensively by environmental scientists. Organic chemists focus their attention on substances that contain carbon and hydrogen in combination with a few other elements. The vast majority of substances are organic chemicals. Inorganic chemists focus on most of the elements other than carbon, though the fields of organic and inorganic chemistry overlap in some ways.

Although chemistry is considered a "mature" science, the landscape of chemistry is dotted with unanswered questions and challenges. Modern technology demands new materials with unusual properties, and chemists must devise new methods of producing these materials. Modern medicine requires drugs targeted to perform specific tasks in the human body, and chemists must design strategies to synthesize these drugs from simple starting materials. Society requires improved methods of pollution control, substitutes for scarce materials, nonhazardous means of disposing of toxic wastes, and more efficient ways to extract energy from fuels. Chemists are at work in all these areas.

1.1.3 Chemical Engineering

Chemical engineering is the profession concerned with the creative application of the scientific principles underlying the transport of mass, energy and momentum, and the physical and chemical change of matter[6]. The broad implications of this definition have been justified over the past few decades by the kinds of problems that chemical engineers have solved, though the profession has devoted its attention in the main to the chemical process industries. As a result chemical engineers have been defined more traditionally as those applied scientists trained to deal with the research, development, design and operation problems of the chemicals, petroleum and related industries. Experience has shown that the principles required to meet the needs of the process industries are applicable to a significantly wider class of problems, and the modern chemical engineer is bringing his established tools to bear on such new areas as the environmental and life sciences[7].

Chemical engineering developed as a distinct discipline during the twentieth century in answer to the needs of a chemical industry no longer able to operate efficiently with manufacturing processes which in many cases were simply larger scale versions of laboratory equipment[8]. Thus, the primary emphasis in the profession was initially devoted to the general subject of how to use the results of laboratory experiments to design process equipment capable of meeting industrial production rates[9]. This led naturally to the characterization of design procedures in terms of the unit operations, those elements common to many different processes. The basic unit operations include fluid flow, heat exchange, distillation, extraction, etc. A typical manufacturing process will be made up of combinations of the unit operations. Hence, skill in designing each of the units at a production scale would provide the means of designing the entire process.

The unit operations concept dominated chemical engineering education and practice until the mid-1950s, when a movement away from this equipment-oriented philosophy toward an *engineering science* approach began[10]. This approach holds that the unifying concept is not specific processing operations, but rather the understanding of the fundamental phenomena of mass, energy and momentum transport that are common to all of the unit operations, and it is argued that concentration on unit operations obscures the similarity of many operations at a fundamental level[11].

Although there is no real conflict between the goals of the unit operations and engineering science approaches, the latter has tended to emphasize mathematical skills and to de-emphasize the design aspects of engineering education. Such a conflict need not exist, and recent educational effort has been directed toward the development of the skills that will enable the creative engineering use of the fundamentals, or a synthesis of the engineering science and unit operations approaches. One essential skill in reaching this goal is the ability to express engineering problems meaningfully in precise quantitative terms. Only in this way can the chemical engineer correctly formulate, interpret, and use fundamental experiments and physical principles in real world applications outside of the laboratory. This skill which is distinct from ability in mathematics, we call analysis[12].

Vocabulary

1. property   ['prɔpəti]   n. 性质
2. greenish   ['gri:niʃ]   a. 浅绿色的
3. silvery   ['silvəri]   a. 银的、银色的、银制的
4. odorless   ['əudəlis]   n. 没有气味的
5. nitrogen   ['naitridʒən]   n. 氮
6. hydrogen   ['haidridʒən]   n. 氢
7. ammonia   [ə'məuniə]   n. 氨、氨水
8. nitric   ['naitrik]   a. 含氮的、硝酸根
9. ingredient   [in'gri:diənt]   n. 组成、组成部分、配料
10. dyestuff   [daistʌf]   n. 染料
11. perfume   [pə'fju:m]   n. 香料、香水、芳香剂
12. endeavor   [in'devə]   n. 努力
13. transistor   [træn'sistə]   n. 晶体管
14. interface   ['intəfeis]   n. 界面、分界面、接触面
15. pharmaceutical   [fa:mə'sju:tikəl]   a. 配药学的
16. pharmacology   [fa:mə'kɔlədʒi]   n. 药理学

6　*English in Applied Chemistry*

17. extract　　　[iks'trækt]　　　n. 萃取物、浓缩物; vt. 榨取、萃取
18. overlap　　　['əuvə'læp]　　　n. vi. 叠盖; vt. 重叠、重复
19. dispose　　　[dis'pəuz]　　　v. 处理
20. toxic　　　　['tɔksik]　　　　a. 有毒的、有害的
21. underlying　　[ʌndə'laiiŋ]　　a. 在下面的、根本的、潜在的
22. momentum　　[məu'mentəm]　　n. 动量
23. implication　　['impli'keiʃən]　n. 牵连、含义、暗示
24. procedure　　[prə'si:dʒə]　　 n. 程序、手续、步骤
25. distillation　　[disti'leiʃən]　　n. 蒸馏
26. unify　　　　['ju:nifai]　　　　v. 使……相同、使……一致、使成一体
27. obscure　　　[əb'skjuə]　　　 v. 使朦胧、遮蔽、使阴暗

Phrases

1. common salt　　食盐
2. resolve into　　分解
3. coal tar　　煤焦油
4. interface of A with B　　A 与 B 交叉
5. extract A from B　　从 B 中提取(萃取)A
6. in combination with　　与……结合,和……联合
7. in the main　　主要地,基本上
8. process industry　　加工业
9. bring ... to bear on　　将……用于,使……受……的影响
10. in answer to　　响应,满足
11. in terms of　　用……术语,按照
12. common to　　对……通用,为……共用

Notes

1. These two sciences are so ... in the other. 句中"that"引导结果状语从句。全句可译为:这两门学科被如此紧密地联系在一起以至于没有一个人能不通过其中一个学科相当多的训练而很好地学习另一学科。
2. Roughly stated, physics is concerned ... physical changes. 句中"roughly stated"

为插入语,译为"一般说来,大致说来";"what are termed"也为插入语,意为"所谓的,所说的"。句中的两个"with"为并列关系。全句可译为:一般说来,物理学研究普遍性质和能量以及由所谓的物理变化产生的事件。

3. Physical changes are those ... at the beginning. 句中"not so ... as to"后接不定式,表示"没有……那么"或"不如……那么";"distinct from ..."为形容词短语作定语。全句可译为:物理变化是指在这些变化中,物质并不彻底变化而转变成为与初始状态截然不同的其它物质。

4. Who but a chemist .... 句中"but"为介词,意为"除了"。句子可译为:除了化学家谁能推测出……。

5. 句中"be concerned with"意为"牵涉及,与……有关"。后面的"in"也与"be concerned"构成短语,"be concerned in"意为"涉及,与……有关"。

6. 此句中"underlying"的动词原形为"underlie",意为"构成(作为)……基础"。全句可译为:化学工程牵涉到作为质量、能量和动量的传递以及物质的物理和化学变化的基础的科学原理在生产中的应用。

7. 此句的主句为"Experience has shown ..."后有两个并列的宾语从句:第一个从"that the principles ..."开始,第二个从"and the modern chemical ..."至句尾。

8. 此句中有一个较长的状语,从"in answer to"至句末。其中从"no longer"开始至句末是修饰"chemical industry"的定语。全句可译为:化学工程是为了满足化学工业的需要而在20世纪作为一个独立的学科发展起来的,在许多情况下实验设备规模的简单扩大,不再能满足生产过程的有效运转。

9. 此句中"how to use ... to"为介词"of"的宾语;"capable of ..."至句末是修饰"process equipment"的定语。

10. 句中"when"前有逗号,此时不是"当……时候"之意,而是"and then"即"那时,在此之后"之意。

11. 这个长句由 and 连接两个并列句,用逗号分开。第一个并列句中的"the unifying concept"有两个表语,由"not ..."和"but rather ..."(不是……,而是……)开始。

12. This skill which is ..., we call analysis. 这是一个半倒装(semi—inversion)句。"This skill"是"call"的宾语,为了强调,将其放在句首(后面又有一个定语从句),因主句的主谓语并未倒装,称为半倒装。

1.2 ACIDS, BASES AND SALTS

The word "acid" is derived from the Latin *acidus*, meaning "sour" or "tart," and is also related to the Latin word *acetum*, meaning "vinegar." Vinegar has been known to mankind since antiquity as the product of the fermentation of wine and apple cider. The sour constituent of vinegar is acetic acid, $HC_2H_3O_2$. Properties commonly associated with water solutions of acids are a sour or tart taste; interaction with metals such as zinc and magnesium to liberate hydrogen gas; interaction with bases to produce a salt and water; ability to change the color of litmus, a vegetable dye, from blue to red; and interaction with carbonates to liberate carbon dioxide[1]. These properties are due to the hydrogen ions, H^+, released by the acid in a water solution.

Classically, a base is a substance capable of liberating hydroxide ions, OH^-, in water solution. Hydroxides of the alkali metals (Group IA) and alkaline earth metals (Group IIA) are the most common inorganic bases. Water solutions of bases are called "alkaline solutions" or "basic solutions." They have the following properties: a bitter or caustic taste; a slippery, soapy feeling; the ability to change litmus from red to blue; and the ability to interact with acids to form a salt and water.

Several theories have been proposed to answer the question, "What is an acid and a base?" One of the earliest, most significant of these theories was advanced by Svante Arrhenius (1859 ~ 1927), a Swedish scientist, who stated that an acid is a hydrogen-containing substance that dissociates to produce hydrogen ions, and a base is a hydroxide-containing substance that dissociates to produce hydroxide ions in aqueous solutions. Arrhenius postulated that the hydrogen ions were produced by the dissociation of an acid in water; and hydroxide ions were produced by the dissociation of bases in water. Thus, an acid solution contains an excess of hydrogen ions, and a base an excess of hydroxide ions.

In 1923, the Brönsted-Lowry proton transfer theory was introduced by J. N. Brönsted, a Danish chemist (1897 ~ 1947), and T. M. Lowry, an English chemist (1874 ~ 1936). This theory states that an acid is a proton donor and a base is a proton acceptor.

Consider the reaction of hydrogen chloride gas with water to form hydrochloric acid.

$$HCl(g) + H_2O(l) \longrightarrow H_3O^+(aq) + Cl^-(aq) \quad (1.1)$$

In the course of the reaction, HCl donates or gives up a proton to form a Cl ion and H_2O accepts a proton to form H_3O^+. Thus, HCl is an acid and H_2O is a base, according to the Brönsted-Lowry theory.

A hydrogen ion, H^+, is nothing more than a bare proton; it does not exist by itself in an aqueous solution. In water, a proton combines with a polar water molecule to form a hydrated hydrogen ion, H_3O^+ or $H(H_2O)^+$, commonly called a hydronium ion. The proton is attracted to a polar water molecule, forming a bond with one of the two pairs of unshared electrons of water. For simplicity of expression in equations, we often use H^+ instead of H_3O^+ with the explicit understanding that H^+ is always hydrated in solution.

Whereas the Arrhenius theory is restricted to aqueous solutions, the Brønsted-Lowry approach has application in all media and has become a more important theory when the chemistry of substances in solutions other than water is studied. Ammonium chloride, NH_4Cl, is a salt, yet its water solution tests to be acidic. From this test we must conclude that NH_4Cl has acidic properties. The Brønsted-Lowry explanation shows that the ammonium ion, NH_4^+, is a proton donor, and water the proton acceptor.

$$NH_4^+ \rightleftharpoons NH_3 + H^+ \quad (1.2)$$
$$\text{Acid} \quad\quad \text{Base} \quad \text{Acid}$$

$$NH_4^+ + H_2O \longrightarrow H_3O^+ + NH_3 \quad (1.3)$$
$$\text{Acid} \quad \text{Base} \quad\quad \text{Acid} \quad\;\; \text{Base}$$

In the reaction of hydrogen chloride and ammonia gases, HCl is the proton donor and NH_3 is the base. ("(g)" after a formula in equations stands for a gas.)

$$HCl(g) + NH_3(g) \longrightarrow NH_4^+ + Cl^- \qquad (1.4)$$
$$\text{Acid} \qquad \text{Base} \qquad \text{Acid} \quad \text{Base}$$

In Equations (1.1), (1.3), and (1.4), a conjugate acid and base are produced as products. The formulas of a conjugate acid-base pair differ by one proton (H^+). In Equation (1.1), the conjugate base of the acid HCl is Cl^-, and the conjugate acid of the base H_2O is H_3O^+. In Equation (1.3), the conjugate acid of the base H_2O is H_3O^+, and the conjugate base of the acid NH_4^+ is NH_3. In Equation (1.4), HCl—Cl^- and NH_4^+—NH_3 are the conjugate acid-base pairs.

A more general concept of acids and bases was introduced by Gilbert N. Lewis (1875 ~ 1946). The Lewis theory deals with the manner in which a substance having an unshared pair of electrons reacts in an acid-base type of reaction. This theory defines a base as any substance that has an unshared pair of electrons (electron-pair donor) and an acid as a substance that will attach itself to or accept a pair of electrons. In the reaction

$$H^+ + :\overset{..}{\underset{..}{N}}:H \longrightarrow H:\overset{..}{\underset{..}{N}}:H^+$$
$$\text{Acid} \quad \text{Base}$$

H^+ is a Lewis acid and $:NH_3$ is a Lewis base.

$$F:\overset{..}{\underset{..}{B}} + :\overset{..}{\underset{..}{N}}:H \longrightarrow F:\overset{..}{\underset{..}{B}}:\overset{..}{\underset{..}{N}}:H$$
$$\text{Acid} \quad \text{Base}$$

According to the Lewis theory, substances other than proton donors (e.g., BF_3) behave as acids. The Lewis and Brønsted-Lowry bases are identical, because to accept a proton, a base must have an unshared pair of electrons. The three theories are summarized in Table 1.1.

These theories explain how acid-base reactions occur. We generally use the theory that best explains the chemical reaction under consideration.

Table 1.1 Summary of acid-base definitions according to Arrhenius, Brönsted-Lowry, and G. N. Lewis theories

Theory	Acid	Base
Arrhenius	A hydrogen-containing substance that produces hydrogen ions in aqueous solution	A hydroxide-containing substance that produces hydroxide ions in aqueous solution
Brönsted-Lowry	A proton(H^+) donor	A proton(H^+) acceptor
Lewis	Any species that will bond to an unshared pair of electrons (electron-pair acceptor)	Any species that has an unshared pair of electrons (electron-pair donor)

Salts are very abundant in nature. Most of the rocks and minerals of the earth's mantle are salts of one kind or another. Huge quantities of dissolved salts also exist in the oceans. Salts may be considered to be compounds that have been derived from acids and bases. They consist of positive metal or ammonium ions (H^+ excluded) combined with negative nonmetal ions (OH^- and O^{2-} excluded). The positive ion is the base counterpart and the nonmetal ion is the acid counterpart. Salts are generally crystalline and have high melting and boiling points.

From a single acid such as hydrochloric acid, HCl, we may produce many chloride salts by replacing the hydrogen with a metal ion (for example NaCl, KCl, RbCl, $CaCl_2$, NH_4Cl, $NiCl_2$). The number of known salts greatly exceeds the number of known acids and bases. Salts are ionic compounds. If the hydrogen atoms of a binary acid are replaced by a nonmetal, the resulting compound has covalent bonding and is therefore not considered to be a salt (for example, PCl_3, S_2Cl_2, Cl_2O, NCl_3, ICl).

Vocabulary

1. base   [beis]   n. 碱
2. tart   [tɑːt]   a. 酸的、尖酸的

3. vinegar ['vinigə] n. 醋
4. antiquity [æn'tikwəti] n. 古代、古老、古代的遗物
5. fermentation [fə:mən'teiʃən] n. 发酵
6. cider ['saidə] n. 苹果酒(汁)
7. acetic [ə'si:tik] a. 醋的、乙酸的
8. magnesium [mæg'ni:ziəm] n. 镁
9. litmus ['litməs] n. 石蕊
10. carbonate ['ka:bəneit] n. 碳酸盐
11. hydroxide [hai'drɔksaid] n. 氢氧化物、羟化物
12. alkali ['ælkəlai] n. 碱; a. 碱性的
13. alkaline ['ælkəlain] a. 碱的、碱性的
14. caustic ['kɔ:stik] a. 腐蚀性的; n. 腐蚀剂
15. dissociate [di'səuʃieit] v. 分离、使分离
16. aqueous ['eikwiəs] a. 水的,水成的
17. postulate [pɔstjuleit] vt. 假定
18. donor ['dəunə] n. 捐赠者、供给者
19. chloride ['klɔ:raid] n. 氯化物
20. donate [dəu'neit] v. 捐赠、赠予
21. hydrated ['haidreitid] a. 含水的、与水结合的
22. hydronium [hai'drəuniəm] n. 水合氢离子
23. explicit [iks'plisit] a. 明白的、明显的
24. conjugate ['kɔndʒugeit] v. 变化,使……变化
25. conjugated ['kɔndʒugeitid] a. 共轭的、结合的、成对的
26. mantle ['mæntl] vt. 覆盖、结皮; n. 覆盖物
27. positive ['pɔzitiv] a. 正的,阳的
28. negative ['negtiv] a. 负的、阴的
29. counterpart ['kauntəpa:t] n. 配对物
30. crystalline ['kristəlain] a. 水晶的、结晶的

			$n.$ 结晶体
31.	ionic	[ai'ɔnik]	$a.$ 离子的
32.	binary	['bainəri]	$a.$ 二元的
33.	covalent	[kəu'veilənt]	$a.$ 共价的

Phrases

1. vegetable dye　植物染料
2. alkali metal　碱金属
   alkaline earth metal = alkali earth metal　碱土金属
3. nothing more (less) than　只不过是,简直是
4. conjugate acid, conjugate base　共轭酸,共轭碱
   conjugate acid-base pair　共轭酸碱对
5. binary acid　二元酸
6. covalent bond　共价键

Note

1. 句中"associated with ... of acids"为分词短语,修饰其前面的"properties"(主句的主语)。主句共有5个宾语,用分号隔开。

1.3 THE PERIODIC LAW

In 1869, Dmitri Ivanovitch Mendeleev (1834 ~ 1907) of Russia and Lothar Meyer (1830 ~ 1895) of Germany independently published their periodic arrangements of the elements. Both of these periodic arrangements were based on increasing atomic weights.

At the time of Mendeleev's periodic table, about 63 elements were known. The brilliance and foresightedness of this work can be seen by the fact that Mendeleev left spaces between certain elements in his original table and predicted that these spaces would be filled by the discovery of new elements. He left a space for an undiscovered element after calcium and called the element eka-boron; another space was left under aluminum, which he called eka-aluminum; and another space under silicon, which he called eka-silicon. The term "eka" comes from Sanskrit meaning "one," which Mendeleev used to indicate that the missing element was one place away in his table from the element indicated. Mendeleev even went so far as to predict, with high accuracy, the physical and chemical properties of those elements yet to be discovered. The three elements above were, in fact, discovered within his lifetime. Scandium (atomic number 21) was discovered in 1879 by Lars F. Nilson (1840 ~ 1899) of Sweden, and was found to correspond in properties to eka-boron; gallium (31) was discovered in 1875 by Lecoq de Baisbaudran (1832 ~ 1912), and was found to correspond to eka-aluminum; and germanium (32) was discovered in 1886 by C.A. Winkler (1838 ~ 1904), and was found to correspond to eka-silicon. The amazing way in which Mendeleev's predictions were fulfilled is illustrated in Table 1.2 which compares the predicted properties of eka-silicon with those of germanium.

Mendeleev constructed his table by arranging the elements in order of increasing atomic weights. The elements were tabulated so that those with similar chemical properties fitted into columns to form family groups. This arrangement left vacant spaces for undiscovered elements.

Several modifications have been made to Mendeleev's table. First, a new family of elements, the noble gases, was discovered and added to the table. Also, it was observed that when the elements were listed according to increasing atomic weights, several discrepancies arose in the table. In the present table, for example, argon appears before potassium, even though the atomic weight of argon is greater than that of potassium. There is no mistaking that potassium should come after argon, because argon is certainly one of the noble gases and potassium behaves like the other alkali metals. There are two other places in the table where this type of deviation occurs.

Table 1.2 Comparison of the properties of eka-silicon predicted by Mendeleev with the properties of germanium

Property	Mendeleev's Predictions in 1871, Eka-silicon (Es)	Observed Properties, Germanium (Ge)
Atomic weight	72	72.6
Color of metal	Dirty gray	Grayish-white
Density	5.5 g·ml^{-1}	5.47 g·ml^{-1}
Oxide formula	EsO_2	GeO_2
Oxide density	4.7 g·ml^{-1}	4.70 g·ml^{-1}
Chloride formula	$EsCl_4$	$GeCl_4$
Chloride density	1.9 g·ml^{-1}	1.89 g·ml^{-1}
Boiling temperature of chloride	Under 100℃	86℃

The correct order of the elements was resolved by the British physicist H.G.J. Moseley (1887~1915), who, while studying the X-ray emission frequencies of the elements, established that the elements should be arranged in order of increasing charge on their nuclei; namely, the atomic number. This correction nullified the above-mentioned discrepancies and led to the current statement of the Periodic Law. This law states that properties of the elements are periodic functions of their atomic numbers. In this sense periodic means recurring in some regular cycle. With the discovery of isotopes for many of the elements, it became more apparent that the atomic number is the correct basis for periodicity.

As one studies the format of the periodic table, it becomes evident that the periodicity in the properties of the elements is due to the recurring similarities of their electron structures.

The most commonly used periodic table is the long form. In this table the elements are arranged horizontally in numerical sequence, according to their atomic numbers; the result is seven horizontal periods. Each period, with the exception of the first, starts with an alkali metal and ends with a noble gas. By this arrangement, vertical columns of elements are formed, having identical or similar outer-shell electron structures and thus similar chemical properties. These columns are known as groups or families of elements.

The heavy zigzag line starting at boron and running diagonally down the table separates the elements into metals and nonmetals. The elements to the right of the line are nonmetallic, and those to the left are metallic. The elements bordering the zigzag line are the metalloids and show both metallic and nonmetallic properties. With some exceptions, the characteristic electronic arrangement of metals is that their atoms have one, two, or three electrons in their outer energy level, while nonmetals have five, six, or seven electrons in their outer energy level.

It is interesting to note that with this periodic arrangement the elements fall into blocks according to the sublevel of electrons that are being filled in their atomic structure. The s block comprising Groups IA and IIA have one or two s electrons in their outer energy level. The p block includes groups IIIA to VIIA and the noble gases (except helium). In these elements electrons are filling the p sublevel orbitals. The d block includes the transition elements of Groups IB to VIIB and Group VIII. The d sublevels of electrons are being filled in these elements. The f block of elements include the inner transition series. In the lanthanide series electrons are filling the 4 f sublevel. In the actinide series electrons are filling the 5 f sublevel.

The periodic table has been used for studying the relationships of many properties of the elements. Ionization energies, densities, melting points, atomic radii, atomic volumes, oxidation states, electrical conductance, and electronegativity are just a few mentioned. However, a detailed discussion of these properties is not practical at this time. The periodic table is not a panacea for the chemist, but it is an important correlating unit for tying together the properties and relationships of the elements. In the past, the use of the periodic table for predicting the existence of certain elements was invaluable to their discovery. Today the periodic table is still being used to predict the probable synthesis of new elements. A thorough understanding of chemical periodicity can lead to a better insight regarding the properties of the elements.

Vocabulary

1. periodic [piəri'ɔdik] a. 周期的
2. foresighted ['fɔːsaitid] a. 能预料的
3. aluminum [ə'luːminəm] n. 铝
4. eka- pref. [化]表示"类","准"
5. scandium ['skændiəm] n. 钪
6. gallium ['gæliəm] n. 镓
7. germanium [dʒəː'meiniəm] n. 锗
8. tabulate [tæbjuleit] vt. 使成平面; a. 平面的
9. vacant [veikənt] a. 空的、空缺的
10. discrepancy [dis'krepənsi] n. 差异
11. argon ['aːgɔn] n. 氩
12. deviation [diːvi'eiʃən] n. 背离
13. emission [i'miʃən] n. 散发、发射、喷射
14. charge [tʃaːdʒ] n. 电荷
15. nuclei [njuːkliai] n. (nucleus 的复数) 核子
16. nullify ['nʌlifai] v. 无效
17. recur [ri'kəː] v. 重现、复发、再来
18. isotope [aisəutəup] n. 同位素
19. periodicity [piəriə'disiti] n. 周期
20. zigzag ['zigzæg] n. Z字形、锯齿形
21. diagonally [dai'ægənli] ad. 对角的
22. metalloid ['metəlɔid] n. 非金属、两性金属; a. 非金属的、类似金属的
23. helium ['hiːliəm] n. 氦
24. lanthanide ['lænθənaid] n. 镧系(原子序数 58~81 间的稀土元素)

25. actinide [ˈæktinaid] n. 锕系
26. radii [ˈreidiai] pl. 半径
27. electronegativity [iˈlektrəuˈnegətiviti] n. 电负性
28. panacea [pænəˈsiə] n. 万能药

Phrases

1. periodic law 周期律; periodic table 周期表
2. atomic weight 原子量; atomic number 原子序数
3. go to far as to 甚至
4. nobble gas 惰性气体, 稀有气体
5. outer-shell electron(s) 外层电子, 价电子
6. energy level 能级
7. inner transition series 内过渡系
8. lanthanide series 镧系

1.4 ELEMENTS, COMPOUNDS AND MIXTURES

All known substances on earth—and most probably in the universe, too—are formed from a sort of "chemical alphabet" consisting of 106 known elements. An element is a fundamental or elementary substance that cannot be broken down, by chemical means, to simpler substances. Elements are the basic building blocks of all substances. The elements are listed in order of increasing complexity beginning with hydrogen, number 1. Of the first 92 elements, 88 are known to occur in nature. The other four—technetium (43), promethium (61), astatine (85), and francium (87)—either do not occur in nature or have only transitory existences resulting from radioactive decay. With the exception of number 94, elements 93 through 106 are not known to occur naturally, but have been synthesized—usually in very small quantities— in laboratories. The discovery of trace amounts of element 94 (plutonium) in nature has been reported recently. Element 103, lawrencium, was first reported in 1961. It was named after the eminent nuclear chemist Dr. Ernest O. Lawrence

(1901 ~ 1958), inventor of the cyclotron. Element 104 was first reported in 1964 by a group of Russian scientists and then again in 1969 by United States scientists at the Lawrence Radiation Laboratory of the University of California. Element 105 was also reported to have been synthesized in 1970 by American scientists. Suggested, but not yet official names are kurchatovium and rutherfordium for element 104 and hahnium for 105. Element 106 was reported to have been synthesized in 1974. No elements other than those on the earth have been detected on other bodies in the universe[1].

Most substances can be decomposed into two or more other distinct substances. We have seen that mercuric oxide can be decomposed into mercury and oxygen and that water can be decomposed into hydrogen and oxygen. Sugar—a common, sweet, crystalline substance—may be decomposed into carbon, hydrogen, and oxygen. Table salt is easily decomposed into sodium and chlorine. An element, however, is a substance that cannot be decomposed into simpler substances by ordinary chemical changes.

If we could take a small piece of an element, say copper, and divide it and subdivide it into smaller and smaller particles, we finally would come to a single unit of copper that we could no longer divide and still have copper. This ultimate particle, the smallest particle of an element that can exist, is called an atom. An atom is also the smallest unit of an element that can enter into a chemical reaction. That atoms are composed of even smaller parts. However, these smaller, subatomic particles no longer have the properties of elements.

Pure substances composed of two or more elements are called compounds. Because they contain two or more elements, compounds, unlike elements, are capable of being decomposed into simpler substances by chemical changes. The ultimate chemical decomposition of compounds produces the elements from which they are made.

The atoms of the elements in a compound are combined in whole-number ratios, not in fractional parts of an atom. Atoms combine with one another to form compounds which exist as either molecules or ions. A molecule is a small, uncharged individual unit of a compound formed by the union of two or more atoms. If we subdivide a drop of water into smaller and smaller particles, we ultimately obtain a single unit of water known as a molecule of water. This water molecule consists of two hydrogen atoms and one oxygen atom bonded together. We cannot subdivide this unit further without destroying the molecule, breaking it up into its elements. Thus, a water molecule is the smallest unit of the compound water.

An ion is a positive or negative electrically charged atom or group of atoms. The ions in a compound are held together in a crystalline structure by the attractive forces of their positive and negative charges. Compounds consisting of ions do not exist as molecules. Sodium chloride is an example of a nonmolecular compound. Although this type of compound consists of large numbers of positive and negative ions, its formula is usually represented by the simplest ratio of the atoms in the compound. Thus, the ratio of ions in sodium chloride is one sodium ion to one chlorine ion.

Compounds exist either as molecules which consist of two or more elements bonded together or in the form of positive and negative ions held together by the attractive force of their positive and negative charges.

The compound carbon monoxide (CO) is composed of carbon and oxygen in the ratio of one atom of carbon to one atom of oxygen. Hydrogen chloride (HCl) contains a ratio of one atom of hydrogen to one atom of chlorine. Compounds may contain more than one atom of the same element. Methane ("natural gas," CH_4) is composed of carbon and hydrogen in a ratio of one carbon atom to four hydrogen atoms; ordinary table sugar (sucrose, $C_{12}H_{22}O_{11}$) contains a ratio of 12 atoms of carbon to 22 atoms of hydrogen to 11 atoms of oxygen. These atoms are held together in the compound by chemical bonds.

Substance	Each Molecule Composed of	Formula
Carbon monoxide	1 carbon atom + 1 oxygen atom	CO
Hydrogen chloride	1 hydrogen atom + 1 chlorine atom	HCl
Methane	1 carbon atom + 4 hydrogen atoms	CH_4
Sugar (sucrose)	12 carbon atoms + 22 hydrogen atoms + 11 oxygen atoms	$C_{12}H_{22}O_{11}$
Water	2 hydrogen atoms + 1 oxygen atom	H_2O

There are over three million known compounds, with no end in sight as to the number that can and will be prepared in the future. Each compound is unique and has characteristic physical and chemical properties. Let us consider in some detail two compounds—water and mercuric oxide. Water is a colorless, odorless, tasteless liquid that can be changed to a solid, ice, at $0°C$ and to a gas, steam, at $100°C$. It is composed of two atoms of hydrogen and one atom of oxygen per molecule, which represents 11.2 percent hydrogen and 88.8 percent oxygen by weight. Water reacts chemically with sodium to produce hydrogen gas and sodium hydroxide, with lime to produce calcium hydroxide, and with sulfur trioxide to produce sulfuric acid. No other compound has all these exact physical and chemical properties; they are characteristic of water alone.

Mercuric oxide is a dense, orange-red powder composed of a ratio of one atom of mercury to one atom of oxygen. Its composition by weight is 92.6 percent mercury and 7.4 percent oxygen. When it is heated to temperatures greater than $360°C$, a colorless gas, oxygen, and a silvery liquid metal, mercury, are produced. Here again are specific physical and chemical properties belonging to mercuric oxide and to no other substance. Thus, a compound may be identified and distinguished from all other compounds by its characteristic properties.

Very few substances in nature are found in the pure state. Small quantities of gold and silver are often found in highly pure form, but most matter occurring in nature is a heterogeneous mixture of several substances.

What are the characteristics of a mixture? How can we distinguish a mixture from a compound? We have seen that a compound contains two or more elements chemically united. Suppose that we mix a quantity of the two elements sulfur and iron. Do we have a compound? No, rather we have a physical mixture of the two elements, for when we mix iron and sulfur, we can, at our discretion, mix together any quantities of the two elements. The iron in the mixture is uncombined and will be attracted by, and cling to, a magnet. In fact, the iron and sulfur of the mixture can be separated by this purely mechanical or physical method. But if the mixture is heated, the iron and sulfur combine chemically to form iron(II) sulfide. This compound(FeS), like any other, must follow the Law of Definite Composition—it contains 63.5 percent Fe and 36.5 percent S by weight. The iron(II) sulfide does not have magnetic properties and therefore is not attracted by a magnet. The iron and sulfur in the compound can be separated only by methods involving chemical changes.

Mixture of Iron and Sulfur	Compound of Iron and Sulfur
Formula: no definite formula; consists of Fe + S	Formula: FeS
Can contain Fe and S in any proportion by weight.	Contains 63.5% Fe and 36.5% S by weight.
Fe and S can be separated by physical means.	Fe and S can be separated only by chemical changes.

Thus, we can see that the properties of the individual elements are lost in forming a compound but are retained in a mixture. A mixture, then, contains two or more substances that do not lose their identity and that can be present in variable concentrations. Several distinct differences between compounds and mixtures are summarized in Table 1.3.

Table 1.3 Comparison of mixtures and compounds

	Mixture	Compound
Composition	May be composed of elements, compounds, or both in variable amounts.	Composed of two or more elements in a definite, fixed proportion by weight.
Separation of components	Separation may be made by simple physical or mechanical means.	Elements can be separated by chemical changes only.
Identification of components	Components do not lose their identities.	A compound does not resemble the elements from which it is formed.

Vocabulary

1. technetium [tek'ni:ʃiəm] n. 锝
2. promethium [prə'mi:θiəm] n. 钷
3. astatine ['æstəti:n] n. 砹
4. francium ['frænsiəm] n. 钫
5. lawrencium [lɔː'rensiəm] n. 铹
6. cyclotron ['saiklətrɔn] n. 回旋加速器
7. rutherfordium [rʌðə'fɔːdiəm] n. 钅卢
8. mercury ['məːkjuri] n. 水银、汞
9. chlorine ['klɔːriːn] n. 氯
10. methane ['miːθein] n. 甲烷、沼气
11. sucrose ['suːkrəuz] n. 蔗糖
12. lime [laim] n. 石灰
13. heterogeneous [hetərə'dʒiːniəs] a. 异种的、异质的、由不同成分形成的
14. discretion [dis'kreʃən] n. 慎重,斟酌
15. cling [kliŋ] vi. 粘紧、附着
16. magnet ['mægnit] n. 磁铁、有吸引力之物
17. sulfide ['sʌlfaid] n. 硫化物

Phrases

1. building block   结构单元
2. radioactive decay   放射性衰变
3. no ... other than M   除 M 外……不
4. whole-number ratio   整数比
5. table sugar = sucrose   蔗糖
6. the Law of Definite composition   固定成分定律

Notes

1. 这句话可译为:除地球上的这些元素外,在宇宙其它星体上还没检测到其它元素。

1.5 METALS, NONMETALS AND METALLOIDS

Elements are classified as metals, nonmetals, or metalloids according to such properties as electrical conductivity and luster.

The periodic table is as important to chemists and chemistry students as good maps are to travelers. The table organizes all sorts of chemical and physical information about the elements and their compounds. It allows us to study systematically the way properties vary with an element's position within the table and, in turn, makes the similarities and differences among the elements easier to understand and remember[1].

Even a casual inspection of samples of the elements reveals that some are familiar metals and that others, equally familiar, are not metals. Most of us are already familiar with metals such as lead, iron, or gold and nonmetals such as oxygen or nitrogen. A closer look at the nonmetallic elements, though, reveals that some of them, silicon and arsenic to name two, have properties that lie between those of true metals and true nonmetals. These elements are called metalloids. Division of the elements into these three categories—metals, nonmetals, and metalloids—is not an even one, however. Most of the elements are metals, slightly over a dozen are nonmetals, and only a handful are metalloids.

You probably know a metal when you see one. Metals tend to have a shine or luster that is easily recognized. For example, the silvery sheen of a freshly exposed surface of sodium would most likely lead you to identify this element as a metal even if you had never seen or heard of it before. We also know that metals conduct electricity. Few of us would poke an iron nail we were holding in our hand into an electrical outlet. In addition, we know that metals conduct heat very well. On a cool day, metals always feel colder to the touch than do neighboring nonmetallic objects because metals conduct heat away from your hand very rapidly. Nonmetals seem less cold because they can't conduct heat away as quickly and therefore their surfaces warm up faster.

Other properties that metals possess, to varying degrees, are malleability—the ability to be hammered or rolled into thin sheets—and ductility—the ability to be drawn into wire. For example, the production of sheet steel for automobiles and household appliances depends on the malleability of iron and steel, and the manufacture of electrical wire is based on the ductility of copper.

Another important physical property that we usually think of when metals are mentioned is hardness. Some, such as chromium or iron, are indeed quite hard; but others, like copper and lead, are rather soft. The alkali metals are so soft that they can be cut with a knife, but they are also so chemically reactive that we rarely get to see them as free elements.

All of the metallic elements, except mercury, are solids at room temperature. Mercury's low freezing point ($-39°C$) and fairly high boiling point ($357°C$) make it useful as a fluid in thermometers. Most of the other metals have much higher melting points, and some are used primarily because of this. Tungsten, for example, has the highest melting point of any metal ($3\,400°C$, or $6\,150°F$), which explains its use as a filament in electric light bulbs. Cobalt and chromium are combined in alloys (mixtures of metals) such as "stellite" to make high-speed, high-temperature cutting tools for industry. This alloy retains its hardness even at high temperatures, and the cutting tools therefore remain sharp.

The chemical properties of metals vary tremendously. Some, such as gold and platinum, are very unreactive toward almost all chemical agents. This property, plus their natural beauty and their rarity, makes them highly prized for use in jewelry. Other metals, however, are so reactive that few people except chemistry students ever have an opportunity to see them. For instance, we all know that the metal sodium reacts very quickly with oxygen or moisture in the air, and its bright metallic surface tarnishes almost immediately. On the other hand, compounds of sodium are quite stable and very common. Examples are table salt (NaCl), baking soda ($NaHCO_3$), lye(NaOH), and bleach(NaOCl).

Many objects we see each day are clearly not metals. Some examples are plastics, wood, concrete, and glass. These aren't elements, though. Most often, we encounter the nonmetallic elements in the form of compounds or mixtures of compounds. There are , however, some nonmetals that are very important to us in their elemental forms. The air we breathe, for instance, contains mostly nitrogen, N_2, and oxygen, O_2. Both are gaseous, colorless, and odorless nonmetals. Since we can't see, taste, or smell them, however, it's difficult to experience their existence. (Although if you step into an atmosphere without oxygen, your body will very quickly tell you that something is missing!) Probably the most commonly observed nonmetallic element is carbon. We find it as the graphite in pencils, as coal, and as the charcoal used for barbecues. It also occurs in the more valuable form of diamonds. Although diamond and graphite differ in appearance, each is a form of elemental carbon.

Properties of nonmetallic elements are almost completely opposite those of metals. Each of these elements lacks the characteristic appearance of a metal. They are poor conductors of heat and, with the exception of the graphite form of carbon, are also poor conductors of electricity. The electrical conductivity of graphite appears to be an accident of molecular structure, since the structures of metals and graphite are completely different.

Many of the nonmetals are solids at room temperature and atmospheric pressure, while many others are gases. All of the Group 0 elements are gases in which the particles consist of single atoms. The other gaseous elements—hydrogen, oxygen, nitrogen, fluorine, and chlorine—are composed of diatomic molecules. Their formulas are H_2, O_2, N_2, F_2, and Cl_2. As we previously learned, a molecule is diatomic if it is composed of two atoms. Bromine and iodine are also diatomic, but bromine is a liquid and iodine is a solid at room temperature.

The nonmetallic elements lack the malleability and ductility of metals. A lump of sulfur crumbles when hammered and breaks apart when pulled on. Diamond cutters rely on the brittle nature of carbon when they split a gem-quality stone by carefully striking a quick blow with a sharp blade.

As with metals, nonmetals exhibit a broad range of chemical reactivity. Fluorine, for instance, is extremely reactive. It reacts readily with almost all the other elements. At the other extreme is helium, the gas used to inflate children's balloons and the Goodyear blimp. This element does not react with anything. Chemists find helium useful when they want to provide a totally inert (unreactive) atmosphere inside some apparatus.

The properties of metalloids lie between those of metals and those of nonmetals. This shouldn't surprise us since the metalloids are located between the metals and the nonmetals in the periodic table. In most respects, metalloids behave as nonmetals, both chemically and physically. However, in their most important physical property, electrical conductivity, they somewhat resemble metals. Metalloids tend to be semiconductors—they conduct electricity, but not nearly so well as metals. This property, particularly as found in silicon and germanium, is responsible for the remarkable progress made during the last decade in the field of solid-state electronics. The newest hi-fi stereo systems, television receivers, and CB radios rely heavily on transistors made from semiconductors. Perhaps the most amazing advance of all has been the fantastic reduction semiconductors have allowed in the size of electronic components[2]. To it, we owe the development of small and versatile hand-held calculators and microcomputers. The heart of these devices is a microcircuit printed on a tiny silicon chip.

Vocabulary

1. luster ['lʌstə] n. 光泽
2. casual ['kæʒuəl] a. 偶然的、临时的
3. arsenic ['aːsnik] n. 砷、砒霜
4. handful ['hændful] n. 一把、少数
5. sheen [ʃiːn] n. 光泽；v. 闪光
6. poke [pəuk] n. 刺；v. 拨开
7. malleability [mæliə'biliti] n. 可展性、可锻性、顺从
8. ductility [dʌk'tiliti] n. 展延性、柔软性
9. chromium ['krəumiəm] n. 铬
10. thermometer [θə'mɔmitə] n. 温度计、寒暑表
11. tungsten ['tʌŋstən] n. 钨、钨锰铁矿
12. cobalt ['koubɔːlt] n. 钴、钴类颜料
13. stellite ['stelait] n. 钨铬钴的合金
14. platinum [plætinəm] n. 铂、白金
15. tarnish ['taːniʃ] n. 晦暗、污点；
    vt. 使失去光泽、玷污；
    vi. 失去光泽、被玷污
16. lye [lai] n. 碱液；v. 用碱液洗涤
17. bleach [bliːtʃ] v. 漂白、变白
18. graphite [græfait] n. 黑铅、石墨

19.	charcoal	['tʃɑːkəul]	n. 木炭
20.	barbecue	['bɑːbikjuː]	n. 烤肉；vt. 烤肉、烧烤
21.	bromine	['brəumiːn]	n. 溴
22.	iodine	['aiədiːn]	n. 碘
23.	lump	[lʌmp]	n. 块；v. 结块
24.	crumble	['krʌmbl]	n. 崩溃、破碎、灭亡； vt. 弄碎
25.	blade	[bleid]	n. 草叶、叶片、刀片、刀刃
26.	blimp	[blimp]	n. 软式小型飞机、肥胖的人
27.	inert	[i'nəːt]	a. 惰性的、迟钝的、无活力的、呆滞的
28.	versatile	['vəːsətail]	a. 多才多艺的、万用的、万向的

Phrases

1. baking soda 小苏打
2. strike a quick blow 快速冲击
3. hi-fi 为 high fidelity（高保真）的缩写
4. CB 为 citizens band（民用电台频带）的缩写

Notes

1. 主句中有两个并列的谓语"allows"和"makes"。句中"study"的宾语"the way"后面接一个定语从句"properties vary with ... the table"修饰它，该定语从句中省略了"in which"，因此，这一部分相当于"the way in which properties vary with ..."。全句可译为：它（周期表）使我们能系统地研究一些性质随元素在周期表内位置变化的方法，同样也使得元素之间的相似性和差别较易于了解和记忆。
2. 句中"semiconductors have allowed in ..."为定语从句，修饰"reduction"。引导定语从句的关系代词"that"在从句中作宾语，被省略了。

1.6 CHEMICAL EQUILIBRIUM

Any system at equilibrium represents a dynamic state in which two or more opposing processes are taking place at the same time and at the same rate. A chemical equilibrium is a dynamic system in which two or more chemical reactions are going on at the same time and at the same rate. When the rate of the forward reaction is exactly equal to the rate of the reverse reaction, a condition of chemical equilibrium exists. The forward reaction rate decreases as a result of decreasing amounts of reactants. The reverse reaction rate starts at zero and increases as the amount of product increases. When the two rates become equal, a state of chemical equilibrium has been reached. So the rates of the forward and reverse reactions become equal at some point in time. The system appears to be at a standstill because the concentration of the products is not changing. The reason the concentration is not changing is that the products are reacting at the same rate they are being produced.

Chemical Equilibrium:
Rate of forward reaction = Rate of reverse reaction

A saturated salt solution is in a condition of equilibrium.

$$NaCl(s) \rightleftharpoons Na^+(aq) + Cl^-(aq)$$

At equilibrium, salt crystals are continuously dissolving, and Na^+ and Cl^- ions are continuously crystallizing. Both processes are occurring at the same rate.

The ionization of weak electrolytes represents another common chemical equilibrium system.

$$HC_2H_3O_2 + H_2O \rightleftharpoons H_3O^+ + C_2H_3O_2^-$$

Weaker acid　　Weaker base　　Stronger acid　　Stronger base

In this reaction, the equilibrium is established in a 1 molar solution when the forward reaction has gone about 1 percent, i.e., when only 1 percent of the acetic acid molecules in solution have ionized. Therefore, only a relatively few ions are present, and the acid behaves as a weak electrolyte. In any acid – base equilibrium system, the position of equilibrium favored is toward the weaker conjugate acid and base. In the ionization of acetic acid, $HC_2H_3O_2$ is a weaker acid than H_3O^+, and H_2O is a weaker

General Knowledge of Chemistry and Chemical Engineering   31

base than $C_2H_3O_2^-$.

At 700 K, the equilibrium mixture

$$H_2 + I_2 \rightleftharpoons 2HI$$

shows that when equal moles of hydrogen and iodine are used, the reaction is 79 percent complete. With 1.00 mole each of H_2 and I_2, the maximum HI that can be formed is 2.00 moles. When the reaction is 79 percent complete, the system is in equilibrium and the amount of HI present is 1.58 moles, 79 percent of the theoretical yield of 2.00 moles. The equilibrium mixture will also contain 0.21 mol H_2 and 0.21 mol I_2, 0.79 mol of each having reacted.

$$\underset{\substack{1.00\\ \text{mol}}}{H_2} + \underset{\substack{1.00\\ \text{mol}}}{I_2} \xrightarrow{700\ K} \underset{\substack{2.00\\ \text{mol}}}{2HI}$$

(This would represent the condition if the reaction were 100 percent complete and 2.00 mol of HI were formed and no H_2 and I_2 were left unreacted.)

$$\underset{\substack{0.21\\ \text{mol}}}{H_2} + \underset{\substack{0.21\\ \text{mol}}}{I_2} \xrightleftharpoons{700\ K} \underset{\substack{1.58\\ \text{mol}}}{2HI}$$

(This represents the actual equilibrium attained starting with 1.00 mol each of H_2 and I_2. It shows that the forward reaction is only 79 percent complete.)

These concentrations in the equilibrium mixture are calculated as follows.

2.00 mol HI × 0.79 = 1.58 mol HI formed
1.00 mol H_2 − 0.79 mol H_2 = 0.21 mol H_2 unreacted
1.00 mol I_2 − 0.79 mol I_2 = 0.21 mol I_2 unreacted

Vocabulary

1. dynamic   [dai'næmik]   *a.* 动态的、有动力的、有力的、动力的；
   n. 动力、动态
2. reactant   [ri'æktənt]   *n.* 反应物
3. standstill   ['stændstil]   *n.* 停止、停顿

4. crystal　　　['kristəl]　　　*a.* 结晶状的、水晶的、透明的；*n.* 晶体、水晶
5. dissolving　　　　　　　　*a.* 毁灭性的，消溶的

Phrases

1. forward reaction　　正(向)反应
   reverse reaction　　逆(向)反应
2. at a standstill　　停顿着，停顿状态

1.7　ENERGY AND CHEMICAL ENERGY

Energy is the capacity of matter to do work. In this respect, matter can have both potential and kinetic energy. Potential energy is stored-up energy or energy an object possesses due to its relative position. Water backed up behind a dam represents potential energy that can be converted into useful work in the form of electrical energy. Gasoline represents a source of stored-up chemical potential energy that can be released during combustion.

Kinetic energy is the energy that matter possesses due to its motion. When the water behind the dam is released and allowed to flow, its potential energy is changed into kinetic energy, which may be used to drive generators and produce electricity. All moving bodies possess kinetic energy. The pressure exerted by a confined gas is due to the kinetic energy of rapidly moving gas particles. We all know the results when two moving vehicles collide—their kinetic energy is expended in the "crash" that occurs.

The common forms of energy are mechanical, chemical, electrical, heat, nuclear, and radiant or light energy.

In all chemical changes, matter either absorbs or releases energy. Chemical changes can be used to produce different forms of energy. Electrical energy to start automobiles is produced by chemical changes in the lead storage battery. Light energy for photographic purposes occurs as a flash during the chemical change in the magnesium flashbulb. Heat and light energies are released from the combustion of fuels. All the energy needed for our life processes—breathing, muscle contraction, blood circulation, etc.—is produced by chemical changes occurring within the cells of the body.

Conversely, energy is used to cause chemical changes. For example, a chemical change occurs in the electroplating of metals when electrical energy is passed through a salt solution in which the metal is submerged. A chemical change also occurs when radiant energy from the sun is utilized by plants in the process of photosynthesis. And, as we saw, a chemical change occurs when heat causes mercuric oxide to decompose. Chemical changes are often used primarily to produce energy rather than new substances. The heat or thrust generated during the combustion of fuels is more important than the products formed.

One type of energy may be transformed into energy of another type. When energy is absorbed during a chemical change, the products contain more chemical or potential energy than do the reactants. Conversely, when energy is released during a chemical change, the products contain less chemical or potential energy than do the reactants. For example, in the electrolysis of water, electrical energy decomposes water, producing hydrogen and oxygen. Energy is absorbed in this decomposition, and the products—hydrogen and oxygen—exist at a higher chemical-energy level than water. This potential energy is released in the form of heat and light when hydrogen and oxygen are burned, again producing water. Figure 1.1 illustrates the energy transformations in the chemical cycle.

$$\text{Water} \longrightarrow \text{Hydrogen + Oxygen} \longrightarrow \text{Water}$$

Water	Electrolysis / Chemical change	Hydrogen and Oxygen	Combustion / Chemical change	Water
Electrical energy	$\longrightarrow$	Chemical energy	$\longrightarrow$	Light and heat energy

Figure 1.1 The transformation of energy from one kind to another

In the process shown in Figure 1.1, electrical energy is transformed into chemical energy which, in turn, is transformed into heat and light energies. From this example we can conclude that energy can be transformed from one form to another or from one substance to another, and therefore is not lost.

The quantities of energy in many different systems have been studied thoroughly. No system has been found to acquire energy except at the expense of the energy possessed by another system[1]. In other words, energy can be neither created nor destroyed, though it may be transformed from one form to another. This is a statement of the Law of Conservation of Energy.

One of the most original concepts ever devised was presented by Albert Einstein(1879 ~ 1955) in 1905. He stated that the quantity of energy (E) equivalent to the mass (m) could be calculated by the equation $E = mc^2$, where m is in grams and c is the velocity of light (3.0×10^{10} cm·s^{-1}). According to Einstein's equation, whenever energy is absorbed or released by a substance, there must be a loss or gain of mass. Although the energy changes in chemical reactions are measurable and may appear to be large, the amounts are relatively small. The difference in mass between reactants and products in chemical changes is so small, in fact, that it cannot be detected by available measuring instruments. According to Einstein's equation 22 million (2.2×10^7) calories of energy are equivalent to 0.000 001 gram (1 microgram) of mass. In a more practical sense, when 2.8×10^3 grams of carbon are burned to carbon dioxide, 2.2×10^7 calories of energy are released. From this large mass of carbon we realize a loss of only 1 microgram of mass (a loss of 3.6×10^{-8} percent of the starting mass). Therefore, in actual practice, we may still treat the substances involved in chemical changes as having constant mass.

Because energy and mass are interchangeable, both the conservation of mass and the conservation of energy laws may be combined indicating that the total mass and energy in a system is constant.

Vocabulary

1. kinetic   [kai'netik]   a. 运动的、动力学的、活动的、活跃的
2. potential   [pə'tenʃəl]   n. 潜在性、可能性、潜力、势、位；
                                         a. 可能的、潜在的、有潜力的
3. radiant   ['reidiənt]   a. 发光的、辐射的
4. electroplating [i'lektroupleit]   vt. 电镀；n. 电镀物品
5. submerge   [səb'mə:dʒ]   vt. 使浸水、使陷入、淹没；
                                         vi. 潜入水中
6. photosynthesis [fəutəu'sinθəsis]   n. 光合作用
7. thrust   [θrʌst]   v. 刺、插、推力
8. electrolysis   [ilek'trɔlisis]   n. 电解
9. transformation [trænsfə'meiʃən]   n. 变化、转化
10. velocity   [vi'lɔsiti]   n. 速度、速率、快速
11. calory   [kæləri]   n. 卡路里

Phrases

1. potential and kinetic energy   势能和动能
2. lead storage battery   铅蓄电池
3. magnesium flashbulb   镁光灯，闪光灯（泡）

Note

1. 句中"at the expense of"意为"以……为代价，在消耗……条件下，靠……"。此句可译为：除了以消耗另一体系的能量为代价外，还没有发现一个体系能从其它方式中得到能量。

1.8 CHEMICAL BONDS

Forces exist within molecules that hold the atoms of a molecule together. These forces are a result of the interaction of electrons between atoms and give rise to chemical bonds. The chemical bond is the glue that holds atoms of a molecule together. The chemical bond and the ramifications of all its forces is a very complex system. We shall study, in their most elementary form, two principal types of bonds: the electrovalent, or ionic bond and the covalent bond.

1.8.1 The Electrovalent Bond

When a sodium atom transfers an electron to a chlorine atom, each atom becomes an electrically charged ion. The Na^+ and Cl^- ions are held or bonded together by the electrostatic attraction of their unlike charges. This bond is called an electrovalent, or ionic bond.

Electrovalent bonds are formed when a complete transfer of an electron (or electrons) takes place from one atom to another. Metallic elements, which have comparatively low electronegativities and little attraction for their valence electrons, tend to form ionic bonds when they combine with nonmetals. Hence, substances like sodium chloride (salts) contain ionic bonds.

1.8.2 The Covalent Bond

The sharing of electrons between atoms results in a chemical bond known as the covalent bond. It is the most predominant chemical bond in nature. The concept of the covalent bond was introduced in 1916 by Gilbert N. Lewis (1875 ~ 1946) of the University of California at Berkeley.

True molecules exist in compounds that are held together by covalent bonds. It is not correct to refer to "a molecule" of sodium chloride or other ionic compounds, since these compounds exist as large aggregates of positive and negative ions. We can refer to a molecule of hydrogen, chlorine, hydrogen chloride, carbon tetrachloride, sugar, or carbon dioxide[1], because these compounds contain only covalent bonds and exist in molecular aggregates.

A study of the hydrogen molecule will give us a better insight into the nature of the covalent bond and its formation. The formation of a hydrogen molecule, H_2, involves the overlapping and pairing of $1s$ electron orbitals from two hydrogen atoms to form a covalent bond between the two atoms. Each atom contributes one electron of the pair that is shared jointly by two hydrogen nuclei. The orbital of the electrons now includes both hydrogen nuclei, but probability factors show that the most likely place to find the electrons (the point of highest electron density) is between the two nuclei.

The tendency for hydrogen atoms to form a molecule is very strong. In the molecule each electron is attracted by two positive nuclei. This attraction gives the hydrogen molecule a more stable structure than the individual hydrogen atoms had. Experimental evidence of stability is shown by the fact that 435.5 kJ are needed to break the bonds between the hydrogen atoms in one mole of hydrogen molecules (2.0 g). The strength of a bond may be determined by the energy required to break it. The energy required to break a covalent bond is known as the bond dissociation energy. The following bond dissociation energies illustrate relative bond strengths. All substances are considered to be in the gaseous state and to form neutral atoms.

The covalent bond is designated by a dash (—) between the two atoms (e.g., H—H; H—Cl; C=O). A single dash means one pair of electrons; two dashes mean two pairs or four electrons are shared between two atoms.

(1) Polar covalent bonds. We have considered bonds to be either covalent or electrovalent, according to whether electrons are shared between atoms or are transferred from one atom to another. In most covalent bonds the pairs of electrons are not shared equally between the atoms. These bonds are known as polar covalent bonds.

The bond between two identical atoms, such as in a hydrogen

molecule, is a nonpolar covalent bond. In a nonpolar covalent bond the shared pair of electrons is attracted equally by the two atoms. The difference between the electronegativities of the two atoms is zero.

When a covalent bond is formed between two atoms of different electronegativities, the more electronegative atom attracts the shared electron pair toward itself. As a result, the atom with the higher electronegativity acquires a partial negative charge and the other atom, a partial positive charge. However, the overall molecule is still neutral. Due to this greater attraction of the electron pair, the bond formed between the two atoms has partial ionic character and is known as a polar covalent bond. The resulting molecule is said to be polar.

A dipole is a molecule that is electrically unsymmetrical, causing it to be oppositely charged at two points. A dipole is often written as $\oplus \ominus$. A hydrogen chloride molecule is polar and behaves as a small dipole. The HCl dipole may be written as H $\longrightarrow$ Cl. The arrow points toward the negative end of the dipole. H_2O, HBr, and ICl are polar; CH_4, CCl_4, CO_2 are nonpolar.

The greater the difference in electronegativity between two atoms, the more polar is the bond between them. When this difference is sufficiently large (greater than 1.7～1.9 electronegativity units), the bond between the two atoms will be essentially ionic (with some exceptions, of course). If the difference is less than 1.7 units, the bond will be essentially covalent. The difference in electronegativity between atoms can also give us a guide to the relative strength of covalent bonds. The greater the difference, the stronger the bond, i.e., the more energy required to break the bond. For example, HF has the strongest bond in the series HF, HCl, HBr, and HI.

(2) Coordinate covalent bond. We saw in the previous sections that a covalent bond was formed by the overlapping of electron orbitals between two atoms. Two atoms each furnish an electron to make a pair that is

shared between them.

Covalent bonds can also be formed by a single atom furnishing both electrons that are shared between the two atoms. The bond so formed is called a coordinate covalent, or semi-polar, bond. This bond is often designated by an arrow pointing away from the electron donor (e.g., A$\longrightarrow$B). Once formed, a coordinate covalent bond has the same properties as any other covalent bond—a pair of electrons shared between two atoms.

The electron-dot structures of sulfurous and sulfuric acids show a coordinate covalent bond between the sulfur and oxygen atoms that are not bonded to hydrogen atoms (see Figure1.1).

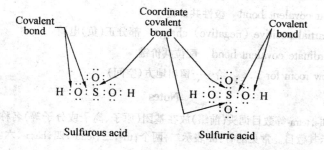

Figure 1.1

The open (unbonded) pair of electrons on the sulfur atom in sulfurous acid allows room for another oxygen atom with six electrons to fit perfectly into its structure in the formation of sulfuric acid[2]. Other atoms with six electrons in their outer shell, such as sulfur, could also fit into this pattern. The coordinate covalent bond explains the formation of many complex molecules.

Vocabulary

1. glue   [gluː]   n. 胶、胶水; vt. 胶合、粘贴
2. ramification [ˌræmifiˈkeiʃən]   n. 分枝、分歧、支流、衍生物
3. electrovalent [iˌlektrəuˈveilənt] a. 电价的

4. iodic [ai'ɔdik] a. 碘的
5. electrostatic [i'lektrə'stætik] a. 静电的,静电学的
6. valence ['veiləns] n. 原子价、化合价
7. aggregate ['ægrigit] n. 合计、总计;a. 合计的、集合的、聚合的;vt.,vi. 聚集、合计

Phrases

1. chemical bond 化学键
   electrovalent bond = ionic bond 离子键,电价键
   covalent bond 共价键
2. probability factor 几率因子
3. bond dissociation energy 键的解离能
4. polar covalent bond 极性共价键
5. a partial positive (negative) charge 部分正(负)电荷
6. coordinate covalent bond 配位共价键
7. allow room for ... 给……留出地方(空间)

Notes

1. 将 di-,tetra- 等数目词头(前缀)放在基团(原子、离子或分子等)名称前面,表示其数目。常见的有:di-表示二,两个;tri-,三;tetra-,四;hexa-,六。更多的此类词头见本书 Unit Five。
2. 句中"fit into ..."意为"适应(于),符合(于)"。这句话可译为:在亚硫酸中硫原子未成键的电子对为另一个有6个电子的氧原子留出了空间,完全适合形成硫酸的结构。

1.9 TYPES OF CHEMICAL EQUATIONS

Part of the problem of writing equations is determining the products formed. There is no sure method of predicting products, nor do we have time to carry out experimentally all the reactions we may wish to consider[1]. Therefore, we must use data reported in the writings of other workers, certain rules to aid in our predictions, and the atomic structure and combining capacities of the elements to help us predict the formulas of the products of a chemical reaction[2]. The final proof, of course, is found by conducting the reaction in the laboratory.

Reactions may be classified into certain types to assist in the formulation of chemical equations. The following four types cover many of the reactions in our studies.

(1) Combination or synthesis reaction. In this type of reaction direct union or combination of two substances produces one new substance. The general form of the equation is

$$A + B \longrightarrow AB$$

where A and B are either elements or compounds and AB is a compound. The formula of the compound in many cases can be determined from a knowledge of the oxidation numbers of the reactants in their combined states. Some reactions that fall into this category are

① Metal/Nonmetal + Oxygen $\longrightarrow$ Metal oxide/Nonmetal oxide

② Metal + Nonmetal $\longrightarrow$ Salt

③ Metal oxide/Nonmetal oxide + Water $\longrightarrow$ Base(Metal hydroxide)/Oxyacid

(2) Decomposition reaction. In this type of reaction a single substance is decomposed or broken down into two or more different substances. The reaction may be considered the reverse of combination. The starting material must be a compound, and the products may be elements or compounds. The general form of the equation is

$$AB \longrightarrow A + B$$

Predicting the products of a decomposition reaction can be difficult and requires an understanding of each individual reaction. Heating of oxygen-containing compounds often results in decomposition.

① Metal oxides. Some metal oxides decompose to yield the free metal plus oxygen. Others give a lower oxide, and some are very stable, resisting decomposition by heating.

② Carbonates and bicarbonates decompose to yield CO_2 when heated.

③ Miscellaneous.

(3) Single replacement or substitution reaction. In this type of reaction one simple substance (element) reacts with a compound substance to form a new simple substance and a new compound, one element displacing another in a compound. The general form of the equation is

$$A + BC \longrightarrow B + AC$$

If A is a metal, it will replace B to form AC, providing A is a more reactive metal than B; if A is a halogen, it will replace C to form BA providing A is a more reactive halogen than C. A brief list of the descending order of reactivity of metals (and hydrogen), known as the Activity Series of Metals, and the relative reactivity of the halogens is given below.

Metals: Li, K, Ba, Ca, Na, Mg, Al, Zn, Sn, Pb, H, Cu, Hg, Ag, Au

Halogens: F_2, Cl_2, Br_2, I_2

Some reactions that fall into this category are

① Metal + Acid $\longrightarrow$ Hydrogen + Salt

② Metal + Water $\longrightarrow$ Hydrogen + Metal hydroxide or Metal oxide

③ Metal + Salt $\longrightarrow$ metal + Salt

④ Halogen + Halogen salt $\longrightarrow$ Halogen + Halogen salt

(4) Double replacement or metathesis reaction. In this type of reaction two compounds react with each other to produce two different compounds. The general form of the equation is

$$AB + CD \longrightarrow AD + CB$$

This reaction may be thought of as an exchange of positive and negative groups, where the positive group (A) of the first compound combines with the negative group (D) of the second compound, and the positive group (C) of the second compound combines with the negative group (B) of the first compound. In writing the formulas of the products, we must take into account the oxidation numbers or charges of the combining groups. Some reactions that fall into this category are

① Neutralization of an acid and a base.

Acid + Base ⟶ Salt + Water

② Formation of an insoluble precipitate.

③ Metal oxide + Acid ⟶ Salt + Water

④ Formation of a gas.

Several factors must be considered as equations are written. ① Write correct formulas for reactants and products. ② All substances that we attempt to react may not react, or the conditions under which they react may not be present. For example, mercuric oxide does not decompose until it is heated; magnesium does not burn in air or oxygen until the temperature is raised to the point at which it begins to react. When silver is placed in a solution of copper (II) sulfate, no reaction takes place; however, when a strip of copper is placed in a solution of silver nitrate, the reaction takes place because copper is a more reactive metal than silver. ③ Predicting the products of a chemical reaction is an area of chemistry that causes students the greatest consternation. The ability to predict products correctly comes with experience. Although you may not be able to predict many reactions at this point, as you continue, you will find that reactions can be categorized, as above, and that prediction of the products thereby becomes easier, if not always certain.

There is a great deal yet to learn about which substances react with each other, how they react, and which conditions are necessary to bring about their reaction[3]. Indeed, these problems are what make chemistry an interesting and fascinating study. But such generalities as the four types of reactions given, the periodic table, atomic structure, oxidation numbers, etc., converge to aid in evaluating chemical reactions.

Vocabulary

1. metathesis   [me'tæθəsis]   n. 转移、置换
2. oxidation   [ɔksi'deiʃən]   n. 氧化
3. oxyacid   [ɔksi'æsid]   n. 含氧酸
4. miscellaneous   [misə'leiniəs]   a. 混杂的

5. halogen   [ˈhælədʒən]   n. 卤素
6. neutralization   [ˌnjuːtrəlaiˈzeiʃən]   n. 中和
7. precipitate   [priˈsipitət]   n. 沉淀物; vt. 使沉淀
8. sulfate   [ˈsʌlfeit]   n. 硫酸盐
9. nitrate   [ˈnaitreit]   n. 硝酸盐
10. consternation   [ˌkɔnstəˈneiʃən]   n. 惊愕、恐怖、惊惶失措

Phrases

1. oxidation number   氧化数
   combined state   化合态
2. single substance   单一物质
3. simple substance   单质(注意与 single substance 意思的不同)
4. reactive metal   活泼金属
5. positive group   阳根,正(性)基
   negative group   阴根,负(性)基
6. combining group   结合的基因
7. at this point   此时(刻)

Notes

1. 句中"nor do we have time to..."为倒装句。
2. 主句的谓语"must use"有三个宾语:"data","certain rules"和"the atomic structure and combining capacities of..."。
3. 主句为"there be"结构,不定式短语"to learn about"作"a large deal"的定语。全句可译为:关于哪些物质相互反应,如何反应,反应需要哪些必要的条件,还有大量的知识需要了解。

1.10 ENTHALPY, ENTROPY AND GIBBS FREE ENERGY

1.10.1 Enthalpy

Because most of the reactions that are of interest to us take place under conditions of constant pressure, rather than constant volume, thermodynamicists invented the quantity that we call enthalpy. It is defined

mathematically by the equation
$$H = E + pV$$
and if we consider only changes at constant pressure, but allow volume changes, then this change in enthalpy, ΔH, is
$$\Delta H = \Delta E + p\Delta V$$

Let's see how this Equation applies to a chemical reaction taking place at constant pressure and during which the system expands against the opposing pressure of the atmosphere. As we saw in the Equation, the work involved for the expanding system is
$$W = -p\Delta V$$
so, once again,
$$\Delta E = q + (-p\Delta V)$$
$$\Delta E = q - p\Delta V$$
Substituting this into the Equation above gives
$$\Delta H = q - p\Delta V + p\Delta V$$
which reduces to
$$\Delta H = q_p$$
We use the subscript p to show that the heat involved this time is the heat of reaction at constant pressure. This equation simply says that the heat of reaction measured under conditions of constant pressure is equal to ΔH.

Another name for enthalpy is heat content. The first of these terms is for the total value of the energy of a system. When it is at constant pressure, this is called the enthalpy of the system. When a system reacts at constant pressure and absorbs or evolves energy, we say that it experiences an enthalpy change, and defined by the equation
$$\Delta H = H_{final} - H_{initial}$$
H_{final} is the enthalpy of the system in its final state and $H_{initial}$ is the enthalpy of the system in its initial state.

For a chemical reaction, the initial state refers to the reactants and the final state to the products, so for a chemical reaction this equation can be rewritten as follows.

$$\Delta H = H_{products} - H_{reactants}$$

After having gone to all this effort to give a formal definition of ΔH, it is perhaps a bit disappointing to learn that we cannot actually calculate it from measured values of H_{final} and $H_{initial}$. This is because the total enthalpy of the system depends on its total kinetic energy plus its total potential energy, and these values can never be determined. We can see why by a little reflection about our universe.

Suppose, for example, that we are interested in the total enthalpy possessed by the reactants in a chemical reaction about to happen in a beaker resting on the laboratory bench[1]. The system might appear to be standing still, and thus it might seem to have no kinetic energy. However, it is on the surface of a possibly shifting continental plate that is on a rotating planet that moves about the sun, which itself moves through a galaxy that moves within the universe. We cannot know all of these velocities, and so the system's total kinetic energy cannot be measured or even calculated. Likewise, we cannot know its total potential energy. We have no way to measure, for example, the potential energies involving the attractions caused by gravity between the beaker and all of the other bodies in the universe.

Our formal definition of H actually serves only one purpose—to define the meaning of positive and negative values of ΔH. We said that when a system absorbs energy from its surroundings, the change is endothermic. This means that the system's final enthalpy, whatever its absolute value is, has to be larger than its initial enthalpy.

In endothermic changes: $H_{final} > H_{initial}$.

Therefore, if we could calculate ΔH from H_{final} and $H_{initial}$, we would subtract a smaller value from a larger one, so the difference (ΔH) would have a positive sign.

General Knowledge of Chemistry and Chemical Engineering 47

On the other hand, if a change is exothermic, the system loses energy and its final enthalpy would have to be less than its initial enthalpy.

In exothermic changes: $H_{final} < H_{initial}$.

Now, to calculate ΔH from $H_{initial}$ and H_{final}, we would subtract a larger number from one that is smaller, so the value of ΔH for all exothermic changes is negative.

Although we cannot measure the total enthalpy, H, the good news is that we really do not need to know its value. We care only about what our system could do for us right here at a particular place on this planet. All we care about is by how much the enthalpy changes, because it is only this enthalpy change that is available to us[2]. In other words, we really have no need for the absolute values of H_{final} and $H_{initial}$. All we need is the difference, ΔH, and this is something we can measure.

1.10.2 Entropy

Entropy is the thermodynamic property related to the degree of disorder in a system, and we designate it by the symbol S. The greater the degree of randomness or disorder in a system, the greater its entropy.

Like internal energy and enthalpy, entropy is a function of state. It has a unique value for a system whose temperature, pressure, and composition are specified. The entropy change, ΔS, is the difference in entropy between two states and also has a unique value.

To represent the mixing of gases symbolically, we can write

$$A(g) + B(g) \longrightarrow \text{mixture of } A(g) \text{ and } B(g)$$

$$\Delta S = S_{\text{mix of gases}} - [S_{A(g)} + S_{B(g)}] > 0$$

In the mixing of gases, disorder and entropy increase and ΔS is a positive quantity, that is, $\Delta S > 0$.

1.10.3 The Gibbs Free Energy

The Gibbs free energy gives us the net effect on spontaneity of the enthalpy and entropy changes.

We have now seen that two factors—enthalpy and entropy—determine whether or not a given physical or chemical event will be spontaneous. Sometimes these two factors work together. For example, when a stone wall crumbles, its enthalpy decreases and its entropy increases. Since a decrease in enthalpy and an increase in entropy both favor a spontaneous change, the two factors complement one another. In other situations, the effects of enthalpy and entropy are in opposition. Such is the case, as we have seen, in the melting of ice or the evaporation of water. The endothermic nature of these changes tends to make them nonspontaneous, while the increase in the randomness of the molecules tends to make them spontaneous. In the reaction of H_2 with O_2 to form H_2O, the enthalpy and entropy changes are also in opposition. In this case, the exothermic nature of the reaction is sufficient to overcome the negative value of ΔS and cause the reaction to be spontaneous.

When enthalpy and entropy oppose one another, their relative importance in determining spontaneity is far from obvious. In addition, temperature is a third factor that influences the direction of a spontaneous change. We have used the melting of ice or freezing of water to illustrate some points about enthalpy, entropy, and spontaneity. But we know that even a slight change in the temperature of an ice-water slush in equilibrium at 0℃ will cause the system to change to all liquid (if raised above 0℃) or all solid (if lowered below 0℃). Thus, we want to know how three factors—enthalpy, entropy, and temperature—interplay in determining spontaneity.

Specialists in thermodynamics, in studying the Second Law, defined a quantity called the Gibbs free energy, G, named to honor one of America's most important scientists, Josiah Willard Gibbs (1839 ~ 1903).

(It's called free energy because it is related, as we will see later, to the maximum energy in a change that is "free" or "available" to do useful work.) The Gibbs free energy is defined as a function of enthalpy, H, entropy, S, and temperature, T, in the following way.

$$G = H - TS \qquad (1.5)$$

Because we cannot know H in an absolute sense, we cannot know G in an absolute sense either. But, as we've often commented, what affects us and what we need most to know about are changes, not absolute values. So we will consider a change in the Gibbs free energy, ΔG, and we will deal with systems at constant temperature and constant pressure. With these conditions in mind, Equation 1.5 becomes

$$\Delta G = \Delta H - T\Delta S \qquad (1.6)$$

Once again we have a state function, a quantity that is independent of the path of the event. This means that

$$\Delta G = G_{final} - G_{initial}$$

What is of particular importance to us is the fact that a change can only be spontaneous if it is accompanied by a decrease in free energy. In other words, for a change to be spontaneous, G_{final} must be less than $G_{initial}$ and ΔG must be negative.

When a change is exothermic and is also accompanied by an increase in entropy, both factors favor spontaneity.

In such a change, ΔG will be negative regardless of the value of the absolute temperature, T (which can only have positive values). Therefore, the change will occur spontaneously at all temperatures.

On the other hand, if a change is endothermic and is accompanied by a decrease in entropy, both factors work against spontaneity. In this case, ΔG will be positive at all temperatures and the change will always be nonspontaneous.

When ΔH and ΔS have the same sign, the temperature becomes critical in determining whether or not an event is spontaneous. If ΔH and ΔS are both positive,

$$\Delta G = (+) - T(+)$$

Only at relatively high temperatures will the value of $T\Delta S$ be larger than the value of ΔH so that their difference, ΔG, is negative. A familiar example is the melting of ice. It is a change that we know is endothermic and occurs with an increase in entropy[3]. At temperatures above 0℃ (when the pressure is 1 atm), ice melts because the $T\Delta S$ term is bigger than the ΔH term. At lower temperatures, ice doesn't melt because the smaller value of T gives a smaller value for $T\Delta S$ and the difference, $\Delta H - T\Delta S$, is positive.

For similar reasons, when ΔH and ΔS are both negative, ΔG will be negative only at relatively low temperatures. The freezing of water is an example. Energy is released as the solid is formed and the entropy decreases. You know, of course, that water freezes spontaneously at low temperatures, that is, below 0℃.

The signs of ΔH and ΔS have effects on ΔG, and hence on the spontaneity of a physical or chemical event.

Vocabulary

1. spontaneity   [spɔntə'ni:iti]   n. 自然发生、自生、自发
2. enthalpy   [en'θælpi]   n. 焓
3. entropy   ['entrəpi]   n. 熵
4. beaker   ['bi:kə]   n. 大口杯、有倾口的烧杯
5. endothermic   [endou'θɔ:mik]   n. 吸热[性]的
6. thermodynamic [θɔ:moudai'næmik]   a. 热力学的、使用热力学的
7. randomness   ['rændəmnis]   n. 随意、无安排
8. slush   [slʌʃ]   n. 烂泥；v. 溅湿
9. interplay   ['intəplei]   n. 互相作用、作用和反作用；
                                         vi. 互相作用

Phrases

1. constant pressure   等压,恒压
   constant volume   等体积,恒容

2. heat content 热焓
3. chemical event 化学过程
4. with ... in mind 考虑到……,把……放在心上

Notes

1. "possessed by..."是过去分词短语作"enthalpy"的定语,"about to happen"(正要发生)作"reaction"的定语。全句可译为:例如,假定我们感兴趣的是放在实验台上的烧杯里正要发生化学反应的反应物所具有的总焓。
2. 句中"All we care about"是主语从句;"by how much...changes"为表语从句,其中"by"为介词,表示各向尺寸,数量变化的差额和倍数;在"because"后是一强调句型。此句可译为:我们关注的一切是焓变化了多少,因为只有这个焓变才是我们可利用的。
3. 这是一个强调句型,句中强调"a change"。另外,该句中"we know"为插入语。

1.11 FRIEDEL-CRAFTS ALKYLATION: PREPARATION OF 4,4′ – Di-*t*-BUTYLBIPHENYL

Alkylation of aromatic compounds is usually accomplished by the reaction of alkyl halides with Lewis acids to generate carbocation intermediates that are the actual alkylating agents. The process involves a typical electrophilic aromatic substitution. The main problem with the reactions are rearrangements of the carbocation intermediates. This type of Friedel-Crafts reaction is often demonstrated by the alkylation of benzene. The cumulative toxicity of benzene has discouraged its use. The same type of chemistry can be demonstrated nicely by the double alkylation of less-volatile biphenyl with the *t*-butyl cation generated from *t*-butyl chloride.

$$(CH_3)_3C-Cl + FeCl_3 \longrightarrow (CH_3)_3C^+$$

$$2(CH_3)_3C^+ + \text{biphenyl} \longrightarrow (CH_3)_3C-\text{—}-C(CH_3)_3$$
4,4′-di-*t*-butybiphenyl

Commercially available *t*-butyl chloride can be used or it can be generated from t-butyl alcohol just before use by the following process.

Cool a 25-to 50 ml round-bottomed flask or Erlenmeyer flask containing 12 ml of concentrated hydrochloric acid in an ice bath in a hood. Add 4 ml of t-butyl alcohol and magnetically stir or swirl the mixture for 20 minutes. Carefully, withdraw the product t-butyl chloride layer with a disposable pipette or transfer the mixture to a small separatory funnel and separate the layers. Dry the halide over $CaCl_2$ in a vial or round-bottomed flask. Distill the product in the hood with a short path still, or with a Hickman still, collecting several fractions boiling between 49 to 50℃. (CAUTION: Avoid contact or inhalation of this alkyl halide!)

Place 1 g(6 mmol) of biphenyl in a 25 ml round-bottomed flask in the hood. As with many aromatic compounds, biphenyl should be handled with extreme care. Although it is not volatile like benzene, contact with the skin should be avoided. Wear rubber gloves. Add 5 ml of dichloromethane and a small stir bar and magnetically stir the mixture until all the biphenyl has dissolved. Add 2 ml (18 mmol) of t-butyl chloride with a syringe or graduated pipette. Continue to stir the solution and add 40 mg of anhydrous ferric chloride. To the top of the flask, immediately attach a T-tube connected to a gas trap containing 10% aqueous sodium hydroxide. Stir the solution for 20 to 30 minutes at room temperature or slightly higher (with the aid of a warm-water bath). Transfer the reaction solution into another small round-bottomed flask or a small separatory funnel containing 4 to 5 ml of 10% HCl. Stir the mixture and separate the layers. If the reaction mixture had been transferred to a flask rather than a separatory funnel, the bottom organic layer can be removed by carefully inserting a disposable pipette through the top aqueous layer and withdrawing the bottom organic layer. Extract the organic layer with several more portions of 10% HCl. Dry the organic layer over anhydrous calcium chloride or sodium sulfate. Gravity filter the solution, or transfer the solution to another tared round-bottomed flask using a disposable pipette containing a cotton plug to remove the drying agent. Remove the solvent with a rotary evaporator and determine the crude yield. Recrystallize the residual 4,4′-di-t-butylbiphenyl from a minimum amount of 95% ethanol. Determine the yield and melting point. The authentic product has a melting point of 127 to 128 ℃. Record the IR spectra (KBr pellet) of the starting material and product. What characteristic bands, typical of a monosubstituted benzene ring, allow you to distinguish the starting material from the product, a p-disubstituted aromatic system? Record the NMR spectra of both the starting material and product in $CDCl_3$ and describe how they can be used to distinguish the two.

Phrases

1. Friedel-Crafts alkylation 弗瑞德－克来福特烷基化
2. alkylating agent 烷基化试剂
3. electrophilic aromatic substitution 亲电芳香取代
4. round-bottomed flask 圆底烧瓶
5. Erlenmeyer flask 锥形瓶
6. ice bath 冰浴
7. separatory funnel 分液漏斗
8. vial = phial 管(形)瓶
9. short path still 短路蒸馏,极窄馏分蒸馏,分子蒸馏
10. graduated pipette 刻度吸量管
11. gas trap 气体收集器,气阱
12. gravity filter 重力滤器
13. cotton plug 棉花塞
14. drying agent 干燥剂
15. rotary evaporator 旋转蒸发器
16. IR spectra = infrared spectra 红外光谱图
17. NMR spectra = nuclear magnectic resonance spectra 核磁共振光谱

1.12 THE PREAPARATION OF FERROCENE [BIS(CYCLOPENTADIENYL) IRON (II)]

$$8KOH + 2C_5H_6 + FeCl_2 \cdot 4H_2O \longrightarrow Fe(C_5H_5)_2 + 2KCl + 6KOH \cdot H_2O$$

Total time required: 1 day (or 5 hours if product dried in vacuo)
Actual working time: 4 hours
Preliminary study assignment:

J. Birmingham, "Synthesis of Cyclopentadiene Metal Compounds", Adv. Organometallic Chem., 2, 365 (1965); G. E. Coates, M. L. H. Green, and K. Wade, Organometallic Compounds, vol. 2, Methuen & Co. Ltd., London, 1968, pp. 90 ~ 115; W. L. Jolly, Inorg. Chem., 6, 1435 (1967); and, in this book, Chap. 5, Solvents, and Appendix 4, Compressed-Gas Cylinders.

Reagents required:

120 ml of 1,2-dimethoxyethane, about 60 g of KOH pellets (protected from moisture), 11 ml of cyclopentadiene—obtained from the thermal cracking of dicyclopentadiene, 13 g of $FeCl_2 \cdot 4H_2O$, 50 ml of dimethyl sulfoxide (DMSO), about 200 g of ice, cylinder of nitrogen or argon.

Special apparatus required:

Fractional distillation apparatus, 300 ml three-necked (standard taper joints) round-bottomed flask, magnetic stirrer and large stirring bar, 100-ml dropping funnel (standard taper), T-tube mercury bubbler with standard-taper connection to flask, Pyrex Petri dish with cover, 150 × 20 mm.

1.12.1 Procedure

The potassium hydroxide pellets are quickly ground with a mortar and pestle until the largest particles are less than 0.5 mm in diameter. Because it is very difficult to pulverize a large quantity of potassium hydroxide at one time, the pulverization should be carried out in batches of 15 g or less. It is important to minimize the exposure of the KOH powder to the atmosphere; therefore it is stored in a tightly capped tared bottle.

The cyclopentadiene is prepared by the thermal cracking of dicyclopentadiene. Dicyclopentadiene is slowly distilled through a fractionating column, collecting only that material which refluxes below 44° (cyclopentadiene boils at 42.5°, and dicyclopentadiene at 170°). If several students plan to prepare cyclopentadiene, it is advised that they cooperate in this step and distill at least 50 ml of dicyclopentadiene from a flask having a capacity twice the volume of the dicyclopentadiene. The freshly distilled cyclopentadiene must be used within 2 or 3 hours, or stored at $-78°$ until use, because slow dimerization occurs at room temperature.

The magnetic stirring bar, 120 ml of 1,2-dimethoxyethane, and 50 g of powdered potassium hydroxide are placed in the three-necked flask. One side neck is stoppered and the other is connected to the T-tube mercury bubbler and the nitrogen cylinder (see Figure 1.2). While the mixture is slowly stirred and the flask is being flushed with a stream of nitrogen, 11.0 ml of cyclopentadiene is added. The main neck is then fitted

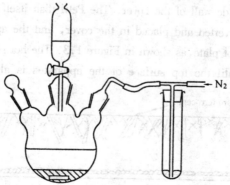

Figure 1.2 Apparatus for the preparation of ferrocene

with the dropping funnel with its stopcock open. After about 99 per cent of the air has been flushed from the flask, the stopcock is closed, and a solution of 13.0 g of $FeCl_2 \cdot 4 H_2O$ in 50 ml of dimethyl sulfoxide is placed in the dropping funnel. The mixture is stirred vigorously. After about 10 min, the T-tube is lifted above the mercury surface (to reduce the pressure in the flask to atmospheric), and drop-by-drop addition of the iron (II) chloride solution is begun. The rate of addition is adjusted so that the entire solution is added in 45 min. Then the dropping funnel stopcock is closed and vigorous stirring is continued for a further 30 min. Finally, the nitrogen flow is stopped, and the mixture is added to a mixture of 180 ml of 6 $mol \cdot L^{-1}$ HCl and about 200 g of crushed ice. Some of the resulting slurry may be used to rinse the reaction flask. The slurry is stirred for about 15 min, and the precipitate is collected on a sintered-glass funnel and washed with four 25 ml portions of water. The moist solid is spread out on a large watch glass and dried in the air overnight. The yield is about 11.5 g of ferrocene. This product should be quite satisfactory as an intermediate for subsequent syntheses. An extremely pure product can be obtained by sublimation. The material to be sublimed is placed in the inverted cover of a Petri dish so that none of the material is within 2 mm of the side wall of the cover. The Petri dish itself (the smaller of the pair) is inverted and placed in the cover, and the apparatus is then placed on a hot plate, as shown in Figure 1.3. The hot plate is gradually warmed up until the top surface of the apparatus is almost too hot to

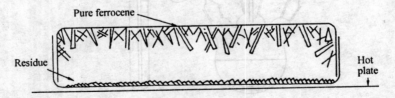

Figure 1.3 Sublimation of ferrocene

touch. After 4 to 10 hours, the ferrocene should be completely sublimed onto the upper glass surface and should be completely separated from the small amount of residue on the bottom by a gap of several millimeters. If any of the ferrocene crystals are touching the residue, the temperature of the hot plate should be increased, and more time allowed for complete sublimation. The yield is about 11.4 g of ferrocene, melting at 173 ~ 174℃.

1.12.2 Characterization

The infrared spectrum of ferrocene may be determined by using either solutions in carbon tetrachloride and carbon disulfide or a KBr pellet. Absorption bands are observed at the following frequencies (in cm^{-1}): 170, 478, 492, 782, 811, 834, 1 002, 1 051, 1 108, 1 188, 1 411, 1 620, 1 650, 1 684, 1 720, 1 758(m), 3 085(s). A complete interpretation of the spectrum has been given by Lippincott and Nelson.

The ultraviolet spectrum in ethanol or hexane shows maxima at 325 mμ (ϵ = 50) and 440 nm(ϵ = 87), and rising short-wavelength absorption (ϵ = 5 250 at 225 nm).

Ferrocene is readily oxidized to the blue ferricinium ion, $Fe(C_5H_5)_2^+$. Ferricinium tungstosilicate may be prepared by the following procedure. Dissolve $\frac{1}{2}$ g of ferrocene in 10 ml concentrated sulfuric acid; allow the solution to stand for 15 min to 1 hour and then pour it into 150 ml of water. Stir the resulting blue aqueous solution for a few minutes and filter to remove the sulfur precipitate. Add a solution of 2.5 g of 12-tungstosilicic acid in 20 ml of water, and filter off, wash with water, and air dry the pale blue tungstosilicate. Ferricinium perchlorate may be prepared by an electrolytic process described by Stranks. Ferrocene may be readily converted to the complex $[(C_5H_5)Fe(C_6H_6)]PF_6$, in which a benzene ring has replaced one of the cyclopentadienide rings.

A molecular orbital treatment of the bonding in ferrocene is described by Cotton.

Phrases

1. compressed-gas cylinder　压缩气体钢瓶
2. fractional distillation　分馏(作用)
3. three-necked round-bottomed flask　三颈(口)圆底烧瓶
4. standard taper joint　标准锥形接头
5. magnetic stirrer　(电)磁搅拌器
6. dropping funnel　滴液漏斗
7. T-tube mercury bubbler　T形管汞鼓泡瓶
8. Pyrex Petri dish　派热克斯陪替氏培养皿
9. tared bottle　配衡瓶(已称过容器皮重的瓶)
10. fractionating column　分馏柱
11. drop by drop = in drops　一滴一滴(地)
12. sintered-glass funnel　烧结玻璃板漏斗
13. watch glass　表(面)玻璃,表面皿

2

Some Basic Chemical Theories

2.1 DALTON'S ATOMIC THEORY

Dalton proposed that each element consists of indestructible particles called atoms that are identical in mass.

The law of definite proportions begs a question also raised by the law of conservation of mass: "What must be true about the nature of matter, given the truth of this law?[1]" In other words, what is matter made of?

At the beginning of the nineteenth century, John Dalton (1766 ~ 1844), an English scientist, used the Greek concept of atoms to make sense out of the emerging law of definite proportions[2]. Dalton, in fact, believed that this law and the law of conservation of mass compelled a belief in atoms as the real and ultimate particles of which all elements and compounds are made.

Dalton reasoned that if atoms really exist, they must have certain properties to account for the two laws of chemical combination. He described such properties, and the list constitutes what we now call Dalton's atomic theory.

2.1.1 Dalton's Atomic Theory

① Matter consists of definite particles called atoms.

② Atoms are indestructible. In chemical reactions, the atoms rearrange but they do not themselves break apart.

③ The atoms of one particular element are all identical in mass (and other properties).

④ The atoms of different elements differ in mass (and other properties).

⑤ When atoms of different elements combine to form compounds, new and more complex particles form. However, their constituent atoms are always present in a definite numerical ratio.

First, let's see how this theory explains the law of conservation of mass. According to the theory, atoms are indestructible and each has its own characteristic mass. Further, a chemical reaction simply involves the reordering of atoms from one combination to another. If there are a certain number of atoms present at the beginning of a reaction, and no atoms are allowed to enter or escape, then the same atoms must also be present after the reaction is over. Since each atom has its own characteristic mass, which does not change during a reaction, the total mass must remain constant. This, of course, is the law of conservation of mass.

The law of definite proportions is equally easy to explain. One of the key parts of the theory is that every atom of a given element has the same mass, which differs from the mass of the atoms of every other element. The theory also states that in a given compound the atoms of the various elements are always combined in a definite numerical ratio. Suppose we let the symbol A stand for one element and B for another. Also suppose that these elements form a compound in which there is one atom of A for each atom of B. In other words, the formula of the compound is AB. Let's also suppose that an atom of B has twice the mass of an atom of A.

This means that, regardless of the size of the sample of the compound, every time we find an atom of A with a certain mass we will also find an atom of B with twice as much mass. In this compound, the ratio of the mass of B to the mass of A must always be 2 to 1. This condition is exactly what the law of definite proportions requires—in any sample of the compound, the elements A and B are always present in the same proportion by mass.

2.1.2 Law of Multiple Proportions

One powerful piece of evidence for Dalton's theory came when Dalton and other scientists studied elements that could combine to give two (or more) compounds. Iron and sulfur, for example, form more than one compound. If we assume simple ratios, a few of the many theoretically possible combinations of iron and sulfur atoms would be as follows.

	Atom Ratio
[FeS]	1 (Fe) to 1 (S)
[FeS$_2$]	1 (Fe) to 2 (S)
[Fe$_2$S]	2 (Fe) to 1 (S)
[Fe$_2$S$_3$]	2 (Fe) to 3 (S)

Suppose that FeS and FeS$_2$ exist. If Dalton is right, then to make FeS$_2$ from a fixed mass of iron —say, 1.000 g of iron— should require exactly twice the mass of sulfur needed to make FeS from 1.000 g of iron. This is because, for a given amount of iron, there are twice as many sulfur atoms in FeS$_2$ as in FeS. In other words, the ratio of the mass of sulfur in FeS$_2$ to its mass in FeS should be 2 to 1— a ratio of simple whole numbers, provided each compound contains the same mass of iron. And this result has been repeatedly observed. Here are the mass ratios in two known compounds of iron and sulfur.

Mineral	Mass Ratio of the Elements
Pyrite (FeS_2)	1.000 g iron to 1.148 g sulfur
Troilite (FeS)	1.000 g iron to 0.574 g sulfur

Compare the masses of sulfur that combine in different ways with 1.000 g of iron.

$$\frac{1.148 \text{ g sulfur}}{0.574 \text{ g sulfur}} = \frac{2}{1}$$

It would be extremely hard to imagine any other view of matter besides Dalton's atomic theory that could account for a simple, whole-number ratio like this[3].

Many other examples of elements forming two or more compounds are known. One example involves the elements tin and oxygen. In one compound, these two combine in a ratio of 1.000 g of oxygen to 7.420 g of tin. In another compound, the mass ratio is 1.000 g of oxygen to 3.710 g of tin. Now compare the two masses of tin that combine with 1.000 g of oxygen.

$$\frac{7.420 \text{ g tin}}{3.710 \text{ g tin}} = \frac{2}{1}$$

The ratio is one of simple whole numbers, 2 to 1. Out of these and many other examples came the third great law of chemical combinations, the law of multiple proportions.

> **Law of Multiple Proportions.** Whenever two elements form more than one compound, the different masses of one element that combine with the same mass of the other element are in the ratio of small whole numbers.

2.1.3 Some Problems with Dalton's Theory

Dalton turned out to be wrong in some ways, but luckily the errors did not affect the basic postulate that atoms do exist. One error was the idea that atoms are indestructible. In this century, scientists have invented "atom-smashing" machines that develop enough energy to split atoms into a number of fragments, with such names as electrons, protons, and neutrons. Dalton's theory worked because in all chemical changes atoms do not fragment into such smaller particles, at least not in any way that observably affects mass relationships. In many chemical reactions atoms do pass some of their tiniest, least massive particles—electrons—back and forth, but as we will see, the total number of electrons is conserved. Therefore, the total mass of a compound is unchanged compared to the sum of the masses of the elements used to make it.

Another incorrect postulate in Dalton's atomic theory was the idea that all of the atoms of an element have identical masses. Actually, most elements are uniformly intermingled mixtures of two or more unique substances called the isotopes of the element. All of the isotopes of the given element have very nearly the same chemical properties, but their atoms have slightly different masses. An iron nail, for example, is made up of a mixture of four isotopes. One contributes 91.66% of the atoms in any iron sample.

The existence of isotopes did not affect the development of Dalton's theory for two important reasons. First, regardless of where on the earth or in the atmosphere an element happens to be found, the relative proportions of its different isotopes are essentially constant. As a result, every sample of a particular element has the same isotopic composition, and the average mass per atom is the same from sample to sample. Second, all the isotopes of a given element have virtually identical chemical properties—all give the same kinds of chemical reactions. For example, any one of the four isotopes of iron can combine with sulfur to give either FeS_2 or FeS. As a result, even though the existence of isotopes makes Dalton's third postulate untrue, an element behaves chemically as if it were true.

Vocabulary

1. indestructible [indis'trʌktəbl] a. 不能破坏的、不可毁灭的
2. numerical [nju:'merikl] a. 数字的、数值的、以数字表示的
3. rearrange [ri:ə'reindʒ] v. 重新整理、重新布置
4. tin [tin] n. 锡、马口铁; vt. 涂锡于
5. fragment ['frægmənt] n. 碎片、破片、断片
6. proton ['proutɒn] n. 质子
7. neutron ['nju:trɒn] n. 中子
8. intermingle [intə'miŋgl] vi. 混入、渗入、掺杂; vt. 使混合
9. nail [neil] n. 钉、指甲; vt. 钉、将……钉牢

Phrases

1. the law of definite proportions   定比定律
2. beg a question   以未证实的假设为依据(来辩论)
3. atom-smashing   以粒子轰击原子核,原子核分裂
   atom-smasher   原子加速器
4. law of multiple proportions   倍比定律
5. turn out (to be) + a. (或 n.)   结果弄清楚是,原来结果是

Notes

1. "given"位于句首时,表示假设,已知,给定,如果,假定等。
2. "… used the Greek concept of atoms to make sense out of the emerging law of definite proportions."其中"make B out of A"意思是用 A 制成 B,使 A 成为 B,在句中作状语。这部分可译为:为了使刚提出的定比定律有意义,应用了希腊的原子概念。
3. 句中"it"是形式主语,动词不定式"to imagine…"为真正的主语。"that"引导的定语从句修饰"atomic theory"。全句可译为:除了运用这个能说明简单整数比的道尔顿原子理论,任何其它的物质观的设想将是相当困难的。

2.2 VALENCE BOND THEORY

The overlap of atomic orbitals provides a way for a pair of electrons to be shared between two nuclei.

We have described the formation of the covalent bond in the hydrogen molecule, H_2. We saw that as two hydrogen atoms approach each other, the electron density shifts toward the region between the two nuclei. In the molecule, both electrons are able to move around both nuclei.

Now, let's look at the formation of this covalent bond in terms of orbitals and electrons. In an isolated hydrogen atom, there is one electron in the atom's 1s atomic orbital. When a hydrogen molecule is formed, the atomic orbitals of two atoms merge so as to allow the electrons to move back and forth between the two nuclei. Valence bond theory gives us a way of describing how this merging of orbitals occurs.

According to VB theory, a bond between two atoms is formed when a pair of electrons is shared by two overlapping atomic orbitals, one orbital from each of the atoms joined by the bond. By overlap of orbitals we simply mean that portions of two atomic orbitals from different atoms share the same space. An important part of the theory is that only one pair of electrons, with their spins paired, can be shared by two overlapping orbitals. This electron pair becomes concentrated in the region of overlap and helps "cement" the nuclei together, so the strength of the covalent bond—and the extent to which the energy of the atoms is lowered—is determined in part of the extent to which the orbitals overlap[1]. Because of this, atoms tend to position themselves so that the maximum amount of orbital overlap occurs. As we will see, this behavior is one of the major factors that control the shapes of molecules.

Next, let's look at a molecule that is just a bit more complex than H_2—hydrogen fluoride, HF. Following the rules described previously, we would write out its Lewis structure and we could diagram the formation of the bond as

$$H\cdot + \cdot \ddot{\underset{\cdot\cdot}{F}}: \longrightarrow H:\ddot{\underset{\cdot\cdot}{F}}:$$

The H—F bond is formed by the pairing of electrons—one from hydrogen and one from fluorine. To explain this according to VB theory, we must have two half-filled orbitals—one from each atom—that can be joined by overlap. (They must be half-filled, because we can't place more than two electrons into the bond.) To see clearly what must happen, it is best to know of the orbital diagrams of the valence shells of hydrogen and fluorine.

The requirements for bond formation are met by overlapping the half-filled 1s orbital of hydrogen with the half-filled 2p orbital of fluorine; there are then two orbitals plus two electrons whose spins can adjust so they are paired.

The overlap of orbitals provides a means for sharing electrons, thereby allowing each atom to complete its valence shell. It is sometimes convenient to indicate this using orbital diagrams. For example, a diagram can show how the fluorine atom completes its 2p subshell by acquiring a share of an electron from hydrogen.

Since the Lewis and VB descriptions of the formation of the H—F bond both account for the completion of the atoms' valence shells, a Lewis structure can be viewed, in a very qualitative sense, as a shorthand notation for the valence bond description of the molecule.

Now, let's turn our attention to a more complicated molecule, hydrogen sulfide, H_2S. (H_2S is the compound that gives rotten eggs their foul odor.) This is a nonlinear molecule, and experiment has shown that the H—S—H bond angle, the angle formed between the two H—S bonds, is about 92°. From the orbital diagram for sulfur's valence shell, we know that sulfur has two p orbitals that each contain only one electron. Each of these can overlap with the 1s orbital of a hydrogen atom. This overlap completes the valence shell of sulfur because each hydrogen provides one electron.

When the 1s orbital of each hydrogen atom overlaps with a p orbital of sulfur, the best overlap occurs when the hydrogen atoms are located along the imaginary y and z axes drawn through the center of the sulfur atom. The angle formed between these two H—S bonds, the predicted bond angle, is 90°. This is very close to the actual bond angle of 92° found by experiment. Thus, the VB theory requirement for maximum overlap quite nicely explains the geometry of the hydrogen sulfide molecule.

The overlap of p orbitals with each other is also possible. For example, we can consider the bonding in the fluorine molecule, F_2, to occur by the overlap of two 2p orbitals. The formation of the other diatomic molecules of the halogens, all of which are held together by single bonds, could be similarly described.

Vocabulary

1. merge   [məːdʒ]   v. 合并
2. spin   [spin]   n. 自旋
3. fluorine   ['fluːəriːn]   n. 氟
4. shell   [ʃel]   n. 贝壳、壳、外形、炮弹；vt. 去壳、脱落、炮轰；vi. 剥落
5. nonlinear   ['nɔn'liniə]   a. 非线性的
6. axes   ['æksiz]   n. 坐标轴、轴

Phrases

1. VB = Valence bond
2. to some extent   在某种程度上
3. in part of   在……方面,就……方面(角度)来讲
4. half-filled orbital   半充满轨道

Note

1. 全句可译为:这个电子对在重叠区域集中出现并有助于把原子核连接起来,因此共价键的强度——和原子能量降低的程度——在某种程度上决定于轨道部分重叠的程度。

2.3　THE KINETIC THEORY OF GASES

The gas laws can be explained by assuming that gases consist of many small particles moving randomly according to the laws of physics.

If you have ever taken something apart to see how it works—a clock, a radio, a car engine—you'll be able to appreciate the impulses in scientists back in the nineteenth century who asked, "How do gases work?"[1] These people were aware of the very predictable behavior of gases, and they knew the gas laws. So they wondered what had to be true about a gas to explain its conformity to the gas laws. The kinetic theory of gases was the answer. This theory begins with a set of postulates that suppose what gases must be like, and it assumes that the laws of physics and statistics hold just as much for the particles of a gas as for any other objects[2].

Postulates of the kinetic theory of gases:

① A gas consists of an extremely large number of very tiny particles that are in constant, random motion.

② The gas particles themselves occupy a net volume so small in relation to the volume of their container that their contribution to the total volume can be ignored.

③ The particles often collide in perfectly elastic collisions with themselves and with the walls of the container, and they move in straight lines between collisions. (This would be true only if the particles exert no forces of attraction or repulsion on each other.)

Physics, the study of matter, energy, and physical changes, has a branch called mechanics that deals with the laws of behavior of moving objects. (Billiard players master the applications of these laws, if not their mathematical forms.) In effect, the kinetic theory treats gases as supersmall, constantly moving billiard balls that bounce off each other and the walls of their container. They are so small, that their individual volume can be ignored. Using mechanics and statistics, physicists were able to show that gas pressure is the net effect of innumerable collisions made by gas particles on the walls. They calculated the relationship between pressure and volume at constant temperature and the result agreed beautifully with the pressure-volume law observed by Boyle.

The biggest triumph of the kinetic theory came with its explanation of gas temperature. Each gas particle has a particular mass and velocity. At any given instant, some particles are momentarily stopped and have zero velocity. Others, after being struck particularly hard, acquire a very high velocity. Since collisions are random, the velocity of any one particle changes constantly. But a gas sample has a huge number of particles, so we could envision an average velocity.

Any object with both mass and velocity has energy of motion called kinetic energy (KE), as we learned previously $KE = \frac{1}{2}(mv^2)$, where m = mass and v = velocity. If we can speak of an average velocity there must therefore also be an average kinetic energy. Individual particles can have momentary energies ranging from zero to high values, but overall there is an average kinetic energy. The various values of kinetic energy are distributed among the particles. The peak of the curve represents the most frequently occurring value of kinetic energy. The average kinetic energy lies at a slightly higher value because the curve is not symmetrical.

When physicists used the postulates of the kinetic theory to find an equation for pressure, they found that the product of pressure and volume, pV, is proportional to the average kinetic energy of the particles.

$$pV \propto \text{average } KE \qquad (2.1)$$

But we know from a source entirely different from mechanics and statistics——the equation for the ideal gas law—that the value of pV is proportional to the Kelvin temperature[3].

$$pV \propto T \quad (2.2)$$

The above equations tell us that pV is proportional both to T and to the average KE. The Kelvin temperature of a gas must, therefore, be directly proportional to the average kinetic energy of its particles.

$$T \propto \text{average } KE \quad (2.3)$$

This extremely important relationship between gas temperature and the average kinetic energy of gas particles gives us a most illuminating glimpse into the realm of atoms and molecules. We can now understand, for example, what happens to gas particles when a gas is heated. Because the absorbed heat energy goes to increasing their average kinetic energy, it must therefore go to increasing their velocities. No change in the mass of a gas particle can occur, so the only way the quantity $\frac{1}{2}(mv^2)$, the KE, can change is by a change in v, velocity. Thus, heating a gas makes its particles move faster, on the average.

Temperature has effect on the kinetic energy distribution in a gas. Notice that at the higher temperature the fractions of molecules with low kinetic energies decrease and the fractions with large kinetic energies increase. This causes the average kinetic energy to shift to a higher value at the higher temperature. The kinetic energy distributions are extremely important because, as it turns out, they apply not only to particles in gases but also to the particles in a liquid or a solid.

The kinetic theory is the theory about gases, and if it is any good it should explain the facts about gases, the gas laws. Let's see how the kinetic theory explains each gas law.

(1) Pressure-Temperature Law. The kinetic theory tells us that an increase in gas temperature increases the average velocity of gas particles.

At higher velocities, the particles must strike the container's walls more frequently and with greater force. But at constant volume, the area being struck is still the same, so the force per unit area—the pressure—must therefore increase. And this is how gas pressure is proportional to gas temperature (under constant volume and mass)—the pressure-temperature law.

(2) Temperature-Volume Law (Charles' Law). Suppose one "wall" of the gas container is a mercury level that can move in or out so that the pressure inside can always equal the pressure outside. When an increase in temperature acts to increase the inside pressure—explained just above in terms of the kinetic theory—the mercury wall moves out instead and so gives the gas more volume. And this is how gas volume is proportional to gas temperature (under constant pressure and mass)—Charles's law.

(3) Pressure-Volume Law (Boyle's Law). Suppose that one wall of the gas container is a movable piston that we can pull out (or push in) and thereby change the gas volume. If we make the gas volume larger without changing its temperature, the gas thins out and there are fewer gas particles per unit volume. The particles still have the same average kinetic energy (since the temperature is the same), but fewer strike any unit area of the walls per second. The pressure, therefore, must decrease. If we made the volume smaller at constant temperature (and constant average kinetic energy), the particles would strike a unit wall area more frequently and create a higher pressure. And this, in a qualitative sense, is how the kinetic theory explains Boyle's law—gas volume and pressure are inversely proportional (under constant temperature and mass).

(4) Dalton's Law of Partial Pressures. The law of partial pressures is actually evidence for the postulate in the kinetic theory that gas particles neither attract nor repel each other—that they act independently. Only if the particles of each gas in a mixture of gases act independently can the partial pressures add up in a simple way to give the total pres-

sure.

(5) Law of Effusion (Graham's Law). The conditions of Graham's Law are that the rates of effusion of two gases with different formula weights must be compared at the same pressure and temperature. When two gases are at the same temperature, their particles must have the same average kinetic energy. This conclusion, you will recall, was deduced from the postulates of the kinetic theory using the laws of physics and statistics. Remember that $KE = \frac{1}{2}(mv^2)$. When two gases have different formula weights, the value of m in this equation is different for the two. So if we change m in going from one gas to the other, the only way KE can still be a constant is for the average v to be different in each gas. When m is small, as for a low-formula-weight gas, v (or rather, v^2) must be larger to compensate. In other words, the gas with the lower formula weight has particles moving around with a greater velocity than the gas with the higher formula weight. And this is how the lower-formula-weight gas effuses more rapidly than the other. Because of the velocity-squared term in the equation for kinetic energy, a square-root term must occur in the equation that compares rates of effusion.

(6) Kinetic Theory and the Ideal Gas Law. The equation for the ideal gas law, $pV = nRT$, nowhere identifies a particular gas. It's true for all gases—a most remarkable fact about our natural world. The kinetic theory accommodates this fact as follows. We will first rearrange the ideal gas law equation.

$$p = \frac{n}{V}RT$$

This form of the ideal gas law tells us that the pressure of any gas is directly proportional to its molar concentration—n/V is moles per liter—when the temperature is constant. The question, then, is why should gases at the same molar concentration and the same temperature have the same pressure regardless of the gas?

Different gases at the same temperature consist of particles with the same average kinetic energy, according to the kinetic theory. It can be shown (and we will not go into the details) that moving particles with the same kinetic energy exert the same individual force per hit as they strike a given wall area. Of course, if we multiply the force per hit by the number of particles hitting the given area, we get the total force on this area, or force per unit area—pressure. But the total number of particles hitting a given area is proportional not to the identities of the particles but to their concentration—their moles per liter, to pick the most useful units. Different gases at identical molar concentrations and temperatures, therefore, ought to exert identical pressures—and they do.

The chief reason why we have gas laws for gases but not comparable laws for liquids or solids is that gases are mostly empty space. Gas molecules do not touch each other except when they collide. There is no other (or nearly no other) interaction between them, so the chemical identity of the gas does not matter. When the pressure on a gas is increased, the molecules are denied some of their "empty space" as the volume shrinks in accordance with Boyle's law. Thus gases are compressible. Liquids and solids, with little if any "empty spaces", cannot respond like gases to compression.

If we think about the relationship of temperature to average kinetic energy, as given by the kinetic theory, $T \propto$ average KE, we can see that if the average KE falls to zero, the temperature must become zero, also. But $KE = \frac{1}{2}(mv^2)$, and we know that m cannot become zero. So the only way KE can become zero is if v goes to zero. A gas particle cannot move any slower than it does at a dead standstill, so if the gas molecules stop moving entirely, the gas is as cold as anything can get—absolute zero.

Vocabulary

1. impulse   ['impʌls] $v.$ 冲击、碰撞

2. conformity　　[kən'fɔ:miti]　　n. 依从、符合、一致
3. collide　　　　[kə'laid]　　　　vi. 碰撞、互撞、抵触
4. mechanics　　 [mi'kæniks]　　　n. 力学、机械学
5. billiard　　　 ['biljəd]　　　　　n. 台球的、弹子戏的
6. innumerable　 [i'nju:mərəbl]　　a. 无数的
7. envision　　　 [in'viʒən]　　　　n. 想象、预见
8. symmetrical　　[si'metrikl]　　　a. 对称的、均匀的
9. illuminating　　[i'lu:mineitiŋ]　　a. 照亮的
10. piston　　　　['pistən]　　　　n. 活塞、瓣
11. effusion　　　 [i'fju:ʒən]　　　n. 流出物、泻出
12. compensate　　['kɔmpenseit]　　v. 补偿、偿还

Phrases

1. take apart　　分开,拆开,剖析
2. back in　　回溯到
3. hold for　　适用于
4. at any given instant　　在任何给定的情况下
5. peak of the curve　　曲线的峰值
6. be proportional to　　与……成比例
   be directly/inversely proportional to　　与……成正/反比
7. from ... source(s)　　从……方面
8. turn out　　证明,结果,产生
9. any good　　适用,有益,有用
10. unit area　　单位面积
11. thin out　　变稀少,稀释
12. partial pressure　　分压
13. formula weight　　(化学)式量,分子量
14. go from M to N　　从M变(转)为N
15. square-root term　　平方根项
16. multiply M by (with) N　　以N乘M
17. little if any = little if anything　　几乎没有

18. as + a. or ad. + as anything = like anything 非常

Notes

1. 全句可译为:如果你已经拆开某件物品,如钟表、收音机、汽车发动机等去看它是如何工作的,你将能理解,在19世纪科学家们提出"气体是如何运动的?"这一问题在科学家中所引起的冲击。
2. 此句子是由"and"连接的并列句。
3. 句中"But"为副词,用于句子开头,意为"可是"。"different from"为形容词短语作后置定语修饰"source"。主句中谓语的宾语为"that"引导的宾语从句。

2.4 COLLISION THEORY AND REACTION MECHANISMS

One of the benefits to be gained from a study of reaction rates is information about the paths, or mechanisms, followed by reactions. For most reactions, the individual chemical steps, called elementary processes that make up the mechanism cannot actually be observed. The mechanism that a chemist writes is really a theory about what occurs step-by-step as the reactants are converted to the products. Since the individual steps in the mechanism can't be observed, arriving at a chemically reasonable set of elementary processes is not at all a simple task. Making reasonable guesses requires a lot of "scientific intuition" and much more chemical knowledge than has presently been provided to you, so you need not worry about having to predict mechanisms for reactions. Nevertheless, to understand the science of chemistry better, it is worthwhile knowing how the study of reaction rates can provide clues to a reaction's mechanism.

We have learned that the basic postulate of the collision theory is that the rate of a reaction is proportional to the number of collisions per second between the reactants. We also learned that, for a given set of conditions, the number of effective collisions is only a certain small fraction of the total. If we could somehow double the total number of collisions, the number of effective collisions would be doubled also.

2.4.1 Predicting the Rate Law for an Elementary Process

Let's suppose, now, that we know for a fact that a certain collision process takes place during a particular reaction. For example, suppose that the decomposition of NOCl into NO and Cl_2 actually involves collisions between NOCl molecules and Cl atoms as one step in the mechanism. In other words, the reaction

$$NOCl + Cl \longrightarrow NO + Cl_2 \qquad (2.4)$$

represents an elementary process. What would happen if we were to double the number of Cl atoms in the container? Since there would then be twice as many Cl atoms with which the NOCl molecules could collide, there should be twice as many NOCl-to-Cl collisions per second. This should, in turn, double the rate of the reaction given by Reaction 2.4. In other words, doubling the Cl concentration would double the rate of this elementary process. The rate law for Reaction 2.4 should therefore contain [Cl] raised to the first power.

Similarly, if we were to double the number of NOCl molecules in the container, there would be twice as many NOCl molecules with which Cl atoms could collide. As a result, the collision frequency should double and the rate of the reaction should double. This means that the NOCl concentration should also be raised to the first power in the rate law for this elementary process. Therefore, the rate law for Reaction 2.4 should be

$$\text{rate} = k[Cl][NOCl]$$

Notice that the exponents in the rate law for this elementary process are the same as the coefficients in the chemical equation for the elementary process.

Let's look at another elementary process, one involving collisions between like molecules.

$$2NO_2 \longrightarrow NO_3 + NO \qquad (2.5)$$

If the NO_2 concentration were doubled, there would be twice as many individual NO_2 molecules and each would have twice as many neighbors with which to collide. The number of NO_2-to-NO_2 collisions per second would therefore be increased by a factor of 4. Increasing the collision frequency by a factor of 4 should also increase the rate by the same factor, so the rate should rise by a factor of 4, which is 2^2. Earlier we saw that when the doubling of concentration leads to a fourfold increase in the rate, the concentration of that reactant is raised to the second power in the rate law. Thus, the rate law for Reaction 2.5 should be

$$\text{rate} = k[NO_2]^2$$

Once again the exponent in the rate law for the elementary process is the same as the coefficient in the chemical equation.

The point that these two examples make is that the rate law for an elementary process can be predicted.

> The exponents in the rate law for an elementary process are equal to the coefficients of the reactants in the chemical equation for that elementary process.

It is very important to understand that this rule applies only to elementary processes. If all we know is an overall equation, we can only be sure of the exponents in the rate law if we determine them by doing experiments!

2.4.2 Predicting Reaction Mechanisms

How does the ability to predict the rate law of an elementary process help chemists predict reaction mechanisms? To answer this question, let's look at two mechanisms. First, consider the gaseous reaction

$$2NO_2 \longrightarrow 2NO + O_2 \qquad (2.6)$$

It has been experimentally determined that this is a second-order reaction having the rate law

$$\text{rate} = k[NO_2]^2$$

The mechanism that has been proposed for the reaction involves two steps with elementary processes. The NO_3 formed from the first step is called a reactive intermediate. We never actually observe the NO_3 because it decomposes so quickly. (If the two reactions are added and NO_3 is canceled from both sides, the net overall reaction, Reaction 2.6, is obtained.)

When a reaction such as this occurs in a series of steps, one step is very often much slower than the others. In this mechanism, for example, it is believed that the first step is slow and that once the NO_3 is formed, it decomposes quickly to give NO and O_2.

The slow step in a mechanism is called the rate-determining step or rate-limiting step. This is because the final products of the overall reaction cannot appear faster than the products of the slow step. In the mechanism for the decomposition of NO_2, then, the first reaction is the rate-determining step because the final products can't be formed any faster than the rate at which NO_3 is formed.

The rate-determining step is similar to a slow worker on an assembly line. Regardless of how fast the other workers are, the production rate depends on how quickly the slow worker does his or her job. The factors that control the speed of the rate-determining step therefore also control the overall rate of the reaction. This means that the rate law for the rate-determining step is directly related to the rate law for the overall reaction.

Because the rate-determining step is an elementary process, we can predict its rate law from the coefficients. The coefficient of NO_2 is 2, so the rate law for the first step would be the same as that found experimentally for the overall reaction. The mechanism, therefore, does not conflict with experimental evidence and could be correct.

Now let's look at the reaction

$$2NO_2Cl \longrightarrow 2NO_2 + Cl_2 \qquad (2.7)$$

which is a first-order reaction that has the experimentally determined rate law

$$\text{rate} = k[NO_2Cl]$$

The rate-determining step for this reaction could not possibly involve collisions of two NO_2Cl molecules because if it did, the reaction would be second order. The actual mechanism here appears to be

$$NO_2Cl \longrightarrow NO_2 + Cl \quad (slow)$$
$$NO_2Cl + Cl \longrightarrow NO_2 + Cl_2 \quad (fast)$$

Note that the sum of the elementary processes (after canceling Cl from both sides) gives the equation for the overall reaction (Reaction 2.7) and that the rate-determining step is first order because the coefficient of NO_2Cl is equal to one.

Although chemists may devise other experiments to help prove or disprove the correctness of a mechanism, one of the strongest pieces of evidence is the experimentally measured rate law for the overall reaction. No matter how reasonable a particular mechanism may appear, if its elementary processes cannot yield a predicted rate law that matches the experimental one, the mechanism is wrong and must be discarded.

Vocabulary

1. elementary ['elimentəri] a. 初步的
2. mechanism ['mekənizm] n. 机械、机构、结构、机理、技巧
3. intuition [intju:'iʃən] n. 直觉、直觉的知识
4. exponent [iks'pounənt] n. 指数、说明者、说明物；
   a. 说明的
5. intermediate [intə'mi:diət] n. 中级；a. 中间的、中级的；
   vi. 起媒介作用
6. coefficient [koui'fiʃənt] n. 系数
7. devise [di'vaiz] v. 设计、发明
8. discard [dis'ka:d] vt. 丢弃、抛弃、解雇；n. 抛弃

Phrases

1. reaction mechanism 反应机理
2. elementary process 基元过程

3. rate law 速率,定律
4. raise to + 序数词 + power 自乘……次幂
5. chemical equation 化学方程式
6. overall equation 总方程式
7. second-order reaction 二级反应
8. reactive intermediate 活性中间体
9. rate(-)determining step 决定速率步骤
   rate(-)limiting step 限速步骤

2.5 REACTIVE INTERMEDIATES OF ORGANIC REACTIONS

A chemical reaction is a process in which one species is converted to another, a process in which a rearrangement in the positions of the nuclei and electrons in the system takes place. To understand a reaction in detail, it is necessary to have information concerning ① the structures of the reactants and products, ② the conditions under which the reaction occurs, and ③ an intimate description of the pathways by which the nuclei and electrons change positions as the reactants change to products. It is the last of these topics that is discussed in the present, which deals with the general outlines of what are called the "mechanisms" of chemical reactions. However, before concerning ourselves with the actual pathways by which chemical conversions take place (i.e., the dynamic aspects), we shall first consider the factors relating to the stabilities of reactants and products (i.e., the static aspects).

The types of chemical equilibria that are probably most familiar to those readers who have had a course in general chemistry involve acids and bases. For example, when phosphoric acid is added to water, an interaction takes place in which a proton is transferred from the phosphoric acid to a water molecule, the latter behaving as a base.

The first of the reactive intermediates to be considered is known as a "carbonium ion" or "carbocation." It is defined as a triply liganded carbon possessing only six electrons in the valence shell. In the majority of cases the carbon atom of a carbonium ion is sp²-hybridized, and the stability of the carbonium ion is greatly diminished if the planar configuration required by the sp² carbon cannot be attained. The stability of a carbonium ion also depends on the attached groups. Phenyl and vinyl groups, for example, have a considerable stabilizing effect, because interaction between the unoccupied p orbital of the cation and the π system of the benzene ring or the vinyl group allows the positive charge to be delocalized over the entire system. The delocalized system of the phenylcarbonium ion is illustrated in Figure 2.1 in orbital and resonance structure representation.

$$\text{Ph}-\overset{+}{\text{C}}R_2 \longleftrightarrow \overset{+}{\text{Ph}}=CR_2 \longleftrightarrow {}^+\text{Ph}=CR_2 \longleftrightarrow \underset{+}{\text{Ph}}=CR_2$$

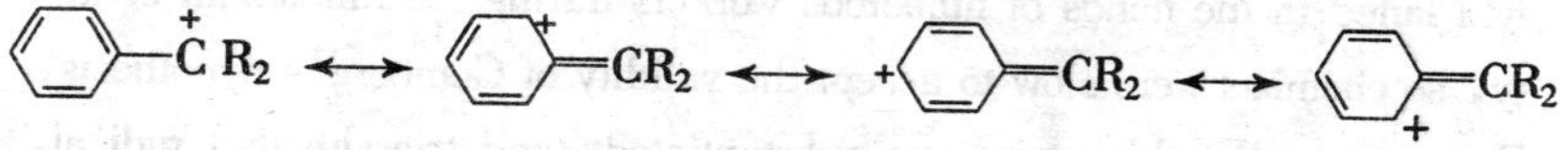

Figure 2.1 Phenylcarbonium ion (benzyl cation)

In addition to the types of resonance structures we have already discussed, still another type is invoked to explain the stabilizing effect of methyl groups on carbonium ions. Numerous experimental data show that the order of stability of alkyl cations decreases from $(CH_3)_3C^+$ to $(CH_3)_2CH^+$ to $CH_3CH_2^+$ to CH_3^+.

The second reactive intermediate to be considered is called a "free radical." It is defined as a triply liganded carbon possessing seven electrons in the outer valence shell. Its structure differs from that of the carbonium ion in having one electron in the p orbital. Thus, whereas the carbonium ion is a positively charged entity, the free radical is neutral. Both are electron-poor species, because they possess fewer than eight electrons in the valence shell. The same groups that stabilize carbonium ions also stabilize free radicals, the order of stability being $Ar\dot{C}H_2$ > $C=C-\dot{C}H_2$ > $R_3\dot{C}$ > $R_2\dot{C}H$ > $R\dot{C}H_2$ > $\dot{C}H_3$. Stabilization by phenyl groups (see Figure 2.2 for an orbital and a resonance structure representation of the delocalized system in the benzyl radical) is so effective,

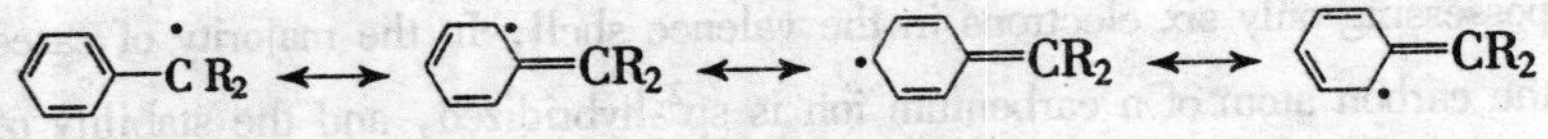

Figure 2.2 Resonance-stabilized benzyl radical

in fact, that when three of them are attached to the radical center, the free radical is stable enough to be isolated and studied. The synthesis of such a compound was accomplished in 1900 by Moses Gomberg during his attempts to prepare hexaphenylethane $((C_6H_5)_3C-C(C_6H_5)_3)$. The properties of the compound that he isolated were so incommensurate with a hexaphenylethane structure that he was forced to postulate the existence of the triphenylmethyl radical, $(C_6H_5)_3C\cdot$. Attempts to make alkyl radicals had failed in the hands of numerous workers during the nineteenth century, so chemists were slow to accept the validity of Gomberg's hypothesis. But, eventually, his claim was substantiated, and triarylmethyl radicals are now represented by many examples. Alkyl free radicals, also, have been established as real entities. They are highly reactive and very short-lived, though, and the abortive efforts of earlier chemists to trap and study them are now readily understood.

A third reactive intermediate to be considered is called a "carbanion," defined as a triply liganded carbon possessing eight electrons in the valence shell. It has two more electrons than the positively charged carbonium ion and one more electron than the neutral free radical, and is, therefore, negatively charged. Whether it has sp^2 or sp^3 geometry depends on the attached substituents. If a phenyl group is attached, the sp^2 geometry is favored, for this allows the negative charge to be delocalized (Figure 2.3). If groups of lesser ability to delocalize the negative charge are attached, the geometry is probably close to that of an sp^3-hybridized carbon, i.e., the electron pair is in a hybrid orbital. Species of this type are configurationally mobile, however, and pass from one tetrahedral form to the other with great facility. The mobility has been demonstrated, for example, by preparing an optically active compound, converting it to a carbanion, and noting the loss in optical activity, which results from the configurational inversion and the eventual production of equal numbers of enantiomers (i.e., racemization)[1].

$$\underset{}{\overset{}{C_6H_5}}-\overset{-}{C}R_2 \longleftrightarrow \cdots \longleftrightarrow \cdots \longleftrightarrow \cdots -CR_2$$

Figure 2.3 Resonance-stabilized phenylcarbanion (benzyl anion)

Vocabulary

1. intimate ['intimeit] a. 亲密的、直接的
2. static [stætik] a. 静态的
3. equilibria [iːkwi'libriə] n. (equilibrium 的复数)均衡、均势
4. phosphoric [fɔs'fɔrik] a. 磷的、含磷的
5. carbonium ['kaːbəuniəm] n. 带正电有机离子
6. cation ['kætaiən] n. 阳离子
7. ligand ['ligənd] n. 配合基、配合体、配体、配位体
8. planar ['plenə] a. 平坦的
9. vinyl ['vainil] n. 乙烯基
10. delocalize [diː'loukəlaiz] vt. 使离开原位
11. phenyl ['fenil] n. 苯基
12. invoke [in'vəuk] v. 调用
13. radical ['rædikl] n. 激进分子、基础；
    a. 激进的、根本的、基本的、根的
14. resonance ['rezənəns] n. 共鸣、回声、反响、谐振、共振
15. incommensurate [inkə'menʃəreit] a. 不相称的、不适应的
16. alkyl ['ælkil] n. 烷基、烃基
17. hypothesis [hai'pɔθəsis] n. 假设
18. validity ['və'liditi] n. 有效性
19. abortive [ə'bɔːtiv] a. 流产的、堕胎的、失败的
20. geometry [dʒi'ɔmitri] n. 几何学
21. carbanion [kaː'bænaiən] n. 负碳离子、阴碳离子

22. hybrid   ['haibrid]   n. 混血儿、杂种、混合物;
                                              a. 混合的、杂种的、混合语的
23. tetrahedral   [tetrə'hi:drəl]   a. 有四面的、四面体的
24. configuration   [kənfigju'reiʃən]   n. 配置、表面配置、形态
25. enantiomer   [i'næntioumə]   n. 对映体、对映异构物
26. racemization   ['ræsimai'zeiʃən]   n. (外)消旋作用

Phrases

1. general outline   概要,大纲
2. general chemistry   普通化学
3. carbonium ion   碳鎓离子
4. triply liganded carbon 三配位碳
5. resonance structure   共振结构
6. free radical   自由基

Note

1. 句中介词"by"后接三个分词短语(作宾语):"preparing","converting"和"noting"。"which"引导的非限性定语从句修饰"the loss in optical activity."

2.6 TYPES OF ORGANIC REACTION

The process in which substances are converted to their constituent atoms is called "cracking" and is common to all compounds, organic or inorganic. Also common to all organic compounds is the reaction with oxygen that yields the oxides of the constituent atoms in a process known as combustion.

The importance of combustion reactions cannot be minimized; most of the heat and power (except that generated by falling water, wind, sun, compressed steam from within the earth, or nuclear sources) comes from the burning of the mineral fuels, oil and coal, and the energy that sustains life is the result of very complicated oxidative processes in which compounds of carbon are ultimately converted to carbon dioxide and water. From the standpoint of the chemist doing basic research, however, cracking and combustion reactions are usually not very interesting. Seldom does he carry them out intentionally, for his fancy is much more likely to be captured by reactions in which only a few bonds in the molecule are changed at a time. In general, it is reactions of this type that are useful to him in his efforts to synthesize new and interesting compounds.

Selective reactions are such reactions in which relatively small changes occur in the transformation of reactants to products. Hundreds of such reactions are now known to the organic chemist, and many of these will be presented in the fields of organic syntheses. To introduce this large and important segment of organic chemistry, let us first focus our attention on some of the reactions that are characteristic of the various types of carbon frameworks that we have discussed previously. To do this, we shall organize the discussion on the basis of four major types of reactions: viz., addition reactions, substitution reactions, elimination reactions, and oxidation reactions.

(1) Addition reactions of alkenes. The most generally useful reactions of alkenes involve the addition of various reagents to the double bond, converting the double bond to a single bond in processes that we can represent as

$$\text{(a)} \quad \diagup\!\!\!\!C=C\!\!\!\!\diagup + AB \longrightarrow A-\overset{|}{\underset{|}{C}}-\overset{|}{\underset{|}{C}}-B \text{, or}$$

$$\text{(b)} \quad \diagup\!\!\!\!C=C\!\!\!\!\diagup + A \longrightarrow \diagup\!\!\!\!\underset{A}{C}\!\!-\!\!C\!\!\!\!\diagup$$

Although the majority of addition processes involving alkenes occur via carbonium ion intermediates, a few follow a different pathway and involve radical intermediates. One of the most interesting of these is the addition of hydrogen bromide in the presence of peroxide reagents. In the absence of peroxides, hydrogen bromide adds to propene via the Markovnikov pathway to yield isopropyl bromide. In the presence of peroxides, however, the order of addition is reversed, and the product is n-propyl bromide; the addition in this case is said to be anti-Markovnikov. This is interpreted in terms of initiation of the addition reaction by bromine atom, $Br\cdot$, rather than by a proton, as in the case for electrophilic addition. The bromine atom, formed by the action of a free radical on hydrogen bromide (e.g., $RO\cdot + HBr \longrightarrow ROH + Br\cdot$), adds to the double bond of propene at either C-1 to form a secondary radical or C-2 to form a primary radical. Since the stabilities of radicals follow the same order as that of carbonium ions, the reaction occurs preferentially via the secondary radical, which subsequently collides with a molecule of hydrogen bromide to form n-propyl bromide and a bromine atom. Once this stage of the reaction sequence is reached, the process becomes self-propagating; every time a carbon radical abstracts a hydrogen from hydrogen bromide, another bromine atom is formed, which adds to propene to form another carbon radical.

(2) Substitution reactions of alkanes. The most characteristic reaction of alkanes, which are much less reactive compounds than alkenes, alkynes, or arenes is a substitution process in which a hydrogen is replaced by some other atom or group, $R-H + Y\cdot \longrightarrow R-Y + H\cdot$. Among the several examples of reactions of this type, the generally most useful ones involve chlorine or bromine. For example, when a mixture of n-butane and chlorine is heated or irradiated, a reaction takes place to produce a mixture of 1-chlorobutane and 2-chlorobutane. A variety of pieces of evidence indicate that this reaction commences

$$CH_3CH_2CH_2CH_3 + X_2 \longrightarrow CH_3CH_2CH_2CH_2X + \begin{array}{c} CH_3CH_2 \\ \diagdown \\ \diagup \\ CH_3 \end{array} CHX$$

n-Butane   28% (X = Cl)   72% (X = Cl)
                 2% (X = Br)   98% (X = Br)

with the dissociation of molecular chlorine to chlorine atoms ($Cl_2 \rightleftharpoons Cl\cdot +$ $Cl\cdot$), and that a chlorine atom then collides with a molecule of n-butane to abstract a hydrogen atom to form hydrogen chloride and a butyl radical. If the collision occurs at the C—H bond of the methyl group, a primary radical is produced ($CH_3CH_2CH_2\dot{C}H_2$); if it occurs at the C—H bond of the methylene carbon, a secondary radical is produced ($CH_3CH_2\dot{C}HCH_3$). Subsequent collisions of these radicals with molecular chlorine then yield 1-chlorobutane and 2-chlorobutane, respectively, along with chlorine atoms. At this stage of the reaction sequence, we have the same situation that we encountered in the peroxide-induced addition of hydrogen bromide to alkenes, viz., a self-propagating or chain reaction. As in that case, here also the chain does not continue indefinitely, for various chain-terminating reactions occur, requiring new chlorine atoms to be constantly supplied to keep the reaction going. However, as many as 10^4 cycles of the self-propagating steps may take place before the chain is broken.

(3) *Elimination Reactions.* Addition reactions, as we have seen, proceed from a less saturated reactant to a more saturated product. Elimination reactions represent the reciprocal case and proceed from a more saturated reactant to a less saturated product. We have already alluded to one type of elimination reaction at the beginning where we discussed the cracking reaction, in which carbon-hydrogen and carbon-carbon bonds are broken to yield, ultimately, atomic carbon and hydrogen. A comparable reaction more generally useful for laboratory synthesis involves the partial removal of hydrogen from six-membered alicyclic compounds to yield aromatic compounds. For example, when cyclohexane is treated at an elevated temperature with platinum, palladium, sulfur, or selenium, hydrogen is removed and benzene is formed (Figure 2.4). In similar fashion, tetralin and decalin can be converted to naphthalene.

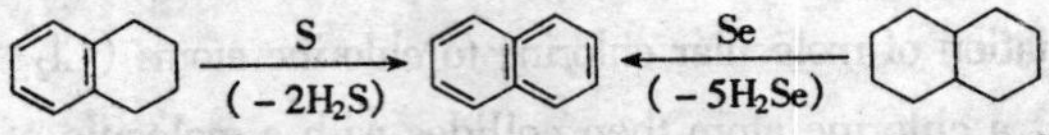

Figure 2.4 Dehydrogenation of cyclohexane, tetralin, and decalin

The process of removing hydrogen from six-membered alicyclic compounds is called dehydrogenation and it has useful application in the determination of the structure of alicyclic compounds. Aromatic compounds are generally crystallizable and less stereochemically complicated than their alicyclic precursors; also, they are more likely to be known compounds or, if unknown, to be accessible by easy synthesis. From the structure of the dehydrogenated product, then, the structural details of the alicyclic compound can be inferred. Caution must be exercised in using such information, however, because rearrangements and loss of functional groups sometimes occur during dehydrogenation.

(4) Oxidation reactions of alkenes and alkynes. For the present discussion, we shall define oxidation reactions as processes in which C—H or C—C bonds are converted to C—O bonds or in which carbon-carbon multiple bonds are converted to C=O bonds.

Alkenes and alkynes are far less resistant to oxidation than alkanes and undergo cleavage of the C=C, or C≡C bond in the presence of a variety of oxidizing agents. A particularly useful oxidizing agent is ozone, which converts alkenes of structure $RCH=CH_2$ or $RCH=CHR'$ to aldehydes, alkenes of structure $R_2C=CR'_2$ to ketones, and alkenes of structure $R_2C=CHR'$ to mixtures of an aldehyde and ketone. Alkynes also react with ozone, yielding carboxylic acids, but the reaction is slower than with alkenes, and may yield mixtures containing other compounds in addition to or instead of carboxylic acids. Ozonolysis provides a most useful way for locating the position of a C=C bond in a compound. For example, suppose that we have a sample of a material that we know to be either compound (a) or compound (b) in Figure 2.5. If it is compound (a), both products of ozonolysis are ketones; if it is compound (b), one of the products of ozonolysis is a ketone and the other is an aldehyde.

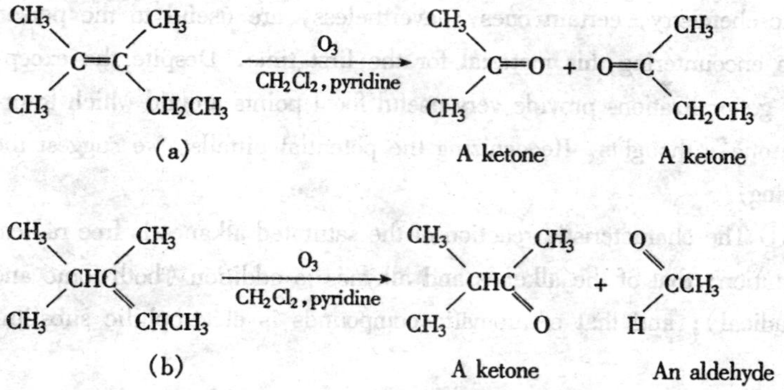

$$\underset{CH_3}{\overset{CH_3}{\diagdown}} C = C \underset{CH_2CH_3}{\overset{CH_3}{\diagup}} \quad \xrightarrow[CH_2Cl_2,\,pyridine]{O_3} \quad \underset{CH_3}{\overset{CH_3}{\diagdown}} C = O \;+\; O = C \underset{CH_2CH_3}{\overset{CH_3}{\diagup}}$$

(a)   A ketone   A ketone

$$\underset{CH_3}{\overset{CH_3}{\diagdown}} CHC = CHCH_3 \overset{CH_3}{\diagup} \quad \xrightarrow[CH_2Cl_2,\,pyridine]{O_3} \quad \underset{CH_3}{\overset{CH_3}{\diagdown}} CHC \overset{CH_3}{\underset{O}{\diagup}} \;+\; \underset{H}{\overset{O}{\diagdown}} CCH_3$$

(b)   A ketone   An aldehyde

Figure 2.5   Ozonolysis of isomeric dimethylpentenes

To be able to use reactions such as those shown above in the design of organic syntheses, we must, of course, remember what kinds of reactants give what kinds of products. For example, we must remember that when bromine reacts with propene the product is 1,2-dibromopropane. Rather than learn each reaction as an isolated fact, however, it is essential that we discern general patterns by means of which we can predict with some certainty how a particular set of compounds will react even though we may not have encountered the specific compounds previously[1]. For example, knowing how propene and bromine react, we can predict, correctly, that cyclohexene will react with chlorine to form 1,2-dichlorocyclohexane. In fact, we can rely on this kind of reaction as a general diagnostic test for the presence of a carbon-carbon multiple bond in a molecule. Thus, if treatment of an unknown compound with a dilute solution of bromine in carbon tetrachloride (pale brown-red color) causes the solution to turn colorless, it is a reasonable inference that the compound contains a carbon-carbon multiple bond to which the bromine in the solution has added. To furnish substance to predictions and extrapolations of this sort, however, it is useful that we have some knowledge of how the reactions take place—i.e., by what mechanisms they occur. Although it is dangerous to place unquestioning faith in many of the generalizations of organic chemistry, certain ones, nevertheless, are useful to the person who is encountering this material for the first time. Despite the exceptions, generalizations provide very useful focal points around which to organize one's thoughts. Recognizing the potential pitfalls, we suggest the following:

① The characteristic reaction of the saturated alkanes is free radical substitution; that of the alkenes and alkynes is addition (both ionic and free radical); and that of aromatic compounds is electrophilic substitution.

② The relative reactivity of C—H bonds with respect to hydrogen atom abstraction decreases in the following order:

$$C=C-\overset{|}{C}-H \approx Ar-\overset{|}{C}-H > R_3CH > R_2CH_2 > RCH_3 > CH_4$$

③ The resistance to oxidation of various carbon frameworks decreases in the following order: arene > alkane > alkyne > alkene. This must be accepted with reservation, however, for the reaction conditions and the oxidizing reagent both play an important role. For example, toluene is oxidized by potassium permanganate at the methyl group rather than at the benzene ring, but ozone attacks the benzene ring preferentially.

The compounds whose reactions are discussed above are those composed solely of carbon and hydrogen. We have already seen, however, that heteroatoms and groups can be attached to carbon frameworks, and it is not surprising to find that the chemistry of the original carbon framework may be changed as a result. For example, chloroform ($CHCl_3$) has a hydrogen that is sufficiently acidic to be removed by strong bases; tetrafluoroethylene ($F_2C=CF_2$) is inert to electrophilic addition and, instead, reacts readily with nucleophiles; benzoic acid ($C_6H_5CO_2H$) is much less susceptible to electrophilic substitution than benzene; and *p*-

hydroxytoluene ($HO-C_6H_4-CH_3$) undergoes oxidation in the aromatic ring rather than the methyl group. Thus, the attached groups (i.e., the functional groups) may alter the properties of the carbon frameworks even to the point of reversing the chemical behavior that is observed in the parent systems.

Vocabulary

1. crack [kræk] n. 裂解; v. (使)破裂
2. combustion [kəm'bʌstʃən] n. 燃烧、氧化
3. sustain [sə'stein] v. 支撑、撑住
4. synthesize ['sinθisaiz] v. 综合、合成
5. viz (拉)即, 也就是(读做 namely)
6. via ['vaiə] prep. 通过、经、经由
7. peroxide [pə'rɔksaid] n. 过氧化物
8. electrophilic [i'lektrəu'filik] a. 亲电子的
9. arene ['æri:n] n. 芳烃
10. commence ['kə'mens] v. 开始、着手
11. dissociation [di,səusi'eiʃən] n. 分裂
12. collision [kə'liʒən] n. 碰撞
13. reciprocal [ri'siprəkl] a. 彼此相反的
14. allude [ə'lu:d] v. 间接提到、暗指、影射
15. alicyclic n. 脂环的、脂环族的
16. palladium [pə'leidiəm] n. 钯
17. selenium [si'li:niəm] n. 硒
18. dehydrogenation [di'haidrɔdʒə'neiʃən] n. 脱氢作用
19. decalin n. 十氢化萘
20. naphthalene ['næfθəli:n] n. 萘
21. cleavage ['kli:vidʒ] n. 劈开、劈开部、分裂
22. ozone ['ouzoun] n. 新鲜的空气、臭氧
23. ozonolysis [əuzə'nɔlisis] n. 臭氧分解
24. discern [di'sə:n] n. 辨别

25. diagnostic [daiəg'nɔstik] a. 诊断的
26. dilute [di:'lju:t] n. 稀释
27. extrapolation [ek͵stræpə'leiʃən] n. 外推法、推断
28. reservation [rezə'veiʃən] n. 保留
29. chloroform ['kɔrəfɔ:m] n. 氯仿; vt. 用氯仿麻醉
30. nucleophile ['nju:kliəfail] n. 亲核试剂
31. benzoic [ben'zəuik] n. 安息香的
32. aromatics [ærou'mætiks] a. 芳香剂、芳族、芳族化合物
33. reagent [ri:'eidʒənt] n. 试药、试剂、反应物

Phrases

1. (be) common to 为……所共用(有),对……通用,common 为形容词
2. secondary radical 二级(仲)自由基
3. self-propagating 自动传播(布)的
4. six-membered 六员(环)的
5. place unquestioning faith in... 不提出疑问就相信……
6. with reservation 保留地
7. functional group 功能(基)团,官能团

Note

1. "Rather than" 出现在句首,表示"不是……,而是……"。全句可译为:然而我们不是把每一个反应作为一个孤立的事件来学习,而是辨明通用的反应类型,这样,即使我们以前还没有遇到某一特殊的化合物,但通过反应类型这一手段,我们确实可以预测一类特定的化合物将如何发生反应。

2.7 FREE RADICAL REACTIONS

A free radical is a very reactive species that contains one or more unpaired electrons. Examples are chlorine atoms that are produced when a Cl_2 molecule absorbs a photon(light) of the appropriate energy

$$Cl_2 + light\ energy \longrightarrow 2Cl\cdot$$

(A dot placed next to the symbol of an atom or molecule indicates that it is a free radical.) The high degree of reactivity of free radicals exists because of the tendency of electrons to pair through the formation of either ions or covalent bonds.

Free radicals are important in many gaseous reactions, including those responsible for the production of photochemical smog in urban areas. In biological systems, they appear to be responsible for the aging process as well as many of the harmful effects of radiation. Reactions involving free radicals have useful applications, too. Many plastics are made by polymerization reactions that occur by mechanisms that involve free radicals. In addition, free radicals play a part in one of the most important processes in the petroleum industry, thermal cracking. This reaction is used to break C—C and C—H bonds of long-chain hydrocarbons to produce the smaller molecules that give gasoline a higher octane rating. A simple example is the thermal-cracking reaction of butane. When butane is heated to 700°C to 800°C, one of the major reactions that occurs is

$$CH_3-CH_2 : CH_2-CH_3 \longrightarrow CH_3CH_2 \cdot + CH_3CH_2 \cdot$$

This reaction produces two ethyl radicals, $C_2H_5 \cdot$.

Free radical reactions tend to have high initial activation energies because chemical bonds must be broken to form the radicals. This can be accomplished by light energy or heat. Once the free radicals are formed, however, the chemical reactions in which they are involved tend to be very rapid. Once a free radical is formed, its reactions with other substances tend to have very low activation energies. In many cases, a free radical reacts with a reactant molecule to give a product molecule plus another free radical. Reactions that involve this step are called chain reactions.

Many explosive reactions are chain reactions involving free radical mechanisms. One of the most studied reactions of this type is the formation of water from hydrogen and oxygen. The reactions involved in the mechanism can be described according to their roles in the mechanism. The chain reaction begins with an initiation step that produces free radicals.

$$H_2 + O_2 \xrightarrow{\text{hot surface}} 2OH\cdot \qquad \text{(initiation)}$$

The chain continues with a propagation step, which produces the product plus another free radical.

$$OH\cdot + H_2 \longrightarrow H_2O + H\cdot \qquad \text{(propagation)}$$

The reaction of H_2 and O_2 is explosive because the mechanism also contains branching steps.

$$\left.\begin{array}{l} H\cdot + O_2 \longrightarrow OH\cdot + O\cdot \\ O\cdot + H_2 \longrightarrow OH\cdot + H\cdot \end{array}\right\} \quad \text{(branching)}$$

Thus, the reaction of one $H\cdot$ with oxygen leads to the net production of two $OH\cdot$ plus another $H\cdot$. Every time an $H\cdot$ reacts with oxygen, then, there is an increase in the number of free radicals in the system. The free radical concentration grows rapidly, and the reaction rate becomes explosively fast.

Chain mechanisms also contain termination steps. In the reaction of H_2 and O_2, the wall of the reaction vessel serves to remove $H\cdot$, which tends to stop the chain process.

$$2H\cdot + \text{wall} \longrightarrow H_2 \qquad \text{(termination)}$$

Direct experimental evidence exists for the presence of free radicals in functioning biological systems. These highly reactive species play many roles, but one of the most interesting is their apparent involvement in the aging process. One theory suggests that free radicals attack protein molecules in collagen. Collagen is composed of long strands of fibers of proteins and is found throughout the body, especially in the flexible tissues of the lungs, skin, muscles, and blood vessels. Attack by free radicals seems to lead to cross-linking between these fibers, which stiffens them and makes them less flexible. The most readily observable result of this is the stiffening and hardening of the skin that accompanies aging (or too much sunbathing).

Free radicals also seem to affect fats (lipids) within the body by promoting the oxidation and deactivation of enzymes that are normally involved in the metabolism of lipids. Over a long period, the accumulated damage gradually reduces the efficiency with which the cell carries out its activities. Interestingly, vitamin E appears to be a natural free radical inhibitor. Diets that are deficient in vitamin E produce effects resembling radiation damage and aging.

Still another theory of aging suggests that free radicals attack the DNA in the nuclei of cells. DNA is the substance that is responsible for directing the chemical activities required for the successful functioning of the cell. Reactions with free radicals cause a gradual accumulation of errors in the DNA, which reduce the efficiency of the cell and can lead, ultimately, to malfunction and cell death.

Vocabulary

1. explosive   [iks'plousiv]   n. 炸药、爆炸物; a. 易爆发的、爆炸的、暴躁的
2. initiation   [iniʃi'eiʃən]   n. 开始、启蒙、入会
3. propagation [prɔpə'geiʃən]   n. 增值、繁殖、广传
4. termination [tə:mi'neiʃən]   n. 终止
5. collagen   ['kɔlədʒən]   n. 胶原质
6. strand   [strænd]   n. 绳、线、串、海滨、河岸; vi. 搁浅; vt. 使搁浅、使落后、弄断、搓
7. lipid   [lipid]   n. 胶质、油脂
8. enzyme   ['enzaim]   n. (生化)酶
9. metabolism   [me'tæblizm]   n. 新陈代谢、变形
10. malfunction ['mæl'fʌŋkʃən]   n. 故障、失灵; vi. 发生故障、不起作用

Phrases

1. photochemical smog　光化学烟雾
2. octane rating = octane number　辛烷值
3. flelxible tissue　软组织
4. cross-linking　交联(作用)

2.8　LAWS OF THERMODYNAMICS

2.8.1　The First Law of Thermodynamics

We have introduced the terms heat and work, and also mentioned internal energy. We need to say more about this concept. First, you need to understand that a system does not contain energy in the form of heat or work. Heat and work are the means by which a system exchanges energy with its surroundings. Heat and work only exist during a change. The energy contained within a system is the internal energy, E. This is the total energy (kinetic and potential) associated with chemical bonds, intermolecular attractions, kinetic energy of molecules, and so on. The relationship involving heat (q), work (W), and changes in internal energy (ΔE) is dictated by the law of conservation of energy, also known as the first law of thermodynamics.

$$\Delta E = q + W$$

In using the above equation keep these important points in mind.

· Any energy entering the system carries a positive sign. Thus, if heat is absorbed by the system, $q > 0$. If work is done *on* the system, $W > 0$.

· Any energy leaving the system carries a negative sign. Thus, if heat is given off by the system, $q < 0$. If work is done by the system, $W < 0$.

2.8.2　The Second Law of Thermodynamics

One of the most far-reaching observations in science is incorporated into the second law of thermodynamics, which states, in effect, that whenever a spontaneous event takes place in our universe, it is accompanied by an overall increase in entropy. Note that the increase in entropy that's referred to here is for the universe, not just the system. This means that a system's entropy can decrease, just as long as there is a larger increase in the entropy of the surroundings so that the net entropy change is positive. Because everything that happens relies on spontaneous changes of some sort, the entropy of the universe is constantly increasing.

In other section we will see how the Second Law translates into concrete mathematical statements that we can use in analyzing the spontaneity of events. But for the moment, let's take a qualitative look at the "big picture" to see how the Second Law influences our daily activities. In particular, let's examine what the Second Law tells us about pollution and our efforts to control it. Can we avoid pollution of our environment entirely? Why are pollutants so easily spread and so difficult to eliminate? These are the kinds of questions we would like to answer.

Much of our time and effort is spent it tidying up our own home and neighborhood. We seem constantly involved in trying to decrease disorder in our lives. We clean up our desks and sweep the floor. We put out the garbage, mow the lawn, and rake leaves in the autumn. Overall our activities are spontaneous because the biochemical reactions driven by the foods we eat and digest are spontaneous and allow us to do these things. Since the total entropy of the universe must be increased by our spontaneous activities, the increased order that we create for ourselves has to be balanced by an even larger increase in disorder somewhere in our surroundings—the environment.

Pollution involves the scattering of undesirable substances through our environment and is accompanied by an increase in entropy. It is a direct result of our efforts to create an orderly world. For example, if we burn our leaves to get rid of them, the smoke from our fire spreads through the neighborhood and disturbs our friends. When we expand our efforts beyond our home and attempt to provide order in our environment, our attempts are met by still more overall disorder. For instance, we use machinery to collect garbage so that we and our neighbors don't have to live in it. But the machinery, by burning fuel, generates air pollution that is even more difficult to clean up. It's a losing game. We are simply faced with the fact that we can never really eliminate pollution. All we can do is to try to keep it from places where it can do us great harm and attempt to confine it to locations where it interferes with life as little as possible[1].

The entropy effect on pollution is important to understand, particularly when we consider the consequences of releasing harmful substances into the environment. Such materials include the highly toxic elements released in trace amounts when coal or nuclear fuels are used, as well as many toxic chemicals such as DDT and dioxin. When these chemicals are released, their spread is unavoidable because of the large entropy increase that occurs as they are scattered about. Eliminating them, once they have had an opportunity to disperse, is an almost impossible task because doing so requires an enormous expenditure of energy and must generate even more disorder around us. Only if the pollutant is extremely hazardous does it warrant the effort necessary to reduce its concentration in the environment, and even if this is done, we can never eliminate the pollutant entirely[2]. We can only reduce our risk of injury. The surest way to overcome pollution, therefore, is not to create it in the first place. It is a trade-off between risk and benefit. For the benefits of activities that pollute (e.g., using fuels to make electricity), we try to reduce the risks as much as we judge possible and endure the rest.

2.8.3 The Third Law of Thermodynamics

The entropy of a substance (i.e., the extent of its thermodynamic disorder) varies with the temperature of the substance—the lower the temperature, the lower the entropy. For example, at a pressure of 1 atm and a temperature above 100°C, water exists as a highly disordered gas with a very high entropy. If confined, the molecules of water vapor will be spread evenly throughout their container, and they will be in constant random motion. When the system is cooled, the water vapor eventually condenses to form a liquid. Although the molecules can still move somewhat freely, they are now confined to the bottom of the container. Their distribution in the container is not as random as it was in the gas and the entropy of the liquid is lower. Further cooling decreases the entropy even more, and below 0°C, the water molecules join together to form ice, a crystalline solid. The molecules are now in a highly ordered state, particularly in comparison to that of the gas, and the entropy of the system is very low.

Yet even in the crystalline form, the order isn't perfect and the water molecules still have some entropy. There is enough thermal energy left to cause them to vibrate or rotate within the general area of their lattice sites. Thus at any particular instant, we would find the molecules near, but probably not exactly at, their equilibrium lattice positions. If we cool the solid further, we decrease the thermal energy and the molecules spend less time away from their equilibrium lattice positions. The order of the crystal increases and the entropy decreases. Finally, at absolute zero, the ice will be in a state of perfect order and its entropy will be zero. This leads us to the statement of the third law of thermodynamics. At absolute zero, the entropy of a pure crystal is also zero.

$$S = 0 \quad \text{at} \quad T = 0 \text{ K}$$

Because we know the point at which entropy has a value of zero, it is possible by measurement and calculation to determine the actual amount of entropy that a substance possesses at temperatures above 0 K. If the entropy of one mole of a substance is determined at a temperature of 298 K (25°C) and a pressure of 1 atm, we call it the standard entropy, $S^{\ominus}$.

Entropy has the dimensions of energy/temperature (i.e., joules per kelvin or calories per kelvin). Once we know the entropies of a variety of substances we can calculate the standard entropy change, $\Delta S^\ominus$, for chemical reactions in much the same way as we calculated $\Delta H^\ominus$ previously.

$$\Delta S^\ominus = (\text{sum of } S^\ominus \text{ of products}) - (\text{sum of } S^\ominus \text{ of reactants})$$

If the reaction with which we are working corresponds to the formation of 1 mol of a compound from its elements, the $\Delta S^\ominus$ that we calculate can be referred to as the standard entropy of formation, $\Delta S_f^\ominus$. Values of $\Delta S_f^\ominus$ are not tabulated; if we need them, we must calculate them from values of $S^\ominus$.

Vocabulary

1. mow [muː] v. 扫除
2. rake [reik] v. 用耙子耙
3. dioxin [daiˈɔksin] n. 二氧(杂)芑
4. expenditure [ekˈspenditʃə] n. 支出
5. hazardous [ˈhæzədəs] a. 危险的
6. warrant [ˈwɔrənt] v. 保证
7. vibrate [vaiˈbreit] v. 振动
8. lattice [ˈlætis] n. 格子、组合格子、格状物；
   vt. 制成格子、使成格子状
9. joule [dʒuːl] n. 焦耳

Phrases

1. incorporate A in(into) B 把 A 加(引、插)到 B
2. as long as 只要……，长达……之久
3. trace amount 痕量
4. lattice site 晶格内位置,(晶)格点
5. at any instant 在任何条件下,随时
6. standard entropy 标准熵

standard entropy of formation　　标准生成熵

Notes

1. 句中有两个不定短语作并列的表语,即"to try to..."和"(to) attempt to..."。全句可译为:我们能做的一切就是设法使它(污染)远离对我们能产生危害的地方和力图把污染限制在对生存妨碍尽可能小的地方。
2. "only"在句首,句中主谓语倒装。"only if..."是"只有当……"的意思;"it"为形式主语,真正的主语为"to reduce..."不定式短语。"warrant the effort necessary"可译作:保证这种工作是必要的。

3

Some Techniques of Chemistry and Chemical Engineering

3.1 FLUID FLOW

Given that a processing plant is a network of pipes and vessels, it is clearly important to be able to size every pump and all of the pipes. Thus techniques to calculate the pressure drop between the ends of each pipe are important. As detailed later, there are two main types of flow which will simply be called 'ordered streamline' and 'chaotic turbulent' for the moment. Detailed knowledge of the flow pattern is unimportant when calculating the pressure drop. Knowledge of pipe size, viscosity, density and velocity of the fluid enable the engineer to select the appropriate equation.

Fluid flow affects the performance of numerous pieces of process equipment and some of the examples mentioned in subsequent chapters are heat transfer, distillation, gas absorption and membrane filtration.

3.1.1 Types of Fluid

Fluids may be classified in two different ways; either according to the behaviour under the action of externally applied pressure, or according to the effects produced by the action of a shear stress.

If the volume of an element of fluid is independent of its pressure and temperature, the fluid is said to be incompressible; if its volume changes it is said to be compressible. No real fluid is completely incompressible but liquids may generally be regarded as such when their flow is considered. Gases have a very much higher compressibility than liquids, and appreciable changes in volume may occur if the pressure or temperature is altered. However, if the percentage change in the pressure or in the absolute temperature is small, for practical purposes a gas may also be regarded as incompressible. Thus, in practice, volume changes are likely to be important only when the pressure or temperature of a gas changes by a large proportion. The relation between pressure, temperature, and volume of a real gas is generally complex, but except at very high pressures the behaviour of gases approximates to that of the ideal gas for which the volume of a given mass is inversely proportional to the pressure and directly proportional to the absolute temperature. At high pressures and when pressure changes are large, however, there may be appreciable deviations from this law and an approximate equation of state must then be used.

The behaviour of a fluid under the action of a shear stress is important in that it determines the way in which it will flow. The most important physical property affecting the stress distribution within the fluid is its viscosity. For a gas, the viscosity is low and even at high rates of shear, the viscous stresses are small. Under such conditions the gas approximates in its behaviour to an inviscid fluid. In many problems involving the flow of a gas or a liquid, the viscous stresses are important and give rise to appreciable velocity gradients within the fluid, and dissipation of energy occurs as a result of the frictional forces set up. In gases and in most pure liquids the ratio of the shear stress to the rate of shear is constant and equal to the viscosity of the fluid. These fluids are said to be Newtonian in their behaviour. However, in some liquids, particularly those containing a second phase in suspension, the ratio is not constant and the apparent viscosity of the fluid is a function of the rate of shear. The fluid is then said to be non-Newtonian and to exhibit rheological properties.

Viscosity is a measure of fluid friction and is proportional to the force required to move a layer of fluid over another layer. Highly viscous materials are those that possess a great deal of internal friction when layers are in relative motion—they cannot be spread or poured as easily as less viscous materials. Consider a model in which two parallel planes of fluid of equal area A, separated by a distance d, are moving in the same direction but at different speeds. The force F which is necessary to maintain this difference is given, for many fluids, by the following equation which was introduced by Sir Isaac Newton

$$\frac{F}{A} = \mu \frac{\Delta v}{d}$$

where Δv is the difference in velocity and μ is a constant for a given material.

The velocity gradient $\Delta v/d$ is a measure of the rate with which velocity changes with distance and it measures the shearing that the fluid experiences. The term is often called the 'rate of shear'. The force per unit area that is required for the maintenance of the shearing action, F/A, is known as the shear stress. A 'stress' is a force per unit area and has the same units as pressure.

At a given temperature, viscosity is, for many fluids, independent of the rate of shear, and if $\Delta v/d$ is doubled then the shear stress F/A is also doubled. Fluids of this type are known as 'Newtonian' and are best exemplified by gases and liquids such as water, petrol, thin motor oils and milk.

For example, as milk is concentrated, the nature of the fluid changes. At low concentrations it behaves as a Newtonian fluid with viscosity being constant for all rates of shear. However, above 25 per cent protein, the material is particularly viscous at low rates of shear. This fact is very important in cheese manufacture and those designing dairy plants have to ensure that a minimum rate of shear is maintained in the processing equipment which will include heat exchangers and maybe membrane units. The above behaviour, known as shear-thinning or pseudoplasticity is typical of polymer solutions and liquids with a second phase in suspension. Their behaviour can often be represented by an equation of the form

$$\frac{F}{A} = C\left(\frac{\Delta v}{d}\right)^n$$

where C is a consistency constant and n is less than unity.

There is a further group of fluids which show time-dependent behaviour. The viscosity of these materials is affected by the amount of shearing applied in the recent past to the material. Often the viscosity will decrease with time during shear but recover, sometimes slowly, when the shear stress is removed. This particular behaviour is termed thixotropic and some paints have this property.

If a fluid subjected to shear suffers an irreversible decrease in viscosity then shear breakdown or rheodestruction is said to have occurred.

Clearly, the flow properties of fluids being processed are very important. Non-newtonian behavior is exhibited by a wide range of industrially important liquids and knowledge of this subject is and will be of increasing importance, not least because many bioprocessing liquids are non-Newtonian.

3.1.2 Nature of Fluid Flow

Consider the apparatus shown in figure 3.1. Will the dyed water that is introduced via the needle form a coloured filament as shown in figure 3.2(a) or will it be dispersed across the whole cross-section of the pipe as shown in figure 3.2(b)? The former corresponds to the ordered streamline state mentioned earlier and is cha-racterised by the absence of convective movement in the radial direction. Thus every fluid element reaches, once any initial disturbances have been dampened, a constant velocity that is parallel to the pipe axis[1].

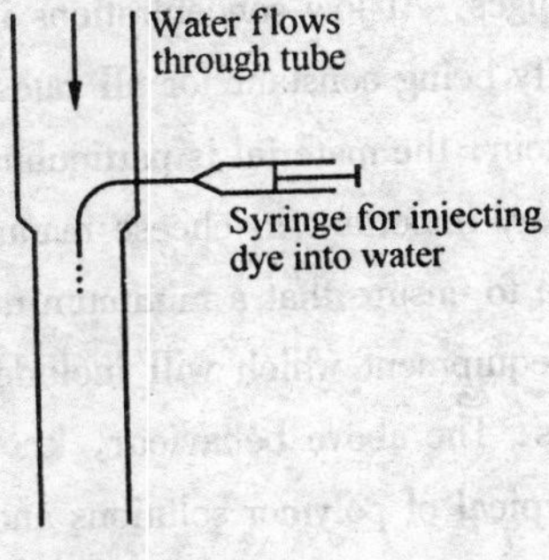

At the pipe wall it is known that there is no slip (that is, the molecules adjacent to the solid surface are stationary) and so the velocity at the wall is zero. Given this boundary condition, the solution of the appropriate equation shows that there is a parabolic variation of velocity u with radial positition

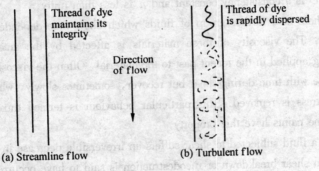

Figure 3.1 Schematic diagram of apparatus for investigating the nature of fluid flow

Figure 3.2 Illustration of the two main types of flow

$$\frac{u}{u_{av}} = 2(1 - \frac{r^2}{(d/2)^2})$$

where u_{av} — volumetric flow rate/cross-section area;

r — radial co-ordinate measured from the axis;

d — pipe diameter.

If the above flow is disturbed by addition of a roughened section of pipe or a valve, it will either revert to the ordered state of streamline or laminar flow, or the induced oscillations become stable and a high degree of radial mixing occurs. In this turbulent state a rapid random motion is imposed on top of the time averaged motion. The latter is given by the following equation.

$$\frac{u}{u_{av}} = 1.22\left(1 - \frac{r}{(d/2)}\right)^{1/7}$$

The turbulent velocity profile is flatter than that for streamline flow and the maximum velocity is only 22 per cent greater than the average velocity. However, the main difference is the rapid random motion which rapidly mixes the fluid elements. If the pipe is part of a heat exchanger in which the fluid in the pipe is being heated by, for example, steam condensing on the outside of the pipe, then the mixing process exchanges hot fluid elements from the wall area with cold fluid elements from the central area. Thus the fluid rapidly receives heat from the condensing steam. The corollary of this desirable state of affairs is that the designer of process equipment often seeks to induce turbulent conditions.

The flow pattern varies with fluid velocity, u, density, ρ, and viscosity, μ, of the fluid and the geometry of the system. For flow in a pipe the characteristic dimension is the diameter d and the viscous forces can be taken to be proportional to $\mu u/d$, which is of the same form as the above equation. By assuming that the flow pattern is determined by the ratio of inertial forces (which are proportional to ρu^2) to viscous forces, it can be shown that the key parameter is $\rho u d/\mu$. This is dimensionless and is known as the Reynolds number after a famous nineteenth century experimentalist who established that this was the key criterion for flow in pipes. Below a Reynolds number of 2 000 the flow is streamline. Above a value of 3 000 it is turbulent.

With regard to pipe flow it is irrelevant whether the pipe is vertical or horizontal, and gravitational forces do not affect the nature of the flow pattern. In mixing vessels on the other hand, gravitational, inertial and viscous forces can all be important and the value of the gravitational constant, g, has to be taken into account. The resulting analysis gives rise to two dimensionless ratios: the Reynolds number introduced above and a second dimensionless number which represents the ratio of inertial to gravitational forces. More dimensionless groups will be met later on and they are favoured by engineers and mathematicians precisely because they are dimensionless and this greatly helps in the formulation of general relationships. For example, the values of 2 000 and 3 000 mentioned above apply to all fluids and all pipe sizes and one would confidently expect the same result in a zero gravity environment.

Vocabulary

1. streamline ['striːmlain] *n.* 流线、流线型；
   vt. 使成流线型、使合理化
2. turbulent ['təːbjulənt] *a.* 狂暴的、吵闹的
3. compressible [kəm'presəbl] *a.* 可压缩的
4. deviation [diːvi'eiʃən] *n.* 背离
5. gradient ['greidiənt] *n.* 梯度；*a.* 倾斜的
6. viscosity [vis'kɔsiti] *n.* 粘质、粘性
7. distillation [disti'leiʃən] *n.* 蒸馏
8. filtration [fil'treiʃən] *n.* 过滤、筛选
9. fluid ['fluː(ː)id] *n.* 流体、液体；*a.* 流动的、流体的、不固定的、易变的、流利的、流畅的
10. inviscid [in'visid] *a.* 非粘(滞)性的、无韧性的、不能延展的
11. dissipation [disi'peiʃən] *n.* 消散、分散、浪费
12. Newtonian [njuː'tounjən] *a.* 牛顿的、牛顿学说的；
    n. 信仰牛顿学说的人
13. rheologic [riːɔ'lɔdʒik] *a.* 流变的
14. parallel ['pærəlel] *n.* 平行、对比；*a.* 平行的、相似的；
    vt. 与……平行、与……相似、相比

15. shear	[ʃiə]	n. 修剪、剪下的东西；
		vt. 修剪、割、剥夺、切
16. pseudo	['psju:dou]	a. 假的、冒充的
17. thixotropic	[θik'sɔtrəpik]	a. 触变的；触变性的
18. irreversible	[iri'və:səbl]	a. 不能倒逆的
19. rheostat	['ri:əstæt]	n. 可变电阻器
20. rheo		[构词成分]表示"流"
21. rheodestruction		n. 流变破坏
22. filament	['filəmənt]	n. 细丝、细线、灯丝单纤维
23. radial	['reidiəl]	a. 放射状的、半径的、径向的；
		n. 光线、射线
24. dampen	['dæmpən]	vt. 弄湿、使衰减；vi. 变湿、丧气
25. axis	['æksis]	n. 坐标轴、轴、轴线、轴心
26. parabolic	[pærə'bɔlik]	a. 抛物线的、抛物线状的
27. laminal	['læminəl]	a. 成薄板[薄片]状的
28. oscillation	[ɔsi'leiʃən]	n. 振动、动摇、变动
29. profile	['proufail]	n. 侧面、轮廓；vt. 描绘……轮廓
30. corollary	['kɔ:rɔləri]	n. 系、推论、必然的结果
31. geometry	[dʒi'ɔmitri]	n. 几何、几何形状、几何学
32. inertial	[i'nə:ʃəl]	a. 不活泼的、惯性的

Phrases

1. fluid flow   液体流动
2. shear stress   剪切作用(力)
3. velocity gradient   速度梯度
4. shear-thinning or pseudoplasticity   假塑性
5. cross-section   横截面
6. convective movement   对流运动
7. co-ordinate   同等者,同等物,坐标(用复数)

Note

1. Thus every fluid element reaches,...to the pipe axis. 句中"a constant velocity"

作"reach"的宾语。

3.2 MOMENTUM, HEAT AND MASS TRANSFER

3.2.1 Introduction

In most of the unit operations encountered in the chemical and petroleum industries, one or more of the processes of momentum, heat, and mass transfer is involved. Thus, in the flow of a fluid under adiabatic conditions through a bed of granular particles, a pressure gradient is set up in the direction of flow and a velocity gradient develops approximately perpendicularly to the direction of motion in each fluid stream; momentum transfer takes place between the fluid elements which are moving at different velocities. If there is a temperature difference between the fluid and the pipe wall or the particles, heat transfer will take place as well, and the convective component of the heat transfer will be directly affected by the flow pattern of the fluid. Here, then, is an example of a process of simultaneous momentum and heat transfer in which the same fundamental mechanism is affecting both processes. Fractional distillation and gas absorption are frequently carried out in a packed column in which the gas or vapour stream rises countercurrently to a liquid. The function of the packing in this case is to provide a large interfacial area between the phases and to promote turbulence within the fluids. In a very turbulent fluid, the rates of transfer per unit area of both momentum and mass are high; and as the pressure rises the rates of transfer of both momentum and mass increase together. In some cases, momentum, heat, and mass transfer all occur simultaneously as, for example, in a water-cooling tower where transfer of sensible heat and evaporation both take place from the surface of the water droplets. It will now be shown not only that the process of momentum, heat, and mass transfer are physically related, but also that quantitative relations between them can be developed.

Another form of interaction between the transfer processes is responsible for the phenomenon of thermal diffusion in which a component in a mixture moves under the action of a temperature gradient[1]. Although these are important applications of thermal diffusion, the magnitude of the effect is usually small relative to that arising from concentration gradients[2].

When a fluid is flowing under streamline conditions over a surface, a forward component of velocity is superimposed on the random distribution of velocities of the molecules, and movement at right angles to the surface occurs solely as a result of the random motion of the molecules. Thus if two adjacent layers of fluid are moving at different velocities, there will be a tendency for the faster moving layer to be retarded and the slower moving layer to be accelerated by virtue of the continuous passage of molecules in each direction. There will therefore be a net transfer of momentum from the fast to the slow-moving stream. Similarly, the molecular motion will tend to reduce any temperature gradient or any concentration gradient if the fluid consists of a mixture of two or more components. At the boundary the effects of the molecular transfer are balanced by the drag forces at the surface.

If the motion of the fluid is turbulent, the transfer of fluid by eddy motion is superimposed on the molecular transfer process. In this case, the rate of transfer to the surface will be a function of the degree of turbulence. When the fluid is highly turbulent, the rate of transfer by molecular motion will be negligible compared with that by eddy motion. For small degrees of turbulence the two may be of the same order.

It was shown in the previous chapter that when a fluid flows under turbulent conditions over a surface, the flow can conveniently be divided into three regions:

① At the surface, the laminar sub-layer, in which the only motion at right angles to the surface is due to molecular diffusion.

② Next, the buffer layer, in which molecular diffusion and eddy motion are of comparable magnitude.

③ Finally, over the greater part of the fluid, the turbulent region in which eddy motion is large compared with molecular diffusion.

In addition to momentum, both heat and mass can be transferred either by molecular diffusion alone or by molecular diffusion combined with eddy diffusion. Because the effects of eddy diffusion are generally far greater than those of the molecular diffusion, the main resistance to transfer will lie in the regions where only molecular diffusion is occurring. Thus the main resistance to the flow of heat or mass to a surface lies within the laminar sub-layer. It has been shown that the thickness of the laminar sub-layer is almost inversely proportional to the Reynolds number for fully developed turbulent flow in a pipe. Thus the heat and mass transfer coefficients are much higher at high Reynolds numbers.

3.2.2 Heat transfer

In the majority of chemical processes, heat is either given out or absorbed, and in a very wide range of chemical plant, fluids must often be either heated or cooled. Thus in furnaces, evaporators, distillation units, driers, and reaction vessels one of the major problems is that of transferring heat at the desired rate. Alternatively, it may be necessary to prevent the loss of heat from a hot vessel or steam pipe. The control of the flow of heat in the desired manner forms one of the most important sections of chemical engineering. Provided that a temperature difference exists between two parts of a system, heat transfer will take place in one or more of three different ways.

① Conduction. In a solid, the flow of heat by conduction is the result of the transfer of vibrational energy from one molecule to another, and in fluids it occurs in addition as a result of the transfer of kinetic energy. Heat transfer by conduction may also arise from the movement of free electrons. This process is particularly important with metals and accounts for their high thermal conductivities.

② Convection. Heat transfer by convection is attributable to macroscopic motion of the fluid and therefore is confined to liquids and gases. In natural convection, it is caused by differences in density arising from temperature gradients in the system. In forced convection, it is due to eddy currents in a fluid in turbulent motion.

③ Radiation. All materials radiate thermal energy in the form of electromagnetic waves. When this radiation falls on a second body it may be partially reflected, transmitted, or absorbed. It is only the fraction that is absorbed that appears as heat in the body.

The three subjects of heat, mass and momentum transfer are interlinked; heat flows down a temperature gradient, mass flows down a concentration gradient and momentum passes down a velocity gradient. These inter-relationships, which are firmly based on fundamental equations, are useful. It is important to realise that the three subjects mentioned above are all interdependent and an umbrella term 'transport phenomena' is often used to group them together.

On a chemical engineering course, the subject of fluid flow (which at a fundamental level is related to momentum transfer) will include not only the flow of gases and the flow of liquids but also two-phase flows such as fluid-solid and gas-liquid flows. An understanding of gas-liquid flow is important if the operation of boilers (and every distillation column has a boiler to provide the upward-flowing vapour) is to be understood.

In food industry the inter-relationship between micro-organism growth, temperature and time is of vital importance. For the destruction of micro-organisms and the inactivation of enzymes, heat treatment may well be selected. Other possibilities include chemicals and ionising radiation.

The importance of heat transfer to chemical engineering will be illustrated by returning briefly to study of sulphuric acid process. In this process there are three vital heat transfer operation:

① removal of heat from the SO_2 rich inlet stream.

② removal of heat from the reactor.

③ removal of the heat of absorption.

In the first two cases, steam is raised by using the high-grade (high-temperature) heat to evaporate water in a boiler. This simultaneously cools the gas stream to reactor temperature and allows the heat to be used elsewhere. This last statement may seem surprising, but steam is an excellent carrier of heat; for every 1 kg condensing, about 2 000 kJ of heat (the exact value is pressure dependent) are released. Thus a steam flow of 1 kg $\cdot$ s^{-1} is equivalent to a useful energy flow of over 2.0 MW. Not only is a unit mass of steam able to store and carry a large quantity of heat, but it does so at a temperature at which it can conveniently be used. It is also worth stating the obvious— it is generated from water, which is harmless and readily available.

The most widely used heat exchangers in the chemical and most other process industries are the various shell and tube units. The advantages of shell and tube heat exchangers are:

① Good mechanical layout; good shape for pressure operation.

② Can be constructed from a variety of materials, including glass and plastic.

③ Uses well-established fabrication techniques.

④ Easily cleaned if straight tubes used.

⑤ Reasonably large surface area per unit volume.

Generally, the materials of construction are metal. However, a few companies offer glass and 'Teflon' exchangers for use with particularly corrosive fluids. Another special type for use with corrosive fluids is the graphite block heat exchanger, which is literally graphite blocks with a matrix of holes. The fluids are generally in cross-flow and although there is a relatively large amount of 'wall' material between the fluid streams, the thermal conductivity of graphite is high and the resultant thermal resistance small.

Plate heat exchangers are of increasing importance. The exchangers consist of a frame, a fixed end plate and a movable plate between which corrugated metal plates, gasketed at the edges, are fixed. A port at each corner is either open or blind, according to need. A variety of gasket materials is used and the temperature range is now −40 to 200°C. Operating pressures of up to 25 bar can also be accommodated.

The modular design enables the exact size and number of plates to be used according to the level of heat transfer required. The heat exchangers are compact and for many duties a plate heat exchanger in stainless steel is cheaper than the equivalent shell and tube unit in carbon steel. While plate heat exchangers were originally developed and used as coolers in the dairy industry, they are now much more versatile. Large types can handle flow rates of 400 t of liquid per hour, while at the other end of the scale, small units recover heat in laundries, heating and ventilation systems and swimming pools.

3.2.3 Mass Transfer

When a concentration gradient exists within a fluid consisting of two or more components, there is a tendency for each constituent to flow in such a direction as to reduce the concentration gradient; this process is known as mass transfer. In a still fluid, or in a fluid flowing under streamline conditions in a direction at right angles to the concentration gradient, the transfer is effected as a result of the random motion of the molecules. In a turbulent fluid this mechanism is supplemented by transference of material by eddy currents.

Mass transfer can take place in either a gas phase or a liquid phase, or in both simultaneously. When a liquid evaporates into a still gas, vapour is transferred from the surface to the bulk of the gas as a result of the concentration gradient; the process continues until the whole of the liquid has evaporated or until the gas is saturated and the concentration gradient reduced to zero. In the absorption of a soluble gas from a mixture with an insoluble gas, mass transfer takes place from the bulk of the gas to the liquid surface and then into the bulk of the liquid. Neither the insoluble gas nor the solvent moves in the direction of mass transfer. In a distillation column, on the other hand, the less volatile component diffuses in the gas phase to the liquid surface and the more volatile material diffuses at an approximately equal molar rate in the opposite direction. In the liquid phase a similar process takes place, with the less volatile material diffusing away from the gas-liquid interface.

The rate of transfer of A in a mixture of two components A and B will therefore be determined not only by the rate of diffusion of A but also by the behaviour of B. The molar rate of transfer of A, per unit area, due to molecular motion is given by

$$N_A = - D_{AB} \frac{dC_A}{dy}$$

where N_A is the molar rate of diffusion per unit area, D_{AB} the diffusivity of A in B, a physical property of the two vapours, C_A the molar concentration of A, and y the distance in the direction of diffusion.

The corresponding rate of diffusion of B is given by

$$N_B = - D_{BA} \frac{dC_B}{dy}$$

where D_{BA} is the diffusivity of B in A and C_B is the molar concentration of B.

The above equation is often referred to as Fick's law. Fick in 1855 derived by analogy with heat transfer that $\partial C_A / \partial t = D(\partial^2 C_A / \partial y^2)$ from which equation is obtained when $\partial C_A / \partial t = 0$.

If the total pressure, and hence the total molar concentration, is everywhere constant, dC_A/dy and dC_B/dy must be equal and opposite, and therefore A and B tend to diffuse in opposite directions. In a distillation process where the two components have equal molar latent heats, condensation of one mole of the less volatile material releases just sufficient heat for the vaporisation of one mole of the more volatile component, and therefore equimolecular counterdiffusion takes place with two components diffusing at equal and opposite rates, as determined by Fick's law. In an absorption process there is net transfer of only one of the components although there is a concentration difference of the other. It is therefore necessary to study the two cases separately.

In many processes B will neither remain stationary nor will it diffuse at an equal and opposite molar rate to A. Exact calculations relating to this type of problem are difficult. A example of this nature is the mass transfer in a distillation column when the two components have unequal molar latent heats.

When the fluid is turbulent, eddy diffusion takes place in addition to molecular diffusion and the rate of diffusion is increased and

$$N_A = -(D_{AB} + E_D)(dC_A/dy)$$

where E_D is known as the eddy diffusivity. E_D will increase as the turbulence is increased and is more difficult to evaluate than the molecular diffusivity.

Vocabulary

1. adiabatic [ˌædiə'bætik] a. 绝热的、隔热的
2. granular ['grænjulə] a. 由小粒而成的、粒状的
3. perpendicularly [ˌpə:pən'dikjuləli] ad. 垂直地
4. convective [kən'vektiv] a. 传达性的、对流的
5. countercurrent ['kauntəˌkʌrənt] n. 反向电流
6. interfacial [ˌintə'feiʃəl] a. 界面的
7. adjacent [ə'dʒeisənt] a. 毗邻的、邻接的、靠近的、贴近的
8. retard [ri'ta:d] n. 阻止、延迟; vt. 妨碍、延

9. eddy	['edi]	n.	逆流、漩涡；
		vt.	(使)起漩涡
10. superimpose	['sjuːpərim'pouz]	v.	重叠、加上去、双重
11. buffer	['bʌfə]	n.	缓冲区；vt. 缓冲
12. drier	['draiə]	n.	干燥工、干燥剂
13. macroscopic	[mækrou'skɔpik]	a.	肉眼可见的
14. electromagnetic		a.	电磁的、由电磁石产生的
15. logarithmic	[lɔːgə'riθmik]	a.	对数的
16. graphical	['græfikəl]	a.	图解的、图表的、生动的
17. watt	[wɔt]	n.	瓦特
18. iterative	['itərətiv]	a.	反复的；n. 反复相
19. matrix	['meitriks]	n.	矩阵、模型
20. corrugate	['kɔrugeit]	vt.	使起皱、成波状；vi. 缩成皱状；a. 有皱的、起皱的、波状的
21. gasket	[gæskit]	n.	束帆索、衬垫、垫圈
22. modular	['mɔdjulə]	a.	模的、有标准组件的
23. versatile	['vəːsətail]	a.	多才多艺的、万用的、万向的
24. laundry	['lɔːndri]	n.	洗衣房、洗衣店、洗好的衣服
25. ventilation	[venti'leiʃən]	n.	通风、空气流通
26. humidify	[hjuː'midifai]	vt.	使潮湿
27. ferroconcrete	['ferou'kɔnkriːt]	n.	钢筋混凝土、钢骨水泥
28. grid	[grid]	n.	格子、栅极
29. hyperboloid	[hai'pəːbəlɔid]	n.	双曲线体
30. equimolecular	[iːkwiməu'lekjulə]	a.	等分子的

Notes

1. Another form of interaction ... the action of a temperature gradient. 句中"be responsible for"意为"对……负责，形成……的原因。"
2. Although these are important... from concentration gradients. 全句可译为：尽管

这些都是热扩散的重要应用,但相对于由浓度梯度引起的增加值,这些效应的增加值通常都是很小的。

3.3 PROCESS DESIGN

A process plant is a network of interconnected vessels joined by pipes. Obviously it is important to calculate the flows through every vessel and pipe so that they can be correctly sized. This must be done not only for the conditions which exist during normal production but also for those conditions which appertain during the start-up period and for those conditions which could occur in the event of an emergency shut-down. Certain pipes and vessels are specified solely for the latter purpose. Although it is hoped that it will not be necessary to use the emergency system, periodic testing of it (particularly the associated controls) is essential.

Knowledge of the mass and volumetric flow rates for each part of the plant results from solution of the material balance and simple cases will be examined shortly[1]. Energy balances are also important since questions such as 'What is the thermal duty?' 'How much steam is required?' and 'Will we have sufficient cooling water?' need to be answered.

Material and energy balances thus lie at the very centre of process design. There is generally no 'right' answer, and while some of the possibilities can be eliminated on the grounds of safety or operability (ease of control), the design group is often left with a degree of choice. Final selection is based on economic considerations and company preferences.

Material and energy balances are also useful when studying plant operation. For example, the task could be to assess performance of a process such as crude oil distillation. Alternatively it might be the location of the cause of excessive material or energy loss. In subsequent sections the fundamentals are introduced together with some simple examples.

3.3.1 Material Balances: General

The balances result simply from the application of the principle of the conservation of matter to a particular unit operation or group of processing units. It is readily understood that

$$\begin{pmatrix}\text{Total amount of}\\ \text{matter into vessel}\\ \text{per second}\end{pmatrix} = \begin{pmatrix}\text{Total amount of}\\ \text{matter out of vessel}\\ \text{per second}\end{pmatrix} + \begin{pmatrix}\text{Rate of accumulation}\\ \text{of total matter}\\ \text{in vessel}\end{pmatrix} \quad (3.1)$$

The last term is a rate or 'per second' term and would typically have units of $kg \cdot s^{-1}$ as would the other terms.

If there is no reaction, the identity of each molecule is maintained and equation (3.1) can be applied to each component in the feed streams. Hence for a general component, i, which might for example, be oxygen, ammonia, ethanol or propane.

$$\begin{pmatrix}\text{Amount of}\\ \text{component } i\\ \text{into vessel}\\ \text{per second}\end{pmatrix} = \begin{pmatrix}\text{Amount of}\\ \text{component } i\\ \text{out of vessel}\\ \text{per second}\end{pmatrix} + \begin{pmatrix}\text{Rate of}\\ \text{accumulation}\\ \text{of component } i\\ \text{in vessel}\end{pmatrix} \quad (3.2)$$

Often the accumulation term can be neglected but sometimes it is vital. To understand when it is of importance, it is necessary to understand that processes fall into two main categories: batchwise processes and continuous processes. The unit operation of distillation can serve as a specific example. Whisky distilling is done on a relatively small scale and in a very traditional manner. A still is charged with a batch at a time and each batch is distilled consecutively. The condensed vapour varies in composition with the passage of time. There is a selective removal of the more volatile components and the concentration of these components in the liquor, from which the vapours arise, decreases with increasing time. The whisky industry traditionally produces the distillate in three distinct and successive stages. The liquor coming over early is known as foreshots, the second phase produces spirits while the last phase yields a liquor known as feints which is low in alcohol.

In an oil refinery and many parts of the chemical industry, the quantities handled are so large that it would be barely feasible and certainly uneconomic to distil one batch at a time. Thus the distillation process is run continuously. After the start-up phase has been completed, the feed rate F kg·s^{-1} will equal the sum of the product rates. Furthermore, once this steady state has been achieved, the compositions at every level in the column and in the product lines will be constant provided the feed composition and rate remain fixed.

Thus for all continuous processes at steady state, equation (3.1) can be written as

Mass flow in per unit time = Mas flow out per unit time   (3.3)

If in addition, there is no reaction, the following holds for each component i

Mass flow of component i in per unit time = Mass flow of component i out per unit time   (3.4)

These common-sense laws are as powerful as they are simple.

3.3.2 Material Balances: With Reaction

Chemical engineering is concerned with the transformation of matter, whether it be the production of a traditional chemical, the manufacture of cement, waste-water treatment or the production of bio-mass. Balances across units in which reactions take place are therefore of great importance. Equation (3.1) is still applicable, since matter can neither be created nor destroyed except in nuclear reactions. Equation (3.2) which referred to a particular component needs to be modified because particular molecules will react or be formed by reactions. An extra 'generation' term allows for this.

| Amount of component i into unit per second | = | Amount of component i out of unit per second | + | Rate of accumulation of component i in unit | + | Net rate of consumption of component i per second |

(3.5)

A material balance equation can be written for every identifiable species present, and so 'component' can be understood to be either element, compound or radical. For example, in the combustion of methane, balances can be made both on the element carbon and the compound methane. When reactions are involved, it is convenient to work in terms of molar rather than mass units.

Stoichiometry is important. The stoichiometry of a particular reaction states unambiguously the ratios in which molecules of different species are consumed or formed[2]. For example, the stoichiometry of the reaction $SO_2 + \frac{1}{2} O_2 = SO_3$ is such that the generation of one mole of SO_3 requires the consumption of one mole of SO_2 and a half a mole of oxygen.

One subject of importance to chemical engineers is combustion. When a fuel is burnt in air there is a minimum of three streams: the fuel, the air and the combustion products. The nitrogen and the inert gases (argon, krypton, etc.) in the air do not react and they can be used as tie components to relate the inlet and outlet compositions. The amount of oxygen is generally in excess of the stoichiometric amount to ensure complete combustion. However, too much excess air creates excessive energy losses and reduces energy.

3.3.3 Material Balances: Recycles and Purge Streams

For a process plant consisting of a mixer, reactor and separator, the balances can be performed unit by unit. The product stream from the mixer is one of the feed streams to the reactor, while the product stream from the reactor is the feed to the separator. The step-by-step approach can be repeated with similar 'left to right' processing schemes, irrespective of the number of units. However, sequential calculations are impossible if a flow stream from a downstream unit is returned (recycled) to a unit upstream. The flow rate and composition of this recycled stream may well not be known because the calculations, for the unit from which the recycle stream came, have yet to be completed[3]. In terms of the example in figure 3.3, the solutions for units 2 and 4 are interdependent. Simultaneous solution of the material balance equations for the units concerned becomes necessary.

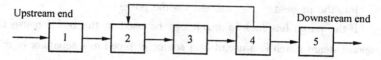

Figure 3.3 A flow scheme with one recycle stream. Which unit is the reactor?

Apart from simple problems, solutions are obtained by using computers. In the 1970s, these would have been large mainframes but now the task is easily handled by desk-top micros.

Recycle problems arise whenever there is a need to recycle unconverted reactant. Even near-equilibrium conversion in the Haber process for ammonia would achieve only 25 per cent conversion in a single pass through the reactor, and so the nitrogen and hydrogen are recycled. The overall conversion of reactants into products can be made to be nearly 100 per cent. The synthesis of methanol from carbon monoxide and hydrogen is similar in that a recycle is essential because the conversion per pass is again around 25 per cent. The production of vinyl chloride (chloroethene) from ethylene (ethene) also involves recycle streams, and many more examples of this sort could be given. More complex processes often involve several recycle loops, and at the process design stage consideration has to be given to the effect they will have not only on steady-state operation but also on the start-up process[4].

Normally it is necessary to bleed off a fraction of the recycle stream to prevent accumulation of unwanted material. In the above example, the argon present in the feed stream was neglected. Unlike the other gases, it is not converted to product and without remedial action it would continue to build up and reach an unacceptable level. A continuous purge is normally used and if under steady-state conditions the inerts are not lost in any other stream, the basic material balance equation can be applied to the inerts to show that

$$\text{Rate of feed of inerts into the process} = \text{Rate of loss of inerts out of process via the purge}$$

If the purge has fuel value, it can be used in the boiler house to generate steam which is a useful heat source or piped to a separator or rejected as waste. On ammonia plants the first was the preferred solution, but with the advent of membrane separators hydrogen is now recovered and recycled to the reactor.

Although the type of work outlined above will have been unfamiliar to most readers it should be clear that the flowrates and compositions of the process streams, which are obtained from the application of equations (3.1) and (3.5), are not only important but also relatively simple to obtain. Material balances in general, with reaction and with recycle have been considered. This covers the full range of possible problems and a complete process can be described by the appropriate number of equations of the same form as those introduced above.

Vocabulary

1. appertain   [æpə'tein] vi. 附属、关系、适合
2. volumetric   [vɔlju'metrik] a. 测定体积的
3. consecutively   [kən'sekjutivli] ad. 连续地
4. distillate   ['distilət] n. 蒸馏物

5. liquor   ['likə]   n. 酒、液体、汁、溶液；vt. 浸水；
                                          vi. 喝酒
6. foreshot   n. 初馏物
7. feint   [feint]   n. 不鲜明、假装
8. refinery   [ri'fainəri]   n. 精炼厂
9. ascertain   [æsə:tein]   vt. 确定、探知
10. stir   [stə:]   n. 骚动、轰动、搅动；vt. 移
                                          动、摇动、激起、惹起、搅和；
                                          vi. 走动；传播
11. stoichiometry   [stɔiki'ɔmitri]   n. 化学计量、化学计算、化学
                                          计算法
12. unambiguously   ['ʌnæm'bigjuəsli]   ad. 明白地、清楚地
13. inert   [i'nə:t]   a. 惰性的、迟钝的、无活力的、呆滞
                                          的
14. argon   ['a:gɔn]   n. 氩
15. krypton   ['kriptɔn]   n. 氪
16. purge   [pə:dʒ]   n. 净化
17. irrespective   [iri'spektiv]   a. 不给……的
18. mainframe   ['meinfreim]   n. 主机
19. vinyl   ['vainil]   n. 乙烯基
20. ethylene   ['eθəli:n]   n. 乙烯
21. remedial   [ri'mi:diəl]   a. 治疗的、补救的、矫正的
22. advent   ['ædvent]   n. 出现、到来

Phrases

1. tie component(s)   结合组分,约束组分
2. desk-top   (计算机)桌面系统
3. bleed off   放出,流出

Notes

1. knowledge of the mass … examined shortly. 全句可译为：这里简要考察物料
平衡时溶液引发的设备中每一部分的质量和体积流动速率(变化)的相关

知识及一些简单的例子。句中"knowledge"和"simple cases"为并列主语。
2. The stoichiometry of ... or formed. 句子的主干结构为："The stoichiometry (...) states (...) the ratio"。
3. The flow rate and composition of ... to be completed. 句中"for"引导的是原因状语从句。
4. More complex processes ... on the start-up processes. 句中含有"effect ... on"结构,意为:对……的影响。

3.4　REACTOR TYPES

3.4.1　The Nature of the Problem

When faced with such kind of industrial design problems as: how to choose the best type of reactor for any particular chemical reaction; how to estimate its necessary size and how to determine its best operating conditions, the chemical engineer must usually regard two things as being fixed beforehand, one of them is the scale of operation (i.e. the required daily output) and the other is the kinetics of the given reaction. Apart from these a chemical engineer has considerable freedom of choice: he can adopt either a batch process or one of the several different kinds of continuous process, he can take whatever values he believes best for the initial concentration of the reagents and also for the operating temperatures and pressures.

3.4.2　Criteria of Choice

In the manufacture of products for sale, the criterion of profitability has usually to be satisfied for the project to go forward. Since no chemical reactor of itself can make a profit, the chemical reaction engineer has to consider the effect of his reactor on the profitability of the project as a whole. There are two main aspects of this issue: ① the cost of the reactor (both capital and running costs); and ② the cost to the rest of the pro-

Some Techniques of Chemistry and Chemical Engineering 127

ject of working up the product of the reactor to its final, saleable form.

3.4.3 Batchwise and Continuous Reaction

Certain chemicals produced in rather small quantities—pharmaceuticals, dyestuffs and so on—are made batchwise. In a typical factory concerned with this sort of production one may be struck by the presence of numbers of autoclaves—each used for producing a ton of one product one day, and a ton of quite a different product the next. Such a system gives great flexibility, especially when the particular factory has a large repertoire of products each produced on a fairly small scale.

A further advantage of batchwise operation is that the capital cost is often less than for a corresponding continuous process when the desired rate of production is low. For this reason it is frequently favoured for new and untried processes which are to be changed over to continuous operation at a more advanced stage of development, when a larger production may be required[1].

The reasons why continuous processes are eventually adopted in almost all large-scale chemical industries are mainly these:

① Diminished labour costs owing to the elimination of operations, such as the repeated filling and emptying of batch vessels.

② The facilitation of automatic control. This also reduces labour costs, although it usually requires considerable capital outlay.

③ Greater constancy in reaction conditions and hence greater constancy in the quality of product.

It will be seen that the decision between batchwise and continuous reaction is very dependent on the magnitude of capital costs in relation to operating costs (of which labour costs may be a very important component). What is best for a highly industrialized country is not necessarily best for one which is less industrialized.

Let us ask now what is the scientific, as distinct from the economic, difference between batchwise and continuous reaction. The kinetics of reactions are usually studied in the laboratory under batchwise conditions, but the application of the results to the design of a continuous process involves no new principles of kinetics since the molecular changes are the same. The difference lies in the existence of a state of flow in the continuous process and this may give rise to important changes of a macroscopic kind. In particular not all molecules passing through the flow system will necessarily have equal residence time, nor will these molecules all undergo the same history of concentration or temperature changes. These factors may cause appreciable differences of yield or of mean reaction rate, as compared to a batch process. This is especially the case when the reaction situation is complicated by the existence of competing side reactions. Here the yield of the desired product may differ considerably as between batchwise and continuous operation, and also as between the two main types of continuous process. Reaction yield is not necessarily lower by continuous process—in some instances it may be higher. However, in examples where it is lower, this factor may so outweigh the normal advantages of continuous operation as to argue in favour of a batchwise system.

It may be remarked that certain processes are neither unambiguously batch nor unambiguously continuous, but should be described as 'semi-batch' or 'semi-continuous'. For example, penicillin is made in large fermenters which are inoculated with the penicillin-producing organism at the start of a production run. After many hours, the contents of the fermenter are emptied, and penicillin recovered from them. This would therefore seem to be a batch process. However, during the run air, and nutrients such as sugar, are continuously added to the fermenter, and gaseous waste products are continuously removed.

3.4.4 Reactor Types

(1) The tubular reactor. In considering continuous reaction equip-

ment, it is often convenient to bear in mind certain simple, or idealized, types of reactor. We shall discuss the general characteristics of these 'model' reactors, leaving the more detailed treatment till later. The first type we shall describe is the tubular reactor.

The tubular reactor is so named because in many of its instances it takes the form of a tube. However, what is meant in general by a tubular reactor is any continuously operating reactor in which there is a steady movement of one or all of the reagents in a chosen spatial direction (the reagents entering at one end of the system and leaving at the other) and in which no attempt is made to induce mixing between elements of fluid at different points along the direction of flow: that is to say, it is the type of continuous reactor for which the most appropriate first approximation useful for predicting its behaviour is the assumption that the fluid moves through it like a plug (the description plug-flow reactor is frequently used)[2]. Some reactors which satisfy this definition and yet which bear no outward resemblance to a tube will be mentioned shortly.

Tubular reactors that actually are tubes are used for many gas reactions and some liquid-phase reactions. As examples we can mention the thermal cracking of hydrocarbons to make ethylene, the oxidation of nitric oxide (one stage in the production of nitric acid from ammonia), and the sulphonation of olefines. For such homogeneous reactions the reactor contains only the reacting fluid.

Tubular reactors are also used extensively for catalytic reactions. Here the reactor is packed with particles of the solid catalyst and for this reason is often referred to as a fixed-bed reactor. Such uses include ammonia synthesis, methanol synthesis and a host of other important heterogeneous reactions. The reactor may be one large-diameter cylinder, or may consist of many tubes in parallel, fixed between two headers as in a tube-and-shell heat exchanger. The tubes are frequently a few centimetres in diameter and may be several metres in length.

In the above reactors the reactants flow axially down a tube, but there are others where the flow is not so simple. In a design of reactor which has been applied to the 'platforming' reaction, the reactant gas flows radially inwards from a perforated cylinder through a catalyst packed between this cylinder and a central perforated tube which collects the product. Though the gas flow is radial, the system satisfies the 'plug-flow' requirement in so far as no attempt is made to cause mixing between points in the direction of flow.

In all such reactors the composition of the reacting fluid necessarily changes in the direction of flow. However, there may also be variations of composition in directions at right-angles to the direction of flow and such variations can be the result of temperature gradients and/or velocity gradients.

Tubular reactors are sometimes operated adiabatically and sometimes with heat transfer through the wall. In the former case the temperature naturally rises along the direction of flow if the reaction is exothermic and falls if it is endothermic. Of course, in many instances, it is necessary to heat the reactants before they enter the reaction zone (e.g. by heat exchange), as otherwise reaction would be too slow. But once reaction is under way, it is often necessary to remove heat through the wall, as otherwise the temperature rise might be excessive and might cause undesirable side reactions to set in.

For efficient removal of heat, the diameter of a cylindrical tubular reactor should clearly be small, in order to reduce the distance over which the heat must be conducted up to the wall. When there are compelling reasons of a different kind for choosing a large diameter, or for placing the catalyst in large trays, it may be necessary to immerse cooling coils in the body of the catalyst.

A striking type of reactor having mercury cooling is that which was used in a process for the production of phthalic anhydride. Naphthalene was vaporized into an air stream and passed through a tubular reactor consisting of as many as 3 000 tubes in parallel, each 1 ~ 2 cm in diameter and up to 3 m long and containing the pelleted catalyst. The construction was similar to that of a tube-and-shell heat exchanger. The heat of reaction was very effectively taken up by the boiling of mercury outside the tubes; the mercury vapour was condensed externally and recirculated. Careful temperature control, at about 350℃, was necessary in this reaction in order to reduce the formation of maleic anhydride and carbon dioxide as by-products.

(2) The continuous stirred tank reactor (C.S.T.R.). This type of reactor consists of a well-stirred tank into which there is a continuous flow of reacting material, and from which the (partially) reacted material passes continuously. It is because such vessels are squat in shape (e.g. cylindrical vessels as wide as they are deep) that good stirring of their contents is essential; otherwise there could occur a bulk streaming of the fluid between inlet and outlet and much of the volume of the vessels would be essentially dead space.

The important characteristic of a C.S.T.R. is the stirring. The most appropriate first approximation to an estimation of its performance is based on the assumption that its contents are perfectly mixed. As a consequence the effluent stream has the same composition as the contents and this demonstrates the important distinction between the C.S.T.R. and the tubular reactor.

A fair approximation to perfect mixing is not difficult to attain in a C.S.T.R., provided that the fluid phase is not too viscous. In general terms, if an entering element of material (e.g. a shot of dye) is distributed uniformly throughout the tank in a time very much shorter than the average time of residence in the tank, then the tank can probably be taken to be 'well-mixed'.

It will be seen later that it is often advantageous to have several C.

S.T.R.s in series, the process stream flowing from one to the next. This results in a stepwise change of composition between successive tanks. It also results in a bypassing loss: a given molecule entering a tank has a significant probability of finding its way into the outflow almost immediately. This is the reason why it is usually necessary to use several tanks in series, especially if high conversions are required. If there were only one or two there would be an appreciable loss of unreacted reagent. Although this loss is, in a sense, the result of the stirring, the loss would usually be even greater in the absence of stirring, due to bulk streaming between inlet and outlet.

A usual consequence of the stepwise change in concentration is that the average reaction rate is lower than it would be in a tubular reactor having the same concentrations of reagents in the feed. The reactor volume, for a given output, must therefore be a good deal larger and this must be allowed for in the design. In view of this it may seem paradoxical that the C.S.T.R. is much used for comparatively slow liquid-phase reactions. The reason is the cheapness of the construction, as compared with a tubular reactor. The greater volume is comparatively unimportant as an economic factor, at any rate in the case of tanks operating at atmospheric pressure and made of an inexpensive material such as mild steel.

One great advantage of the C.S.T.R., apart from simplicity of construction, is the ease of temperature control. The reagents entering the first vessel plunge immediately into a large volume of partially reacted fluid and, because of the strring, local hot spots do not tend to occur. Also the tanks of the C.S.T.R. offer the opportunity of providing a very large area of cooling surface. In addition to the external surface of the vessels themselves, a large amount of internal surface, in the form of submerged cooling coils, can be provided. Sometimes, in place of colis, a calandria is used, as in the example of the Schmid nitrator for nitroglycerine.

A further advantage, as compared to the tubular reactor, is the openness of the construction. This makes it easy to clean the internal surfaces and this is important in the case of reactions where there is a tendency for solid matter to be deposited, e.g. polymerization processes and reactions in which tarry material is formed as a by-product.

For these various reasons the typical fields of application of the C.S.T.R. are continuous processes of sulphonation, nitration, polymerization, etc.. It is used very extensively in the organic chemical industry and particularly in the production of plastics, explosives, synthetic rubber and so on. The C.S.T.R. is also used whenever there is a special necessity for stirring; for example, in order to maintain gas bubbles or solid particles in suspension in a liquid phase, or to maintain droplets of one liquid in suspension in another as in the nitration of benzene or toluene. The rate of such reactions can be very dependent on the degree of dispersion, and therefore on the vigour of agitation.

Certain gas-phase reactions, e. g. combustion, chlorination of gaseous hydrocarbons, etc., are also sometimes carried out in reaction vessels which approximate to single-stage C.S.T.R.s, although without having any mechanical stirring. The shape of the vessels and the position of the gas inlet jets (e.g. tangential entry) may be such as to cause fairly complete mixing. The occurrence of hot spots can thereby be avoided.

(3) The fluidized-bed reactor. This can be used for reactions involving a solid and fluid (usually a gas). The earliest and perhaps best-known example is the 'cat-cracker', in which the solid is a catalyst for the cracking of hydrocarbon vapours. A similar catalytic process has been used for the oxidation of naphthalene with air to give phthalic anhydride.

Examples of processes in which the solid actually reacts with the gas are the reaction of alumina with hydrogen fluoride to produce aluminium fluoride, and the reactions

$$UO_3 \xrightarrow{H_2} UO_2 \xrightarrow{HF} UF_4$$

which are carried out in fluidized-bed reactors.

In such reactors the solid material in the form of fine particles is contained in a vertical cylindrical vessel. The gas stream is passed upwards through the particles at a rate great enough for them to lift, but not so great that they are prevented from falling back into the fluidized phase above its free surface by carry-over in the gas stream[3]. The bed of particles in this condition presents the appearance of boiling; bubbles of the upflowing gas may be seen bursting at the upper surface.

The effect of the rapid motion of the particles is a high degree of uniformity of temperature and thus an avoidance of the hot spots which occur with fixed-bed tubular reactors. This is often a considerable advantage in the case of reactions which can be allowed to proceed adiabatically, finding their own temperature as determined by the heat of reaction. If heat must be removed, in order to keep the temperature down to some prescribed level, use can be made of the fact that heat transfer to cooling tubes is more easily brought about in a fluidized bed than in a fixed bed[4]. Care must be taken to ensure that the heat-transfer surface does not interfere with the efficiency of fluidization. A disadvantage of this type of reactor is attrition of the solid, resulting in, for example, loss of catalyst and dust problems in the effluent gas stream.

The fluidized reactor is difficult to treat theoretically because of the complicated nature of the fluid flow. With gas-fluidized beds much of the gas may pass through the bed as bubbles. The reaction only occurs on the surface of the solid particles and the flow of bubbles through the bed acts as a bypass stream. Gaseous reactant can move from the bubble phase to the particulate phase by diffusion and by convection (there is a convective flow through the bubbles, which do not have 'skins'). The relative importance of reaction, diffusion and convection depends upon the fluid mechanics, which are not precisely known.

It is particularly difficult to scale up fluidized-bed reactors from laboratory experiments. The behaviour of bubbles in relatively shallow fluidized beds in small-diameter tubes can be very different from their behaviour in deep, large-diameter beds. For example, slugging behaviour may occur in narrow tubes, and bubble coalescence may be important in deep beds.

(4) Other types of reactor. There are many other, diverse, types of reactor which have been found useful in particular cases. Brief mention can be made of four.

① Bubble-phase reactors. Here a reagent gas is bubbled through a liquid with which it can react, because the liquid contains either a dissolved involatile catalyst or another reagent. The product may be removed from the reactor in the gas stream. Mass transfer is clearly important, and may control the rate of 'reaction'. An example is the Hoechst-Wacker process for the production of acetaldehyde by the oxidation of ethylene.

② Slurry-phase reactors. These are similar to bubble-phase reactors, but the 'liquid' phase consists of a slurry of liquid and fine solid catalyst particles. They are used in the catalytic carbonylation of hydrocarbons to produce alcohols. 'Three-phase fluidized beds' are of this type.

③ Trickle-bed reactors. In these reactors the solid catalyst is present, not as fine fluidized particles, but as a fixed bed. The reagents, which may be two partially-miscible fluids, are passed either co-currently or counter-currently through the bed. An example is the high-pressure hydration of propylene to give isopropyl alcohol.

④ Moving-burden bed reactors. Here a fluid phase passes upwards through a packed bed of solid. Solid is fed to the top of the bed, moves down the column in a closely plug-flow manner, and is removed from the bottom. This has been used in the catalytic isomerization of xylenes, and in continuous water treatment by ion-exchange.

Vocabulary

1. autoclave　　　　['ɔːtəkleiv]　　n. 压热器、高压消毒锅、高压

			罐
2. repertoire	['repətwa:]	n.	全部剧目、全部技能、所有组成部分
3. fermenter	['fə(:)mentə]	n.	发酵罐
4. inoculate	[i'nɔkjuleit]	vt.	接种、嫁接
5. tubular	['tju:bjulə]	a.	管状的
6. spatial	['speiʃəl]	a.	空间的、空间性的、场所的
7. sulphonate	['sʌlfouneit]	n.	磺酸盐
8. olefine	['əuləfin]	n.	链烯烃
9. homogeneous	[houmə'dʒi:niəs]	a.	同种的、同质的、均质的
10. heterogeneous	[hetərə'dʒi:niəs]	a.	异种的、异质的、由不同成分形成的
11. axially	['æksiəli]	ad.	向轴的方向
12. perforate	['pə:fəreit]	vt.	穿孔于、刺空;
		vi.	穿孔、穿过;
		a.	有孔的、穿孔的
13. adiabatic	['ædiə'bætik]	a.	绝热的、隔热的
14. phthalic(anhydride)	[θælik]	n.	邻苯二甲酸(酐)
15. maleic(anhydride)	[mə'li:ik]	n.	马来酸(酐)
16. paradoxical	[pærə'dɔksikl]	a.	似非而是的、矛盾的、诡论的
17. calandria	[kə'lændriə]	n.	(物)加热体、加热器、(化)排管式
18. nitrator		n.	硝化器
19. nitroglycerine	['naitrouglisə'ri:n]	n.	硝化油
20. toluene	['tɔljui:n]	n.	甲苯[染料、火药的原料]
21. agitation	[ædʒi'teiʃən]	n.	搅动、鼓动、煽动、激动、骚动、焦虑

22. chlorination [klɔ:rə'neiʃən] n. 氯化、用氯化理
23. alumina [ə'lju:minə] n. 矾土、氧化铝
24. slug [slʌg] n. 金属小块;
vt. 插嵌片于
25. coalescence [kouə'lesns] n. 接合、结合、合并
26. acetaldehyde [æsi'tældəhaid] n. 乙醛、醋醛
27. slurry ['slə:ri] n. 泥浆、浆
28. carbonyl ['ka:bənil] n. 羰基
29. propylene ['proupəli:n] n. 丙烯
30. isomerization [aisəu'məraizeiʃən] n. 异构化(作用)
31. xylene ['zaili:n] n. 二甲苯

Phrases

1. give rise to 提高,加大
2. residence time 停留时间

Notes

1. For this reason it is ... may be required. 全句可译为:由于这个原因,间歇操作通常对新的和未试过的过程是有利的,这些过程是在开发的更进一步阶段将要变成连续操作的过程。

2. However, what is meant in general... through it like a plug. 句中"that is to say"后的意思为:(这就是说),这是一种连续反应器,对这种反应器行为进行预测的最恰当、首要的近似处理就是假定液体像一个塞子一样流过它。

3. The gas stream is passed upwards ... in the gas stream. 全句可译为:气流以远大于使粒子升起的速度向上通过粒子,使粒子上移。但速度不致过大从而使粒子通过滞留在气流中从而阻碍其落回到自由层以上的流动相中。句中"be prevented from"意为"阻止、妨碍";"carry-over"意为"遗留,遗留物"。

4. If heat must be removed, in order to keep... than in a fixed bed. 此句中"use can be made of ..."原形为"(we) can make use of ..."。全句可译为:如果为了保证温度降至一定的规定值,一定要移走热量,人们可以利用这样一个事实:移动床比固定床更易将热量移至冷却管中。

3.5　MEMBRANES AND MEMBRANE PROCESSES

A membrane is a selective barrier between two homogeneous phases (gas or liquid). The selectivity is based on the fact that the different components in the feed will diffuse through the barrier at different rates. It is worth remembering that, in general, a membrane is not a complete semi-permeable barrier and some of the solute, which should ideally be retained in the concentrate, passes into the permeate.

The first synthetic membranes were prepared at the beginning of the century and developed for laboratory scale use after the First World War. The membranes were composed of rather thick layers of cellulose nitrate or cellulose nitrate/cellulose acetate. Although technically feasible, separations were not economically feasible on the industrial scale at this time. The necessary equipment would have been excessively large because the amount flowing through the membrane per unit of membrane area per unit time (the 'flux') was very low. The major breakthrough came in the 1960s with the development of asymmetric membranes, divides membranes into three main groups and indicates the nominal pore-diameters of the various classes of membrane.

A thin barrier layer is essential for gas separations and should be dense, non-porous and selective. The passage of components through this layer is often modelled on the basis that the components dissolve in the top surface, diffuse across the membrane and desorb on the downstream side. One the other hand, the free space within microporous membranes is sufficiently large for fluids to flow through in a conventional manner. These membranes are generally still asymmetric in order to have only a small area of high resistance but lack the extra 'skin'[1]. Their pore diameters are in the range $0.1 \sim 10$ μm, which means that they can be pictured by simple electron microscopes. The resolution of scanning electron microscopes is, however, insufficient to elucidate the structure of ultrafiltration membranes, which nevertheless exhibit porous behaviour. Figure 3.4 includes hyperfiltration (reverse osmosis) for which the main use has been in desalination.

An advantage of a microfilter is that they enable sterility to be maintained at the separation stage and so contamination problems are minimised. In the beverage industry, they can be used for the recovery of yeast from brewery waste and so reduce the effluent problem, and in the sugar industry to clarify glucose syrup and beet sugar juice. These filters can be ceramic which conveys obvious advantages.

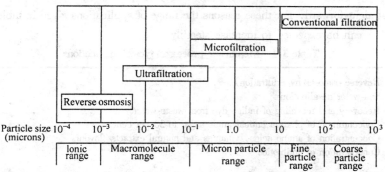

Figure 3.4 Size range of various classes of membrane

The application of ultrafiltration membranes, which are currently all of a polymeric nature, is more widespread because it is possible to separate smaller molecules such as sugars from larger molecules such as proteins. The main attraction for ultrafiltering cheese milk is the increased yield that results from the incorporation of whey protein into the cheese. In a traditional process these proteins remain in the waste liquid whey. The waste stream from a membrane unit still contains lactose (milk sugar) and this can be used for alcohol production, as an animal feed or as a feed to an anaerobic digester which will produce methane. The technology is being applied to both hard and soft cheese; examples includes Cheddar and Camembert.

For yoghurt manufacture, the starter solution must be sufficiently concentrated and skim milk powder is often added back into skim milk prior to incubation. A cheaper alternative is to concentrate skim milk by ultrafiltration which has the added advantage that the lactose passes into the permeate.

Apart from the dairy industry, the main food and beverage areas of application are in fruit juice clarification and effluent pre-concentration. Unlike thermal processes, ultrafiltration allows ambient temperature separation and concentration, avoiding both high energy costs and degradation of heat-sensitive compounds. The units are compact and process control is straight-forward. For these reasons the range of applications given in table 3.1 can be expected to increase steadily.

Table 3.1 Membrane processes: typical applications

Reverse osmosis(hyperfiltration)
Sea-water desalination
Recovery and recycling of indigo dye from waste-water
Concentration of pharmaceutical-grade sugar to 30 per cent
Concentration of amino acids, vitamins and natural extracts/flavours
Concentration of antibiotics

Ultrafiltration
Removal of pectin hazes and suspensions from fruit juices
Recovery of lignosulphonate and vanillin from pulp/paper effluent
Recovery of proteins, for example, from whey
Concentration of milk solids prior to cheese manufacture
Recovery of protein and enzymes from food processing effluents
Fractionation of blood plasma
Removal of pyrogens from water
Hydrogen recovery (special membrane required)
Electropaint recovery
Recovery of oil from oil-water emulsions
Waste-water treatment

Microfiltration
Concentration and separation of products from fermentation broths
Filtration of solvents
Clarification and cold sterilisation of juices, wine and beer
Recovery of beer from yeast sediment

Vocabulary

1. permeable ['pə:miəbl] a. 有浸透性的、能透过的
2. cellulose ['seiljuləus] n. 纤维素
3. asymmetric [æsi'metrik] a. 不均匀的、不对称的
4. desorb [di:sɔ:b] v. 使……放出吸收之物、解吸、脱附
5. elucidate [i'lu:sideit] vt. 阐明、说明
6. hyperfiltration n. 反渗透、反渗透法、超(过)滤
7. desalination [di:sæli'neiʃən] n. 除去盐分
8. brewery ['bruəri] n. 酿造[啤酒等]所
9. glucose ['glu:kəus] n. 葡萄糖
10. sterility ['steriliti] n. 消毒、除菌
11. syrup ['sirəp] n. 糖浆、果汁
12. beet [bi:t] n. 甜菜
13. whey [hwei] n. 乳浆
14. yoburt ['jɔgət] n. 酸乳酪
15. lactose ['læktous] n. 乳糖
16. anaerobic [æneiə'roubik] a. 没有空气而能生活的、厌氧性的
17. ambient ['æmbiənt] a. 周围的

Note

1. These membranes are generally ... lack of the extra 'skin'. 句中 "small area of high resistance" 指 "(有)很小的高阻力区域"; "lack of the extra 'skin'" 意为 "缺少额外的壳(层)"。

3.6 PROCESS SELECTION

There is generally a number of technically feasible processes for any given separation. For example, a solution of common salt can be concentrated by freezing, by evaporation or by reverse osmosis. The skill of a process design team is to select an economically feasible process. The separation of xylenes is a classic example.

Para-xylene (1,4-dimethylbenzene) is an important petrochemical which is an intermediate for the manufacture of chemicals used in the production of polyester fibres, for example, terephthalic acid (benzene-1,4-dicarboxylic acid), $HOOC\text{—}C_6H_4\text{—}COOH$. It is one of the three xylene isomers. The term 'mixed xylenes' refers to these isomers and ethyl benzene:

ortho-xylene meta-xylene para-xylene ethyl benzene

They have similar vapour pressures and their respective normal boiling points differ by less than 8.5℃. A consequence is that the production of individual components of high purity by distillation is difficult, and with regard to one separation (para-meta), almost impossible[2].

	Ethyl benzene	p-xylene	m-xylene	o-xylene
Boiling point (℃) at 1 atm	136.19	138.35	139.10	144.41
Freezing point (℃)	−94.98	13.26	−47.87	−25.18

The process that has evolved includes crystallisation; para-xylene has a significantly higher freezing point. Figure 3.5 gives the overall flow diagram. The ortho-xylene is removed in column 1, together with the C_{9+} aromatics from which it is subsequently separated in another distillation column. The 5.3℃ difference in boiling points between ortho-xylene and meta-xylene is not large and over 100 stages are required. After removal of ethyl benzene, a feedstock for styrene production, the meta-xylene and para-xylene are cooled and para-xylene starts to crystallise out. Its recovery is limited because once the eutectic is reached, mixed crystals will form. When the liquor reaches a concentration of 87 per cent meta-xylene, 13 per cent para-xylene, the practical limit has been reached and over 35 per cent of the para-xylene in the original feed stream is still uncrystallised.

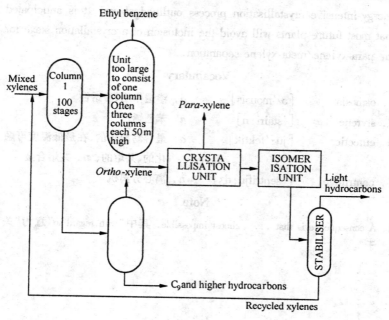

Figure 3.5 Flow diagram of a xylene plant

Filtration of the mixture yields wet crystals which are further purified by removal of the mother liquor in a centrifuge. A final purity of 99.5 per cent is achieved. The filtrate which contains the unrecovered para-xylene is by weight 87 per cent meta-xylene, for which there is very little demand. This stream is isomerised in a reaction stage which results in the production of a mixture of all four C_8 compounds. These are recycled, mixed with fresh feed and reprocessed.

The above example, which completes the overview on the range and choice of separation processes, involves a wide range of unit operations and indicates how different classes of separation processes can be combined to produce an economic plant.

However, technology has developed in competing areas and newer methods of separation such as adsorption now compete with the traditional energy-intensive crystallisation process outlined above. It is anticipated that most future plants will avoid the inclusion of a crystallation stage for the para-xylene/meta-xylene separation.

Vocabulary

1. osmosis　　　[ɔz'mousis]　　　n. 渗透作用、渗透性
2. styrene　　　['stairi:n]　　　n. 苯乙烯
3. eutectic　　　[ju:'tektik]　　　a. 最容易溶解的、在最低温度可融化的、共熔的; n. 共熔合金
4. centrifuge　　　['sentrifju:dʒ]　　　n. 离心分离机

Note

1. A consequence is that ... , almost impossible. 其中"with regard to"意为"关于"。

4

Instrumental and Chemical Analysis Methods

4.1 INTRODUCTION TO INSTRUMENTAL METHODS

Most instrumental techniques fit into one of three principal areas: spectroscopy, electrochemistry, and chromatography (Table 4.1). Although several important techniques, including mass spectrometry and thermal analysis, do not fit conveniently into these classifications, these three areas do provide a basis for a systematic study of chemical instrumentation.

Advances in both chemistry and technology are making new techniques available and expanding the use of existing ones. Photoacoustic spectroscopy is an example of an emerging analytical technique. A number of existing techniques have been combined to expand the utility of the component methods. Gas chromatography-mass spectrometry (GC-MS) and inductively coupled plasma-mass spectrometry (ICP-MS) are examples of successful "hyphenated" methods. The distribution of computer power to individual instruments has led to the widespread use of methods such as the Fourier transform to produce new techniques: Fourier transform infrared (FTIR) and pulsed nuclear magnetic resonance (carbon-13) spectroscopy.

Although the instrument is often the most visible and exciting element of the analytical method, it is only one component of the total analysis. Before focusing on the role of instrumentation in an analytical method, the analyst should consider other steps important to the determination. The following discussion reviews the steps common to analytical methods and thus helps to place the role of instrumentation in proper perspective. A detailed discussion of the topics will not be attempted; they are covered in greater depth in the bibliographies and literature citations.

The first task is to define the analytical problem. When possible, this is done through direct interaction with the person(s) who desires the analysis. The analyst should determine the nature of the sample, the end use of the analytical results, the species to be analyzed, and the information required. Qualitative information may include elemental composition, oxidation states, functional groups, major components, minor components, and complete identification of all species present in the sample. Quantitative data include the required accuracy and precision, range of expected analyte (substance being determined) concentrations and detection limits for the analyte. Other considerations are the unique physical and chemical properties of the analyte, properties of the sample matrix, the presence of interferences that are likely to eliminate the use of certain analyte properties as measurement indicators, and finally an estimated cost of the analysis. A major component of the cost is the time required for the analysis, both the instrument time and the labor required to perform the analysis. When appropriate, the costs of manual versus automated methods should be compared.

Table 4.1 Principal Types of Chemical Instrumentation

Spectroscopic techniques
Ultraviolet and visible spectrophotometry
Fluorescence and phosphorescence spectrophotometry
Atomic spectrometry (emission and absorption)
Infrared spectrophotometry
Raman spectroscopy
X-ray spectroscopy
Radiochemical techniques including activation analysis
Nuclear magnetic resonance spectroscopy
Electron spin resonance spectroscopy

Table 4.1 Principal Types of Chemical Instrumentation (Continued)

Electrochemical techniques
Potentiometry (pH and ion selective electrodes)
Voltammetry
Voltammetric techniques
Stripping techniques
Amperometric techniques
Coulometry
Electrogravimetry
Conductance techniques
Chromatographic techniques
Gas chromatography
High-performance liquid chromatographic techniques
Miscellaneous techniques
Thermal analysis
Mass spectrometry
Kinetic techniques
Hyphenated techniques (see Table 4.2)
GC-MS (gas chromatography-mass spectrometry)
ICP-MS (inductively coupled plasma-mass spectrometry)
GC-IR (gas chromatography-infrared spectroscopy)
MS-MS (mass spectrometry-mass spectrometry)

Table 4.2 The state of the art in hyphenated techniques

Initial technique \ Subsequent technique	Gas chromatography	Liquid chromatography	Thin layer chromatography	Infrared	Mass spectrometry	Ultraviolet (visible)	Atomic absorption	Optical emission spectroscopy	Fluorescence	Scattering	Raman	Nuclear magnetic resonance	Microwaves	Electrophoresis
Gas chromatography	■	□		■	■	□		□				□		
Liquid chromatography	□	■	□	■	■	■	□	□	■			□		
Thin layer chromatography		□	■	□	■	□			□					
Infrared					□					□	□			
Mass spectrometry					■									
Ultraviolet (visible)					□	■			■	□	□			
Atomic absorption							□			□				
Optical emission spectroscopy								□						
Fluorescence					■	□			■	□		□		
Scattering									□	□				
Raman					□						□			
Nuclear magnetic resonance					□							■		
Microwaves														
Electrophoresis														■

NOTE. After personal communication from Tomas Hirschfeld.

■ Established
□ Feasible in the state of the art

Vocabulary

1. spectroscopy [spek'trɔskəpi] n. 光谱学、波谱学
2. spectrometry [ˌspektrə'metri] n. 光谱测定(法)
3. instrumentation [ˌinstrumen'teiʃən] n. 仪器使用、测试设备、仪器制造学
4. acoustic [ə'kuːstik] a. 听觉的、声学的
5. photoacoustic [ˌfəutəuə'kuːstik] a. 光声学的
6. plasma ['plæzmə] n. [物]等离子体、等离子区
7. hyphenate ['haifəneit] vt. 用连字号连接；联用
    a. 用连字号连接的
8. bibliography [bibli'ɔgrəfi] n. 文献学
9. phosphorescence [ˌfɔsfə'resns] n. 磷光(现象)
10. voltammetric [vɔltæ'metrik] a. 伏安测量的
11. amperometric ['ӕmpɛərou'metrik] a. 测量电流的
12. coulometry [kuː'lɔmitri] n. 电量测量、库仑测量
13. chromatography [krəumə'tɔgrəfi] n. 色谱法

Instrumental and Chemical Analysis Methods 149

Phrases

1. mass spectrometry 质谱分析,质谱学
2. thermal analysis 热分析
3. photoacoustic spectroscopy 光声光谱法
4. gas chromatography 气相色谱法
5. inductively coupled plasma 感应耦合等离子体
6. Fourier transform infrared spectroscopy 傅立叶变换红外光谱(法)
7. pulsed nuclear magnetic resonance spectroscopy 脉冲核磁共振光谱(法)
8. end use 最终用途,主要用途
9. qualitative information 定性信息
   quantitative data 定量数据
10. detection limit 检测限,测定极限
11. ultraviolet and visible spectrophotometry 紫外和可见分光光度法
12. fluorescence and photophorescence spectrophotometry 荧光和磷光分光光度法
13. atomic absorption spectrometry 原子吸收光谱法
    atomic emission spectrometry 原子发射光谱法
14. infrared spectrophotometry 红外分光光度法
15. Raman spectroscopy 喇曼光谱法(学)
16. X-ray spectroscopy X-射线光谱法(学)
17. radiochemical techniques including activation analysis 放射化学技术包括活化分析(放射化分析)
18. electron spin resonance spectroscopy 电子自旋共振光谱法(学)
19. ion selective electrode 离子选择(性)电极
20. stripping technique 溶出技术
21. high-performance liquid chromatographic technique (HPLC) 高效液相色谱技术
22. thin layer chromatography 薄层色谱法
23. optical emission spectroscopy 发射光谱法

4.2 ELECTROMAGNETIC RADIATION

By far the largest class of instrumental methods of chemical analysis depends on the use of electromagnetic radiation as a probe to determine the identity, structure, and quantity of chemical species. The most obvious example of electromagnetic radiation is light (visible electromagnetic radiation), but light occupies only a small region in a spectrum of electromagnetic radiation that covers a range of at least 15 orders of magnitude in energy.

Instrumentation and measurement techniques differ widely in the various regions of the electromagnetic spectrum employed for chemical analysis, and at first glance the various spectroscopic methods of analysis appear to have little in common. In fact, the properties of the radiation and the types of experiments performed are basically the same whatever the region of the spectrum, and it is therefore to our advantage to attempt to understand, at least in a qualitative fashion, the properties of electromagnetic radiation and how it interacts with matter before attempting to examine the individual spectroscopic methods of analysis[1].

Light has been studied for centuries and much is known about its properties, but there is no straightforward answer to the simple question. What is light? In some of its aspects, as, for example, in the focusing of light by lenses and its reflection by mirrors, light is best described as a wave phenomenon. In other aspects, such as the emission or absorption of light by atoms or molecules, light can only be considered as a stream of discrete energy packets. Thus, light is said to have a dual nature, exhibiting both wave and particle characteristics. This duality is not confined to the visible portion of the electromagnetic spectrum, but can be demonstrated for all electromagnetic radiation; nor is it limited only to radiation. Wave properties can be demonstrated for all subatomic particles. The diffraction of electrons by crystals is an obvious example of the wave properties of particles. Electromagnetic radiation differs from other particles in that the energy packets, the photons, have zero rest mass.

We shall briefly review phenomena pertinent to the application of electromagnetic radiation for analytical spectroscopy.

Every elementary system—nucleus, atom, molecule—when reasonably isolated, appears to have a number of discrete energy levels. To change from one to another, a system must be exposed to radiation or bombarded with particles having energies at least equal to the energy differences of the states. Emission and absorption of radiation arise from changes in the energy of such systems. There are two important exceptions to this type of behavior:

① the translational motion of atoms, molecules, and ions in free space.

② the "blackbody" emission and absorption of radiation exhibited by solids.

The energy levels of a reasonably isolated nucleus, atom, or molecule are uniquely characteristic, and the interest of the chemist in electromagnetic radiation is based chiefly on his ability to obtain information about the energy levels by measurement of the emission and absorption of electromagnetic radiation. If the intensity of emission or absorption is measured as a function of the energy of the electromagnetic radiation (or some quantity related to the energy, such as frequency, wave number, or wavelength), then one obtains a unique signature of the elementary system, which we call its spectrum. Further, with proper calibration, it is possible to relate the intensity of a particular energy transition to the concentration of interacting species.

The instrumentation required to carry out the measurements described above varies greatly as one varies the frequency of the electromagnetic radiation, but basically there are only two types of experiments—measurement of the intensity of either the emission or absorption of the electromagnetic radiation.

Figure 4.1 summarizes the types of transitions observed in the various regions of the electromagnetic spectrum and the approximate ranges of energy involved.

Gamma radiation is emitted or absorbed as a result of transitions between nuclear energy levels. A nucleus may be produced in an excited state in the course of radioactive decay or by capture of a particle. Gamma radiation may then be emitted as the nucleus deactivates to the ground state. The measured energy spectrum of gamma rays, emitted in transitions between the energy states of a nucleus, is used to determine the energies of these states—the connection being made by the same trial-and-error procedures as were used in the very early days of the study of atomic spectra and energy states. The energies of the nuclear states give important information about nuclear structure, but from the point of view of the chemist a more interesting type of information is obtained in a gamma-ray absorption experiment of a type that has come to be known as Mössbauer spectroscopy.

(SOURCE: R. Mavrodineanu and H. Boiteux, Flame Spectroscopy, Wiley, New York, 1965, p.218)

If the gamma ray emitted by an excited nucleus is allowed to interact with a ground-state nucleus of the same element, the gamma ray may be reabsorbed. Although this process was a well-recognized possibility, it was believed to be a rather improbable process for the following reason: Because of the very high energy of the gamma ray, a small but appreciable fraction of the energy provided by the nuclear transition should go into recoil of the nucleus upon emission, and thus the energy of the gamma ray should be somewhat less than the energy change of the transition. Moreover, in the absorption process the gamma ray should impart some of its energy to the nucleus as kinetic energy. Consequently, for identical emitting and absorbing nuclei, it would be expected that the emitted gamma

Instrumental and Chemical Analysis Methods 153

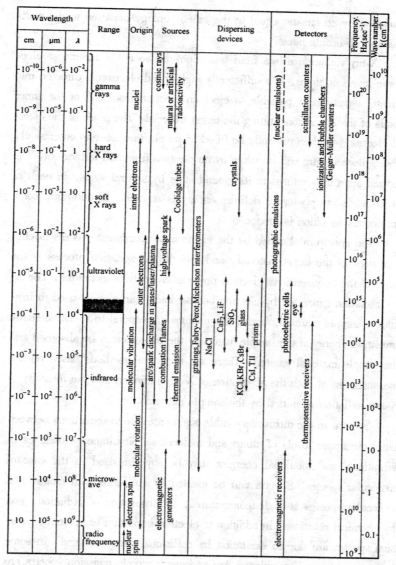

Figure 4.1 Survey of the electromagnetic spectrum

ray would fall short of supplying the energy required for the absorption transition by an amount equal to the kinetic energy imparted to the emitting and absorbing nuclei[2].

X-ray radiation arises from transitions between the innermost electronic levels of atoms. If a sufficiently high-speed electron is directed into a metal target, it is possible to eject an electron from one of the inner shells of the atoms constituting the target material. The place of the ejected electron will then promptly be filled by an electron from an outer shell, whose place in turn will be taken by an electron from still farther out. The ionized atom thus returns to its ground state by several steps, in each of which an X-ray photon of definite energy is emitted. In addition, some continuous radiation is produced.

The spectrum obtained by the electron-bombardment process is characteristic of the target material, and this direct excitation process is the basis of the very sensitive electron-probe microanalysis technique. The X-ray radiation generated by electron bombardment can also be used to irradiate a sample material, and if the photons correspond to a characteristic transition energy of the sample material, the photons may be absorbed and the sample material excited. There are analytical methods based on the measurement of both the absorption of X-ray radiation and the intensity of X-ray radiation reemitted by the sample material.

Spectra in the ultraviolet-visible region are due to transitions between outer electronic levels of atoms and molecules. Simultaneous changes in vibrational and rotational energies may be superimposed in the case of molecular species. Spectra can be excited by collisions with atomic and molecular species at high temperatures, by absorption of radiation, and by chemical reactions, in addition to electron impact. Flame atomic emission spectra are due to excitation by collisional processes and, in some cases, by chemical reactions. Arc and spark atomic emission spectra are due to collisional and electron-impact excitation. Atomic absorption and atomic fluorescence spectrometry depend on the absorption of radiation at characteristic frequencies and measurement, respectively, of the intensity of radiation absorbed and reemitted. Molecular spectra are largely observed as absorption spectra, but molecular fluorescence and phosphorescence are based on measurement of the reemission of absorbed radiation.

Spectra in the infrared region are mostly associated with changes in vibrational energy levels of molecules, although some pure rotational energy transitions for heavier molecules are found in the low-energy end of this region. Changes in vibrational energy may also be accompanied by changes in rotational energy. Note that lower vibrational energy levels may be appreciably populated at room temperature.

Almost all infrared spectra are observed as absorption spectra. The spectra obtained in this region are highly characteristic, and infrared spectroscopy is today an almost indispensable tool in the characterization of organic compounds.

Spectra in the microwave region arise from two distinct sources: changes in rotational energies of molecules, and changes in the energies associated with the alignment of unpaired electrons in an external magnetic field. In both cases excited levels may be appreciably populated, and in both cases absorption measurements are experimentally more convenient in most situations.

Spectra in the radio frequency region are associated with changes in nuclear orientation in an external magnetic field. At room temperature, all levels are almost equally populated. Again, absorption measurements prove experimentally most convenient in the majority of cases.

The energy required for the nuclear orientational transition is dependent on the molecular environment of the nucleus, and the technique of nuclear magnetic resonance has become an indispensable tool in the characterization of complex chemical species.

Radiation from a heated solid does not exhibit discrete lines or bands but rather is continuous; that is, the intensity of the radiation varies smoothly with wavelength over an extended range of wavelengths. Moreover, to a large extent the distribution of radiation with wavelength does not depend on the material of which the emitter is composed, but depends only on the absolute temperature of the source. If the solid has the property of absorbing all the radiation incident on it, then the source is said to be an ideal blackbody. A close approximation to an ideal blackbody can be achieved by an arrangement. An evacuated hollow inside any solid may be maintained at equilibrium with respect to both emission and absorption. If the only radiation that escapes from the hollow comes through an opening that is small compared to the total surface area of the cavity, then the spectrum obtained is that of an ideal blackbody and does not depend on the nature, size, or shape of the solid, but only on the temperature[3].

Vocabulary

1. spectrum ['spektrəm] n. 光谱、范围、系列
2. magnitude ['mægnitju:d] n. 大小、数量
3. duality [dju:'æliti] n. 二元性
4. pertinent ['pə:tinənt] a. 相关的、有关的、中肯的、贴切的
5. bombard [bɔm'ba:d] vt. 炮击、攻击、轰击；n. 射石炮
6. deactivate [di:'æktiveit] vt. 使无效、使不活动、去活化
7. calibration [kæli'breiʃən] n. 校准、刻度、标准化
8. innermost ['inəmoust] a. 最里面的、最深处的；n. 最深处
9. bombardment [bɔm'ba:dmənt] n. 炮击
10. superimpose ['sju:pərim'pouz] vt. 重叠、添加、附加
11. arc [a:k] n. 弧形、弓形、弧光、弧
12. fluorescence [fluə'resns] n. 荧光、发荧光、荧光性
13. alignment [ə'lainmənt] n. 队列、结盟、校直
14. resonance ['rezənəns] n. 共鸣、回声、反响、谐振、共振
15. discrete [dis'kri:t] a. 离散的、不连续的
16. evacuate [i'vækjueit] vt. vi. 疏散、撤出、排泄

17. cavity ['kæviti] n. 洞、空穴、腔

Phrases

1. electromagnetic radiation　电磁辐射
2. to (one's) advantage　(对……)有利,有益处
3. a stream of discrete energy packets　不连续的能束流,一束不连续的能量流
4. pertinent to　与……有(相)关的
5. translational motion　平移运动
6. radioactive decay　放射性衰变
7. excited state　受激(状)态,激发态
   ground state　基态
8. trial and error　试错法,反复试验
9. Mössbauer spectroscopy　穆斯堡尔谱(学)
10. impart A to B　把A给与(传给)B
11. fall short of　不足,达不到,不符合
12. electron probe　电子探针
13. vibrational energy　振动能
    rotational energy　转动能
14. flame atomic emission spectrum　火焰原子发射光谱
15. arc and spark atomic emission spectrum　电弧和火花原子发射光谱
16. vibrational energy level　振动能级

Notes

1. In fact, the properties of the radiation... methods of analysis. 句中"and it is therefore to ..."中"it"是形式主语,其指代的是"to attempt to understand..."至句尾。其中"understand"有两个宾语:"the properties"和"how"引导的从句。

2. Consequently, for identical emitting ... and absorbing nuclei. 句中"it would be expected that ..."的"it"为形式主语,"that"开始到句末为真正主语,"that"引导主语从句,其中"supplying"为分词,作其前面"of"的宾语。"required"和

"imparted"均为后置定语,分别修饰其前面的名词。句中"by an amount equal to","by"表示相差,"equal to"为形容词短语修饰"an amount"。
3. If the only radiation that ... on the temperature. 句首为条件状语从句,其中"that escapes from the hollow"为定语从句修饰"the only radiation"。"that is small... the cavity"为定语从句修饰"an opening"。主句中有两个并列的谓语动词"is"和"does not depend on ..."。

4.3 MOLECULE SPECTROSCOPY

Molecules, even diatomic species, exhibit spectra that are much more complex than atomic spectra. The additional complexity occurs because the molecular energy is composed of quantized components associated with rotation and vibration of the molecule in addition to the electronic transitions observed in atoms.

The energy of a molecule may be treated as though it consisted of additive quantized components due to ① rotation of the molecule, ② vibration of constituents of the molecule with respect to one another, and ③ intramolecular electronic transitions. The three types of motion can be separated because of the great differences in their frequency of occurrence. The time required for an electronic transition is of the order of 10^{-15} or 10^{-16} s. In this time, the components of a molecule, vibrating with a frequency of approximately 10^{12} s^{-1}, will have covered a distance of only one-thousandth or less of the amplitude of the vibration and can therefore be regarded as fixed during the time required for the electronic transition. In the same fashion, rotations, with a frequency of the order of 10^{10} s^{-1}, are much slower than vibrations.

The energy levels for the rotation of a molecule are relatively close to each other and, if a rotational transition is permitted, the radiation associated with the transition is of long wavelength. Pure rotational spectra are observed in the region of the electromagnetic spectrum extending from the far infrared into the microwave, with wavelengths of the order of 1 cm.

The separation of the vibrational levels is greater, and the corresponding spectra are of shorter wavelength. Vibrational spectra are most often observed in the IR region between 1 and 30 μm. The spacing of electronic levels is usually large, and electronic transitions are, of course, most frequently observed in the UV and visible portions of the electromagnetic spectrum.

In gases, vibrational transitions are generally accompanied by changes in the rotational state, so that the spectrum corresponding to a transition between two vibrational levels, instead of being a single line, contains a number of lines close together representing the rotational transitions superposed on the vibrational transition[1]. Similarly, electronic spectra of gaseous molecules may assume a very complicated structure because of the superposition of vibrational and rotational changes on the electronic transition.

The numbers 0, 1, 2, and so on, assigned to the rotational levels, indicate the values of the rotational quantum number J and, similarly, the vibrational levels are characterized by a vibrational quantum number v. The spacing between rotational levels tends to increase with increasing J, while the spacing between vibrational levels tends to decrease with increasing v. The dashed horizontal lines associated with the electronic levels A and B indicate the energy that the molecule would possess at the absolute zero of temperature if it were possible to suppress all vibratory motion within the molecule. The energy represented by the separation between the dashed line and the level with $v = 0$ and $J = 0$ is the residual energy of the molecule at the absolute zero by virtue of the vibratory motion. This quantity is called the zero-point energy.

There are a great many permitted transitions between the vibrational and rotational levels of A and B, a few of which are indicated as absorption transitions by the vertical arrows. Thus, it is clear that a very complicated pattern may arise in the absorption spectrum corresponding to an electronic transition in a molecule, whereas in an atom an electronic transition gives rise to a single line.

When observed with a spectrometer of sufficiently high resolving power, the visible and UV spectra of diatomic molecules and of some relatively simple polyatomic molecules are indeed found to consist of a large number of fine lines constituting the rotational structure of the spectrum. The rotational structure is, however, seldom observed in analytical spectroscopy. Various circumstances are responsible for this failure to see the rotational structure:

① The lines may be so close together as to be unresolvable with the available instrumentation.

② Most spectra are measured for liquid samples, and molecular collisions occur before a rotation can be completed. When this occurs, the energy levels are perturbed to varying degrees, and the concept of quantized rotations breaks down.

③ The spacing between rotational levels for large molecules may be so small that even in the gas phase the rotational structure is lost, because the width of the lines imposed by Doppler and collisional broadening may be greater than the spacing between consecutive lines.

Unlike rotational structure, vibrational structure of electronic spectra is often seen. The vibrational spectrum associated with an electronic transition consists of a series of more or less broad bands, usually with an asymmetric distribution of intensity determined by the concentration of unresolved rotational lines. To illustrate, the UV absorption spectrum of benzene is associated with a single electronic transition, and the band structure is determined by the vibrational changes which accompany the electronic transition. Vibrational bands persist in the spectra of many species even for liquid and solution samples.

Vocabulary

1. amplitude   ['æmplitju:d] $n.$ 广阔、充足、增幅、振幅

2. superposition [sju:pəpə'ziʃən] n. 重叠、重合
3. quantum ['kwɔntəm] n. 量、总量、总额、量子
4. dash [dæʃ] n.v. 冲撞
5. spectrometer [spek'trɔmitə] n. 分光计
6. polyatomic [pɔliə'tɔmik] a. 多原子的
7. perturb [pə'tə:b] vt. 烦(干)扰、扰(搅)乱、使紊乱

Phrases

1. (be) of the order of 数量级为,大约,约为
2. amplitude of vibration 振幅
3. superpose on 叠加……之上
4. quantum number 量子数
   rotational quantum number 旋转量子数
5. by virtue of 由于,基于,借助于
6. zero-point energy 零点能(量)
7. resolving power 分辨率,分辨能力
8. Doppler broadening 多普勒(谱线)变宽

Note

1. In gases, vibrational transitions are ... on the vibrational transition. "so that"引导的结果状语从句中,主语"the spectrum"后面有一个修饰它的分词短语"corresponding to...","close together"和"representing"(动名词)都是修饰从句宾语"a number of lines"的。"superposed"短语修饰其前名词。

4.4 IR AND RAMAN SPECTROSCOPY

Spectroscopy based on transitions between the vibrational levels of molecules and polyatomic ions forms a powerful and well-developed tool for identification, structure determination, and quantitative estimation of these species. Two rather different experimental approaches to vibrational spectroscopy exist. These are the direct observation of the absorption of radiant energy in the IR region of the spectrum, and the indirect observation of vibrational transitions using a scattering phenomenon known as the Raman effect. Both IR and Raman spectroscopy will be discussed in this chapter.

Vibrational transitions are, in general, of much greater energy than the thermal energy ($KT = 4 \times 10^{-14}$ erg/molecule at 298 K). A typical (5 μm) vibrational transition requires 4×10^{-13} erg; as a result, only the ground vibrational level is appreciably populated at normal temperatures as noted previously. There will, therefore, usually be only one absorption band corresponding to each mode of vibration of a molecule. The vibrational transitions of organic molecules that are most often used in analysis lie in the wavelength region 2 ~ 15 μm. This region is referred to as the "ordinary" infrared. The spectral region between 2 μm and the visible is known as the near infrared (near to the visible), while the region from approximately 50 μm to the microwave region is termed the far infrared (far from the visible). These regions are utilized, but less so than the ordinary infrared.

Infrared radiation was discovered by William Herschel in 1800, but because of the difficulty in producing suitable detectors, it was not until the experiments of W. W. Coblentz in the period 1900 ~ 1910 that the chemical value of IR radiation even began to be recognized. Although Coblentz actually measured the IR absorption spectra of a number of organic liquids, the structure-spectra correlations known now were not perceived and IR spectroscopy remained confined to the realm of the research of academic physicists.

In the 1930s, organic chemists began to consider IR as a possible analytical method, and some of the larger chemical companies undertook the construction of spectrometers. It was World War II, however, that provided the stimulus for the commercial development of IR. At that time, it was realized that IR provided a rapid, accurate method of analyzing the C_4 hydrocarbon fraction required for the production of much-needed synthetic rubber. The combination of instrumental availability with the recognition of the chemical potentialities of the method has since made IR a ubiquitous and well-exploited field.

Although the Raman effect was known during this period of explosive development in IR, Raman spectrometry has lagged behind IR in its chemical applications because of instrumental difficulties that are only now being overcome. This "exploitation gap" seems certain to narrow in the future.

In this section, we shall review briefly some future applications of vibrational spectroscopy beyond the identification and structure determination applications in organic chemistry emphasized in the preceding section. This review is by no means exhaustive, but rather is intended to give the reader some feeling for the potentialities of other aspects of this branch of spectroscopy.

Resolved bands in the ordinary IR region are usually found to obey Beer's law and may form the basis of rapid quantitative analyses. In the past, when the majority of IR spectrometers were of the single-beam type, more use was made of quantitative IR analysis than is done in current practice. There are two reasons for the declining use of IR in quantitative procedures. The convenient, scanning double-beam instruments now in general use are largely of the optical-null design and have lower photometric accuracy than their single-beam predecessors; and other techniques, principally gas chromatography, have replaced IR for quantitative measurements. Nonetheless, IR should not be overlooked as a quantitative tool when an accuracy of a few percent is satisfactory.

The wavelength region between 0.7 and 2.5 μm is customarily regarded as the near infrared. The near IR is distinguished from the NaCl region by instrumental differences: Quartz is a suitable optical material for cells, and double-quartz monochromators using photoconductive detectors are employed. As a result of the differences in instrumentation, high-resolution, ratio-recording spectrometers are generally available, making the near IR suitable for quantitative measurements with precisions in the 0.004 ~ 0.5 percent range.

With few exceptions, all of the absorption bands in this region arise from overtones ($\Delta v > 1$) and combinations involving hydrogen stretching vibrations of molecular fragments X—H. In practice, the majority of spectra studied involve C—H, N—H, and O—H stretching vibrations. In addition to the fundamental spectroscopic interest in these bands, in particular the study of the interactions from which combination bands arise, useful analytical procedures have been developed for the determination of water, alcohols, amines, organic acids, and olefins. For example, the amine content of hydrocarbons can be assayed with good precision and a detectability limit of the order of 0.04 percent using a near IR absorption at 1.5 μm. Studies of hydrogen bonding are also well suited to the near IR.

The far infrared is, as one worker in the field observed, "the region beyond where your instrument stops." While in the recent past, anything beyond 15 μm was considered "far," the routine availability of grating instruments that cover the 2 ~ 50 μm region has caused the far IR to mean the region from 50 ~ 1 000 μm (200 ~ 1 cm^{-1}). This region is instrumentally difficult to study, since no source is available that provides the major fraction of its energy output at the desired frequencies[1]. The source employed is the high-pressure mercury arc which has, of course, its greatest power output in other spectral regions. Only the Golay and pyroelectric detectors are suitable for this very low-energy radiation. Grating instruments function to about 700 μm, although difficulties are encountered in filtering out unwanted orders; the longer-wavelength region is accessible only by interferometry.

Far IR has been of interest to inorganic chemists for some time as a method of studying crystal lattice vibrations. More recently, it has been applied to measurement of the very low molecular vibrational frequencies. Knowledge of these frequencies is needed to complete normal-mode analyses and for the calculation of thermodynamic properties. Other applications of far IR include the study of chain and lattice vibrations in polymers, studies of hindered intramolecular rotations, and the detection of very weak intermolecular interactions. In the last category, studies of hydrogen bond vibrations in the $200 \sim 100$ cm^{-1} region have provided insight on molecular interactions of the liquid phase. Bands below 90 cm^{-1} in spectra of nitriles and nitro compounds have been observed and assigned to vibrations of dipole-dipole complexes. Spectra of organic salts in benzene solution have revealed bands due to interionic vibrations and the vibrations of ion aggregates.

The far IR appears to be exclusively the province of spectroscopists and physical chemists; no routine analytical procedures in this region are known.

Since Raman spectra consist of scattered radiation shifted in wavelength, they appear as emission, rather than absorption spectra. In this spectrum, the intense line with zero shift is the Rayleigh line; the Raman line shifted about 200 cm^{-1} appears symmetrically on both sides of the Rayleigh line with the high-frequency (anti-Stokes) line noticeably less intense than the Stokes line, as anticipated.

In many instances, the magnitudes of the Raman shifts correspond exactly to the frequencies of IR absorptions. This is true for the doublet in the CCl_4 spectrum centered near 770 cm^{-1}: The most intense bands in the IR spectrum of this molecule occur at 782 and 755 cm^{-1}. If, of course, the same information were always available from both techniques, one of the two would be redundant. However, as discussed previously, the two methods represent very different fundamental processes, and some vibrational motions interact only with IR radiation while others produce only Raman effects. For such cases, the IR and Raman spectra complement each other, and the complete vibrational spectrum can be obtained only by using both techniques. From theoretical considerations, more vibrational modes should be active in the Raman than in the IR, so intrinsically more information should be contained in the Raman spectrum.

As noted above, for some structures a rule of mutual exclusion applies, and none of the IR-active modes gives rise to a Raman effect. In such cases, of which polyethylene is one example, the differences in the vibrational frequencies obtained from the two spectra may be used to great advantage in determination of symmetry and structure.

One notes in the discussion of IR functional-group frequencies that most of the organic functions that absorb strongly are highly polar[2]. This fact is, of course, related to the requirement that a change in molecular dipole moment is associated with an IR-active vibration. On the contrary, Raman effects are associated with a change in polarizability, and nonpolar groupings tend to be the most easily polarized; thus these give strong Raman scattering. Application of vibrational spectroscopy to the study of polymeric materials is divided along these lines: IR investigations usually involve study of polar substituents on the polymer chain, whereas Raman spectra give information about the carbon skeleton itself.

Perhaps the single most exploited advantage of Raman spectroscopy is its ability to deal with aqueous solutions, which are anathema to commonly used IR-transparent materials. Much use has been made of the Raman effect in studying polyatomic inorganic ions and metal complexes[3]. Information regarding the symmetry of these species, types of bonding, and the degree of covalency of metal-ligand bonds is obtainable from Raman spectra. It is also possible to make quantitative measurements of concentrations of species in solution from the integrated intensities of Raman bands. On this basis, ionic equilibria in solution can be investigated.

One can say with certainty that the perfection of laser-source Raman instruments will increase the utilization of this method. It is likely to become effective competition for IR in organic analysis and structure determination for the reasons of larger information content of the spectra, and because the entire vibrational frequency range can be examined with one instrument rather than the three separate instruments required to cover the near, medium, and far IR. It can be seen that it would be easy to resolve Raman bands with shifts as small as 100 cm^{-1}; the limitation on observation of low-frequency bands is simply the ability of the instrument to limit the width of the Rayleigh line. The source and detector problems of far IR are circumvented.

Biochemistry is a field in which Raman spectroscopy has obvious applications. The ability to study conformations of polymeric materials combined with the ease of handling aqueous samples means that biopolymers may be studied with relative ease under conditions approximating those in living organisms. An example is the comparison of the Raman spectrum of solid poly-L-proline with that obtained from an aqueous solution of this biopolymer. It was possible to conclude that the polymer retains its helical conformation even in solution, in spite of the fact that it lacks the intramolecular hydrogen bonds usually believed to be the basis of the stability of such helical conformations.

Vocabulary

1. radiant ['reidiənt] a. 发光的、辐射的
2. scatter ['skætə] v. 散射
3. stimulus ['stimjuləs] n. 刺激物、促进因素
4. ubiquitous [juː'bikwitəs] a. 到处存在的、普遍存在的
5. lag [læg] vt. 落后于、滞后
6. declining [di'klainiŋ] a. 倾斜的、衰退中的
7. predecessor ['priːdisesə] n. (被取代的)原有事物
8. monochromator [mɔnəu'krəumeitə] n. 单色光镜

9. precision [pri'siʒən] n. 精度、精密度
10. overtone ['əuvətəun] n. 折射的光彩
11. assay [ə'sei] n.v. 化验
12. grating ['greitiŋ] n. 光栅
13. pyroelectric [paiərəui'lektrik] a. (焦)热电的
14. filtering ['filtəriŋ] n. 滤波器
15. interferometry [ˌintəfə'rɔmitri] n. 干涉测量(法)
16. nitrile ['naitrail] n. 腈
17. redundant [ri'dʌndənt] a. 多余的、过多的、冗长的
18. polarizable ['pouləraizəbl] a. 可极化的
19. proline ['prəuli:n] n. 脯氨酸

Phrases

1. near(far) infrared 近(远)红外
2. single(double)-beam 单(双)光束
3. optical null method 光平衡法
4. stretching vibration 伸缩振动
5. crystal lattice 晶格
6. normal-mode 简正模式
7. integrated intensity 积分强度

Notes

1. This region is ... at the desired frequencies. 注意句中从句"that provides..."修饰"no source",二者被谓语动词和表语分隔开了。
2. One notes in the discussion ... are highly polar. 句中 note 为及物动词,其宾语为 that 引导的从句"that most of ..."。"note"和宾语从句之间插入了一个"in"开头的介词短语(做状语)。
3. Much use has been made ... and metal complexes. 此句为被动语态,应用了短语"make use of"。由于缺少主语,句子相应变为被动语态,由 use 做主语。

4.5 INTRODUCTION TO ELECTROANALYTICAL METHODS OF ANALYSIS

Analytical techniques based on electrochemical principles make up one of the three major divisions of instrumental analytical chemistry. Each basic electrical measurement of current, resistance, and voltage has been used alone or in combination for analytical purposes. If these electrical properties are measured as a function of time, many additional electroanalytical methods of analysis are possible (Table 4.4).

Table 4.4   Selected Electroanalytical Methods

Quantity measured	Variable controlled	Name of method
E	$i=0$	1. Ion selective potentiometry
		2. Null-point potentiometry
E versus volume of titrant	$i=0$	Potentiometric titrations
Weight of separated phase	E	Controlled potential electrodeposition
i versus E	Concentration	Voltammetry
	t	Linear potential sweep stripping chronoamperometry
	t	Linear potential sweep voltammetry
i versus volume of titrant	E	Amperometric titrations
Coulombs (current × time)	E	Coulometry
$1/R$ (conductance)	Concentration	Conductance measurements
$1/R$ versus volume of titrant		Conductometric titrations

The individual techniques are best recognized by their excitation-response characteristics. Less confusion arises when each technique is described by an operational nomenclature that consists of an independent-variable part followed by a dependent-variable part. An example is voltammetry. The name is often preceded by system-specific modifiers such as cyclic voltammetry or square-wave voltammetry.

There are highly refined ways of making reliable electrical measurements in the submicroampere and microvolt range. Reliable analyses at the picogram range are possible. By contrast, selectivity is one of the weakest aspects of electrochemical methods due to poor resolution. However, the combination of electroanalytical methods with chromatography is a powerful tool for both qualitative and quantitative work. Chromatography provides the selectivity and electroanalytical methods provide the sensitivity.

Electroanalytical methods are conveniently divided into two categories: steady state and transient. The steady-state or static methods, such as potentiometry, are firmly rooted in the basic concepts of equilibrium and mass action, and the rigor of their presentation in introductory analytical and physical chemistry courses is considered adequate. Steady-state methods entail measurements of the potential difference at zero current. The system defined by the solid-solution interface is not disturbed and equilibrium is maintained. Time is effectively eliminated as a variable, and equilibrium is assured by vigorously stirring the solution with the indicator electrode held stationary, or vice versa. Any concentration gradients at the solution-electrode interface are completely or nearly completely eliminated. In such cases, the indicator electrode potential is related to bulk and surface concentration.

Voltammetry is concerned with the current-potential relationship in an electrochemical cell and, in particular, with the current-time response of an electrode at a controlled potential. In a typical voltammetric experiment, the amount of material actually removed or converted to another form is quite small. Polarography is the name applied to dc voltammetry at the dropping mercury electrode. Although the term polarography is acceptable in this context, its more general usage is undesirable.

The integrated current (current multiplied by time), or charge (coulombs), is a measure of the total amount of material converted to another form. In controlled-potential coulometry and in controlled-potential electroanalysis, this corresponds to the total removal of the reactant solute species.

There is yet another group of methods in which electron-transfer reactions and diffusional transport are unimportant. Charge transport by migration forms the basis for conductometry and conductometric titrations.

Vocabulary

1. nomenclature [nəu'menklətʃə] n. 命名
2. precede [pri'siːd] v. 领先于，在……之前
3. vigor ['vigə] n. 严格、严厉、苛刻、严密、精确
4. polarography [ˌpəulə'rɔgrəfi] n. 极谱法
5. titration [ti'treiʃən] n. 滴定

Phrases

1. cyclic voltammetry 循环伏安法
   square-wave voltammetry 方波伏安法
2. null-point 零(电流)点
3. potentiometric titration 电位滴定(法)
   amperometric titration 电流滴定法
4. indicator electrode 指示电极
5. dropping mercury electrode 滴汞电极
6. integrated current 积分电流

4.6 THERMAL ANALYSIS

Thermal analysis includes a group of techniques in which specific physical properties of a material are measured as a function of temperature. The production of new high-technology materials and the resulting requirement for a more precise characterization of these substances have increased the demand for thermal analysis techniques. Current areas of application include environmental measurements, composition analysis, product reliability, stability, chemical reactions, and dynamic properties.

Thermal analysis has been used to determine the physical and chemical properties of polymers, electronic circuit boards, geological materials, and coals. An integrated, modern thermal analysis instrument can measure transition temperatures, weight losses, energies of transitions, dimensional changes, modulus changes, and viscoelastic properties.

Thermal analysis is useful in both quantitative and qualitative analyses. Samples may be identified and characterized by qualitative investigations of their thermal behavior. Information concerning the detailed structure and composition of different phases of a given sample is obtained from the analysis of thermal data. Quantitative results are obtained from changes in weight and enthalpy as the sample is heated. The temperatures of phase changes and reactions as well as heats of reaction are used to determine the purity of materials.

Table 4.5 Summary of Thermal Analysis Techniques

Technique	Quantity measured	Typical application
Differential scanning calorimetry (DSC)	Heats and temperatures of transitions and reactions	Reaction kinetics, purity analysis, polymer cures
Differential thermal analysis (DTA)	Temperatures of transitions and reactions	Phase diagrams, thermal stability
Thermogravimetric analysis (TGA)	Weight change	Thermal stability, compositional analysis
Thermomechanical analysis (TMA)	Dimension and viscosity changes	Softening temperatures, expansion coefficients
Dynamic mechanical analysis (DMA)	Modulus, damping, and viscoelastic behavior	Impact resistance, mechanical stability
Evolved gas analysis (EGA)	Amount of gaseous products of thermally induced reaction	Analysis of volatile organic components of shale

The use of microprocessors has both enhanced and simplified the techniques of thermal analysis. The sample is heated at a programmed rate in the controlled environment of the furnace. Changes in selected properties of a sample are monitored by specific transducers, which generate voltage signals. The signal is then amplified, digitized, and stored on a magnetic disk along with the corresponding direct temperature responses from the sample. The data may also be displayed or plotted in real time. The microcomputer is used to process the data with a library of applications software designed for thermal analysis techniques. The multitasking capabilities of some computer systems allow a single microcomputer to operate several thermal analyzers simultaneously and independently.

A major advantage of microcomputer systems in thermal analysis is that the operator seldom, if ever, needs to repeat an analysis because of an improper choice of ordinate scale sensitivity. The software does this rescaling after all the data have been collected. In some systems both axes are automatically rescaled after the last data point has been received. When the amount of time necessary to obtain thermal data is considered, the advantage is obvious. For example, a differential scanning calorimetric run at 10°C/min from room temperature to 1 100°C takes 100 min, and thus the rerun of the sample would take longer than 2 h including the time required for cooling and sample reloading.

Differential scanning calorimetry (DSC) has become the most widely used thermal analysis technique. In this technique, the sample and reference materials are subjected to a precisely programmed temperature change. When a thermal transition (a chemical or physical change that results in the emission or absorption of heat) occurs in the sample, thermal energy is added to either the sample or the reference containers in order to maintain both the sample and reference at the same temperature. Because the energy transferred is exactly equivalent in magnitude to the energy absorbed or evolved in the transition, the balancing energy yields a direct calorimetric measurement of the transition energy. Since DSC can measure directly both the temperature and the enthalpy of a transition or the heat of a reaction, it is often substituted for differential thermal analysis as a means of determining these quantities except in certain high-temperature applications.

In differential thermal analysis (DTA), the difference in temperature between the sample and a thermally inert reference material is measured as a function of temperature (usually the sample temperature). Any transition that the sample undergoes results in liberation or absorption of energy by the sample with a corresponding deviation of its temperature from that of the reference. A plot of the differential temperature, ΔT, versus the programmed temperature, T, indicates the transition temperature(s) and whether the transition is exothermic or endothermic. DTA and thermogravimetric analyses (measurement of the change in weight as a function of temperature) are often run simultaneously on a single samle.

Thermogravimetry (TG) or thermogravimetric analysis (TGA) provides a quantitative measurement of any weight changes associated with thermally induced transitions. For example, TG can record directly the loss in weight as a function of temperature or time (when operating under isothermal conditions) for transitions that involve dehydration or decomposition. Thermogravimetric curves are characteristic of a given compound or material due to the unique sequence of physical transitions and chemical reactions that occur over definite temperature ranges. The rates of these thermally induced processes are often a function of the molecular structure. Changes in weight result from physical and chemical bonds forming and breaking at elevated temperatures. These processes may evolve volatile products or form reaction products that result in a change in weight of the sample. TG data are useful in characterizing materials as well as in investigating the thermodynamics and kinetics of the reactions and transitions that result from the application of heat to these materials. The usual temperature range for TG is from ambient to 1 200℃ in either inert or reactive atmospheres.

The analysis of the purge gas exit stream from differential thermal analysis, differential scanning calorimetry, and thermogravimetric analyzers is useful in establishing mechanisms and stoichiometric relationships of thermal decompositions. In evolved gas analysis (EGA) the absolute identities of the gaseous components are determined, whereas in evolved gas detection (EGD) the presence of only a single, preselected component of the evolved gas is sensed. An appropriate analyzer may be coupled to a thermogravimetric system for performing either EGA or EGD. The resulting hyphenated methods are powerful analytical tools. Two analyzers that have been successfully coupled to TG systems are mass spectrometers (MS) and flame ionization detectors (FID). The TG-MS or TG-MS-MS combination is used for evolved gas analysis, whereas the TG-FID combination provides evolved gas detection. These hyphenated methods are used in studies of the volatile organic pyrolysis products of oil shales.

Vocabulary

1. modulus   ['mɔdjuləs]   n. 率、系数、模数
2. digitize   ['didʒitaiz]   vt. 使数字化
3. rescaling   [ri'skeiliŋ]   n. 尺度改变、改比例
4. stoichiometric   [ˌstɔikiə'metrik]   a. 化学数量的、化学计算的
5. pyrolysis   [pai'rɔlisis]   n. 高温分解
6. transducer   [trænz'dju:sə]   n. 变换器、转换机构、传感器

Phrases

1. transition temperature   转变温度
2. viscoelastic property   粘弹性能
3. differential scanning calorimetry(DSC)   差示扫描量热法
   differential thermal analysis (DTA)   差热分析
   thermogravimetric analysis (TG)   热重分析
   thermomechanical analysis (TMA)   热机械分析
   dynamic mechanical analysis(DMA)   动态机械分析
4. evolved gas analysis   逸出气分析

5. reference material   参比物,标准物质
6. programmed rate   程序控制的速率
7. programmed temperature   程序控制温度
8. thermal decomposition   热分解
9. flame ionization detector   火焰离子(化)检测器
10. oil shale   油页岩

5

Nomenclature

5.1 NOMENCLATURE OF INORGANIC COMPOUNDS

5.1.1 Common or Trivial Names

Chemical nomenclature is the system of names that chemists use to identify compounds. When a new substance is formulated, it must be named in order to distinguish it from all other substances. Before chemistry was systematized, a substance was given a name that generally associated it with one of its outstanding physical or chemical properties. For example, nitrous oxide, N_2O, used as an anesthetic in dentistry, has been called "laughing gas," because it induces laughter when inhaled. The name "nitrous oxide," however, is now giving way to the more systematic name "dinitrogen oxide." Nonsystematic names are called common, or trivial names.

Common names for chemicals are widely used in many industries, since the systematic name frequently is too long or too technical for everyday use. For example, CaO is called "lime," not "calcium oxide", by plasterers; nurserymen called $CCl_3CH(C_6H_4Cl)_2$ by the abbreviation "DDT," not "dichlorodiphenyltrichloroethane." These common names are chemical nicknames, and, as the DDT example shows, there is a practical need for short, common names. Table 5.1 lists the common names, formulas, and chemical names of some familiar substances.

Table 5.1 Common Names, Formulas, and Chemical Names of Some Familiar Substances

Common Name	Formula	Chemical Name
Acetylene	C_2H_2	Ethyne
Lime	CaO	Calcium oxide
Slaked lime	$Ca(OH)_2$	Calcium hydroxide
Galena	PbS	Lead(II) sulfide
Alumina	Al_2O_3	Aluminum oxide
Baking soda	$NaHCO_3$	Sodium hydrogen carbonate
Cane or beet sugar	$C_{12}H_{22}O_{11}$	Sucrose
Blue stone, blue vitriol	$CuSO_4 \cdot 5H_2O$	Copper(II) sulfate pentahydrate
Calcite, marble, limestone	$CaCO_3$	Calcium carbonate
Grain alcohol	C_2H_5OH	Ethyl alcohol, ethanol
Litharge	PbO	Lead(II) oxide
Washing soda	$Na_2CO_3 \cdot 10H_2O$	Sodium carbonate decahydrate
Wood alcohol	CH_3OH	Methyl alcohol, methanol

5.1.2 Systematic Chemical Nomenclature

The trivial name is not entirely satisfactory to the chemist who wants a name that will tell the composition of the substance. Therefore, as the number of known compounds increased, it became necessary to develop a scientific, systematic method of identifying compounds by name. The systematic method of naming inorganic compounds considers the compound to be composed of two parts, one positive and one negative. The positive part is named and written first. The negative part, generally nonmetallic, is named second. Names of the elements are modified with suffixes and prefixes to identify the different types or classes of compounds. Thus, the compound composed of sodium ions and chloride ions is named sodium chloride; the compound composed of iron(II) and chloride ions is named iron(II) chloride (read as iron-two chloride).

5.1.3 Binary Compounds

Binary compounds contain only two different elements. Their names consist of two parts, the name of the more electropositive element followed by the name of the electronegative element, which is modified to end in ide.

(1) Binary compounds in which the electropositive element has a fixed oxidation state. The chemical name is composed of the name of the metal, which is written first, followed by the name of the nonmetal, which has been modified to an identifying stem to which is added the suffix ide. For example, sodium chloride, NaCl, is composed of one atom each of sodium and chlorine. The compound name is sodium chloride.

Elements: Sodium (metal)
    Chlorine (nonmetal)
    name modified to the stem chlor + ide
Name of compound: Sodium chloride

Stems of the more common nonmetals are shown in the following table.

Symbol	Element	Stem	Binary Name Endings
B	Boron	Bor-	Boride
Br	Bromine	Brom-	Bromide
Cl	Chlorine	Chlor-	Chloride
F	Fluorine	Fluor-	Fluoride
H	Hydrogen	Hydr-	Hydride
I	Iodine	Iod-	Iodide
N	Nitrogen	Nitr-	Nitride
O	Oxygen	Ox-	Oxide
P	Phosphorus	Phosph-	Phosphide
S	Sulfur	Sulf- or Sulfur-	Sulfide

Compounds may contain more than one atom of the same element, the name follows the rule for binary compounds.

Examples:

$CaBr_2$ Mg_3N_2

Calcium bromide   Magnesium nitride

Table 5.2 shows more examples of compounds with names ending in ide.

Table 5.2 Examples of Compounds with Names Ending in ide

Formula	Name
$MgBr_2$	Magnesium bromide
Na_2O	Sodium oxide
CaC_2	Calcium carbide
$AlCl_3$	Aluminum chloride
PbS	Lead(II) sulfide
LiI	Lithium iodide
NH_4F	Ammonium fluoride
NaOH	Sodium hydroxide
KCN	Potassium cyanide

(2) Binary compounds containing metals of variable oxidation numbers and nonmetals. Two systems are commonly used for compounds in this category. The official system, designated by the International Union of Pure and Applied Chemistry (IUPAC), is known as the Stock System. In the Stock System, when a compound contains a metal that can have more than one oxidation number, the oxidation number of the metal in the compound is designated by a Roman numeral in parentheses (e.g., (II)) written immediately after the name of the metal. The Roman numerals are used only in the name, not in the formula of the compound. The nonmetal is treated in the same manner as in the previous case.

Examples:

$FeCl_2$	Iron(II) chloride	(Fe^{2+})
$FeCl_3$	Iron(III) chloride	(Fe^{3+})
CuCl	Copper(I) chloride	(Cu^+)
$CuCl_2$	Copper(II) chloride	(Cu^{2+})

When a metal has only one possible oxidation state there is no need to distinguish one oxidation state from another, so Roman numerals are not needed and are not used.

In classical nomenclature, when the metallic ion has only two oxidation numbers, the name of the metal is modified with the suffixes ous and ic to distinguish between the two. The lower oxidation state is given the ous ending and the higher one the ic ending.

Examples:

$FeCl_2$	Ferrous chloride	(Fe^{2+})
$FeCl_3$	Ferric chloride	(Fe^{3+})
CuCl	Cuprous chloride	(Cu^+)
$CuCl_2$	Cupric chloride	(Cu^{2+})

Notice that the ous-ic system does not indicate the oxidation state of an element but merely that at least two oxidation states exist. The Stock System clearly indicates the oxidation number and avoids the uncertainty of the oxidation numbers for ous and ic compounds.

(3) Binary compounds containing two nonmetals. The most electropositive element is named first. In a compound between two nonmetals, the element that occurs earlier in the following sequence is written and named first in the formula: B, Si, C, P, N, H, S, I, Br, Cl, O, F. To each element is attached a Latin or Greek prefix indicating the number of atoms of that element in the molecule. The second element still retains the modified binary ending. The prefix mono is generally omitted except when needed to distinguish between two or more compounds. Common prefixes and their numerical equivalences are the following.

Mono = 1   Hexa = 6
Di = 2   Hepta = 7
Tri = 3   Octa = 8
Tetra = 4   Nona = 9
Penta = 5   Deca = 10

Examples of compounds illustrating this system are shown below.

CO	Carbon monoxide
CO_2	Carbon dioxide
PCl_3	Phosphorus trichloride
PCl_5	Phosphorus pentachloride
N_2O	Dinitrogen oxide
N_2O_4	Dinitrogen tetroxide
NO	Nitrogen oxide
N_2O_3	Dinitrogen trioxide

(4) Exceptions using ide endings. Three notable exceptions using the ide ending are hydroxides (OH^-), cyanides (CN^-), and ammonium (NH_4^+) compounds. These polyatomic ions, when combined with another element, take the ending ide, even though more than two elements are present in the compound.

NH_4I	Ammonium iodide
$Ca(OH)_2$	Calcium hydroxide
KCN	Potassium cyanide

(5) Acids derived from binary compounds. Certain binary hydrogen compounds, when dissolved in water, form solutions that have acid properties. Because of this property, these compounds are given acid names in addition to their regular ide names. Binary acids are composed of hydrogen and one other nonmetallic element. However, not all binary hydrogen compounds are acids. To express the formula of a binary acid, it is customary to write the symbol of hydrogen first, followed by the symbol of the second element (e.g., HCl, HBr, H_2S).

To name a binary acid, place the prefix *hydro* in front of, and the letters *ic* after, the stem of the nonmetal. Then add the word "acid."

Examples:

HCl

Hydro chlor/ic acid(hydrochloric acid)

H_2S

Hydro sulfur/ic acid (hydrosulfuric acid)

Table 5.3 shows examples of other binary acids.

Table 5.3 Names of Selected Binary Acids

Formula	Acid Name
HF	Hydrofluoric acid
HCl	Hydrochloric acid
HBr	Hydrobromic acid
HI	Hydroiodic acid
H_2S	Hydrosulfuric acid
H_2Se	Hydroselenic acid

5.1.4 Ternary Compounds

Ternary compounds contain three elements: an electropositive group, which is either a metal or hydrogen, combined with a polyatomic negative ion. We will consider the naming of compounds in which one of the three elements is oxygen.

In general, in naming ternary compounds the name of the positive group is given first, followed by the name of the negative ion. The negative group usually contains two elements: oxygen, and a metal or a nonmetal. To name the polyatomic negative ion, add the endings ite or ate to the stem of the element other than oxygen. Thus, SO_4^{2-} is called sulfate, and SO_3^{2-} is called sulfite. Note that oxygen is not specifically included in the name. The suffixes ite and ate represent different oxidation states of the element other than oxygen in the polyatomic ion.

ite	ate
$CaSO_3$	$CaSO_4$
Calcium sulfite	Calcium sulfate
ite ending indicates	ate ending indicates
lower oxidatioin state	higher oxidation state

When an element has only one oxidation state, such as C in carbonate, the ate ending is used. Examples of ternary compounds and their names are given in Table 5.4.

Table 5.4 Names and Formula of Selected Ternary Compounds

Formula	Name
K_2CO_3	Potassium carbonate
$Al_2(SO_4)_3$	Aluminum sulfate
$KClO_3$	Potassium chlorate
$AlPO_4$	Aluminum phosphate
$FeSO_4$	Iron(II) sulfate or ferrous sulfate
$Fe_2(SO_4)_3$	Iron(III) sulfate or ferric sulfate
$PbCrO_4$	Lead(II) chromate
H_2SO_4	Hydrogen sulfate
HNO_3	Hydrogen nitrate
Li_3AsO_4	Lithium arsenate
$NaNO_2$	Sodium nitrite
$ZnMoO_4$	Zinc molybdate

Inorganic ternary compounds containing hydrogen, oxygen, and one other element are called oxy-acids. The ous-ic system is used in naming ternary acids. The suffixes ous and ic are used to indicate different oxidation states of the element other than hydrogen and oxygen. The ous ending again indicates the lower oxidation state and the ic ending, the higher oxidation state.

To name these acids, we place the ending ic or ous after the stem of the element other than hydrogen and oxygen, and add the word "acid." If an element has only one usual oxidation state, the ic ending is used. Hydrogen in a ternary oxy-acid is not specifically designated in the acid name, but its presence is implied by use of the word "acid."

Examples:

$$H_2SO_3 \qquad\qquad H_2SO_4$$
Sulfur/ous acid sulfur/ic acid

In cases where there are more than two oxy-acids in a series, the ous-ic names are further modified with the prefixes per and hypo. Per is placed before the stem of the element other than hydrogen and oxygen when the element has a higher oxidation number than in the ic acid. Hypo is used as a prefix before the stem when the element has a lower oxidation number than in the ous acid. The use of per and hypo is illustrated in the oxy-acids of chlorine.

Formula	Name	Oxidation Number of Chlorine
HClO	Hypochlorous acid	+1
HClO$_2$	Chlorous acid	+3
HClO$_3$	Chloric acid	+5
HClO$_4$	Perchloric acid	+7

Examples of other ternary oxy-acids and their names are shown in Table 5.5. The endings ous, ic, ite and ate are part of classical nomenclature; they are not used in the Stock System to indicate different oxidation states of the elements. These endings are still used, however, in naming many common compounds.

Table 5.5 Names and Formulas of Selected Ternary oxy-acids

Formula	Acid Name	Formula	Acid Name
H$_2$SO$_3$	Sulfurous acid	HNO$_2$	Nitrous acid
H$_2$SO$_4$	Sulfuric acid	NHO$_3$	Nitric acid
H$_3$PO$_2$	Hypophosphorous acid	HBrO$_3$	Bromic acid
H$_3$PO$_3$	Phosphorous acid	HIO$_3$	Iodic acid
H$_3$PO$_4$	Phosphoric acid	H$_3$BO$_3$	Boric acid
HClO	Hypochlorous acid	H$_2$C$_2$O$_4$	Oxalic acid
HClO$_2$	Chlorous acid	HC$_2$H$_3$O$_2$	Acetic acid
HClO$_3$	Chloric acid	H$_2$CO$_3$	Carbonic acid
HClO$_4$	Perchloric acid		

5.1.5 Salts

The same rules given above are used in naming the positive part of a salt. In binary compounds the usual ide ending is given to the negative part of the salt name. In ternary compounds, the ous and ic endings of the acids become ite and ate, respectively, in the salt names, but the names of the stems remain the same.

Ternary oxy-acid		Ternary oxy-salt
ous ending of acid	becomes	ite ending in salt
ic ending of acid	becomes	ate ending in salt

Acid		Salt	
H_2SO_4	Sulfur/ic acid	Na_2SO_4	Sodium sulf/ate
H_2SO_3	Sulfur/ous acid	$CaSO_3$	Calcium sulf/ite
HClO	Hypochlor/ous acid	LiClO	Lithium hypochlor/ite
$HClO_4$	Perchlor/ic acid	$NaClO_4$	Sodium perchlor/ate

Other examples of ternary acids and salts are given in Table 5.6

Table 5.6 Comparison of Acid and Salt Names in Ternary oxy-Compounds

Acid	Salt	Name of Salt
HNO_3	KNO_3	Potassium nitrate
Nitric acid	$HgNO_3$	Mercury(I) nitrate or mercurous nitrate
	$Hg(NO_3)_2$	Mercury(II) nitrate or mercuric nitrate
	$Fe(NO_3)_2$	Iron(II) nitrate or ferrous nitrate
	$Al(NO_3)_3$	Aluminum nitrate
HNO_2	KNO_2	Potassium nitrite
Nitrous acid	$Co(NO_2)_2$	Cobalt(II) nitrite or cobaltous nitrite
	$Mg(NO_2)_2$	Magnesium nitrite
HClO	NaClO	Sodium hypochlorite
$HClO_2$	$NaClO_2$	Sodium chlorite
$HClO_3$	$NaClO_3$	Sodium chlorate
$HClO_4$	$NaClO_4$	Sodium perchlorate

5.1.6 Salts with More Than One Positive Ion

Salts may be formed from acids containing two or more acid hydrogen atoms by replacing only one of the hydrogen atoms with a metal or by replacing both hydrogen atoms with different metals. Each positive group is named first and then the appropriate salt ending is added.

Acid	Salt	Name
H_2CO_3	$NaHCO_3$	Sodium hydrogen carbonate or sodium bicarbonate
H_3PO_4	$MgNH_4PO_4$	Magnesium ammonium phosphate
H_2SO_4	$NaKSO_4$	Sodium potassium sulfate

Note the name "sodium bicarbonate." The prefix bi is commonly used to indicate a compound in which one of two acid hydrogen atoms has been replaced by a metal. Table 5.7 shows examples of other salts containing more than one positive ion.

Table 5.7 Names of Selected Salts Containing more than One Positive Ion

Acid	Salt	Name of Salt
H_2SO_4	$KHSO_4$	Potassium hydrogen sulfate or potassium bisulfate
H_2SO_3	$Ca(HSO_3)_2$	Calcium hydrogen sulfite or calcium bisulfite
H_2S	NH_4HS	Ammonium hydrogen sulfide or ammonium bisulfide
H_3PO_4	$MgNH_4PO_4$	Magnesium ammonium phosphate
H_3PO_4	NaH_2PO_4	Sodium dihydrogen phosphate
H_3PO_4	Na_2HPO_4	Disodium hydrogen phosphate

Note that prefixes are also used in chemical nomenclature to give special clarity or emphasis to certain compounds as well as to distinguish between two or more compounds. Examples follow.

Na_3PO_4	Trisodium phosphate
Na_2HPO_4	Disodium hydrogen phosphate
NaH_2PO_4	Sodium dihydrogen phosphate

5.1.7 Bases

Inorganic bases contain the hydroxyl group, OH^-, in chemical combination with a metal ion. These compounds are called hydroxides. The OH group is named as a single ion and is given the ending *ide*. Several common bases are listed below.

KOH	Potassium hydroxide
NH_4OH	Ammonium hydroxide
$Ca(OH)_2$	Calcium hydroxide

5.1.8 Coordination Compounds

In formulae the usual practice is to place the symbol for the central atom(s) first (except in formulae which are primarily structural), with the ionic and neutral ligands following and the formula for the whole complex enclosed in square brackets.

In names the central atom(s) should be placed after the ligands.

(1) Numeral Prefixes

The presence of identical ligands in a complex is shown by using prefixes di, tri, tetra, penta, hexa, octa, etc. for 2, 3, 4, 5, 6, 8, etc. respectively. For organic ligands, prefixes *bis*, *tris*, etc. are used to avoid confusion. Thus, bispyridine means presence of two pyridine ligands, dipyridine is the ligand $C_5H_5N-NC_5H_5$. Trismethylamine means the presence of three methyl amine as ligands.

(2) Indication of oxidation number and proportion of constituents

The names of coordination entities always have been intended to indicate the charge of the central atom (ion) from which the entity is derived. Since the charge on the coordination entity is the algebraic sum of the charges of the constituents, the necessary information may be supplied by giving either the STOCK number (formal charge on the central ion, i. e., oxidation number) or the EWENS-BASSETT number:

$K_3[Fe(CN)_6]$ potassium hexacyanoferrate(III)
                        potassium hexacyanoferrate(3−)
                        tripotassium hexacyanoferrate
$K_4[Fe(CN)_6]$ potassium hexacyanoferrate(II)
                        potassium hexacyanoferrate(4−)
                        tetrapotassium hexacyanoferrate

Structural information may be given in formulae and names by prefixes such as *cis*, *trans*, *fac*, *mer*, etc..

Anions are given the termination -ate. Cations and neutral molecules are given no distinguishing termination.

The ligands are listed in alphabetical order regardless of the number of each. The name of a ligand is treated as a unit. Thus, "diammine" is listed under "a" and "dimethylamine" under "d". [Until the present Rules were published in 1971, this rule required the listing of negative ligands first (e.g., dichlorodiammineplatinum(II) rather than the proposed diamminedichloroplatinum(II)). The older usage is found in much of the literature cited.]

(3) The names of anionic ligands, whether inorganic or organic, end in -o

In general, if the anion name ends in -ide, -ite, or -ate, the final -e is replaced by -o, giving -ido, -ito, and -ato, respectively. Enclosing marks are required for inorganic anionic ligands containing numerical prefixes, as (triphosphato), and for thio, seleno, and telluro analogues of oxo anions containing more than one atom, as (thiosulfato). Examples of organic anionic ligands which are named in this fashion [are]:

CH_3COO^- acetato
$(CH_3)_2N^-$ dimethylamido

The anions listed below do not follow exactly the above rule and modified forms have become established:

Ion		Ligand
F⁻	fluoride	fluoro
Cl⁻	chloride	chloro
Br⁻	bromide	bromo
I⁻	iodide	iodo
O^{2-}	oxide	oxo
H⁻	hydride	hydrido(hydro)①
OH⁻	hydroxide	hydroxo
O_2^{2-}	peroxide	peroxo
CN⁻	cyanide	cyano

The letter in each of the ligand names which is used to determine the alphabetical listing is given in boldface type in the following examples to illustrate the alphabetical arrangement. For many compounds, the oxidation number of the central atom and/or the charge on the ion are so well known that there is no need to use either a Stock number or a Ewens-Bassett number. However, it is not wrong to use such numbers and they are included here.

$Na[B(NO_3)_4]$ sodium tetranitratoborate(1−)
        sodium tetranitratoborate(III)

$K_2[Os(Cl_5N)]$ potassium pentachloronitridoosmate(2−)
        potassium pentachloronitridoosmate(VI)

$[Co(NH_2)_2(NH_3)_4]OC_2H_5$ diamidotetraamminecobalt(1+) ethoxide
        diamidotetraamminecobalt(III) ethoxide

$[CoN_3(NH_3)_5]SO_4$ pentaammineazidocobalt(2+) sulfate
        pentaammineazidocobalt(III) sulfate

$Na_3[Ag(S_2O_3)_2]$ sodium bis(thiosulfato)argentate(3−)
        sodium bis(thiosulfato)argentate(I)

$NH_4[Cr(NCS)_4(NH_3)_2]$ ammonium diamminetetrakis(isothiocyana-

---

① Both hydrido and hydro are used for coordinated hydrogen, but the latter term usually is restricted to boron compounds.

	to)chromate(1 −)
	ammonium diamminetetrakis(isothiocyanato)chromate(III)
Ba[BrF$_4$]$_2$	barium tetrafluorobromate(1 −)
	barium tetrafluorobromate(III)

Although the common hydrocarbon radicals generally behave as anions when they are attached to metals and in fact are sometimes encountered as anions, their presence in coordination entities is indicated by the customary radical names even though they are considered as anions in computing the oxidation number.

K[B(C$_6$H$_5$)$_4$]	potassium tetraphenylborate(1 −)
	potassium tetraphenylborate(III)
K[SbCl$_5$C$_6$H$_5$]	potassium pentachloro(phenyl)antimonate(1 −)①
	potassium pentachloro(phenyl)antimonate(V)

The name of the coordinated molecule or cation is to be used without change. All neutral ligands are set off with enclosing marks. (Four exceptions are listed below.)

cis-[PtCl$_2$(Et$_3$P)$_2$]	cis-dichlorobis(triethylphosphine)platinum
	cis-dichlorobis(triethylphosphine)platinum(II)
[CuCl$_2$(CH$_3$NH$_2$)$_2$]	dichlorobis(methylamine)copper
	dichlorobis(methylamine)copper(II)
[Pt(py)$_4$][PtCl$_4$]	tetrakis(pyridine)platinum(2 +) tetrachloroplatinate(2 −)
	tetrakis(pyridine)platinum(II) tetrachloroplatinate(II)
[Co(en)$_3$]$_2$(SO$_4$)$_3$	tris(ethylenediamine)cobalt(3 +) sulfate
	tris(ethylenediamine)cobalt(III) sulfate
K[PtCl$_3$(C$_2$H$_4$)]	potassium trichloro(ethylene)platinate(1 −)
	potassium trichloro(ethylene)platinate(II) or

---

① Normally phenyl would not be placed within enclosing marks. They are used here to avoid confusion with a chlorophenyl radical.

potassium trichloromonoethyleneplatinate(II)

[Ru(NH$_3$)$_5$(N$_2$)]Cl$_2$ pentaammine(dinitrogen)ruthenium(2 +) chloride
pentaammine(dinitrogen)ruthenium(II) chloride

Water and ammonia as neutral ligands in coordination complexes are called "aqua"(formerly "aquo") and "ammine", respectively. The groups NO and CO, when linked directly to a metal atom, are called "nitrosyl" and "carbonyl", respectively. In computing the oxidation number these ligands are treated as neutral.

[Cr(H$_2$O)$_6$]Cl$_3$ hexaaquachromium(3 +) chloride
hexaaquachromium trichloride

Na$_2$[Fe(CN)$_5$NO] sodium pentacyanonitrosylferrate(2 -)
sodium pentacyanonitrosylferrate(III)

K$_3$[Fe(CN)$_5$CO] potassium carbonylpentacyanoferrate(3 -)
potassium carbonylpentacyanoferrate(II)

(4) Alternative modes of linkage of some ligands

The different points of attachment of a ligand may be denoted by adding the italicized symbol(s) for the atom or atoms through which attachment occurs at the end of the name of the ligand. Thus the dithiooxalato anion

$$\begin{array}{c} ^-S \quad\quad S^- \\ \diagdown\quad\quad\diagup \\ C-C \\ \diagup\quad\quad\diagdown \\ O \quad\quad\quad O \end{array}$$

conceivably may be attached through S or O, and these are distinguished as dithiooxalato-S, S' and dithiooxalato-O, O', respectively.

In some cases different names are already in use for alternative modes of attachment as, for example, thiocyanato(—SCN) and isothiocyanato(—NCS), nitro(—NO$_2$), and nitrito(—ONO).

Na$_3$[Co(NO$_2$)$_6$] sodium hexanitrocobaltate(3 -)
sodium hexanitrocobaltate(III)

Co(ONO)(NH$_3$)$_5$]SO$_4$ pentaamminenitritocobalt(2 +) sulfate
pentaamminenitritocobalt(III) sulfate

$[Co(NCS)(NH_3)_5]Cl_2$ pentaammineisothiocyanatocobalt(2+) chloride
pentaammineisothiocyanatocobalt(III) chloride

Vocabulary

1. nomenclature ['noumənkleitʃə] n. 命名法、专门语、术语
2. anesthetic [ænəs'θetik] n. 麻醉剂、麻药、麻醉的
3. plasterer ['plæstərə] n. 泥水匠、石膏师
4. thiosulfate [θaiou'sʌlfeit] n. 硫代硫酸盐
5. phenyl ['fenil] n. 苯基
6. suffix ['sʌfiks] n. 后缀、下标; vt. 添后缀
7. prefix ['pri:fiks] n. 字首、前缀
8. electropositive a. 阳电的、带阳电的、阳性的
9. electronegative a. 阴电的、带阴电的、阴性的
10. parentheses [pə'renθisi:s] n. pl. 括弧、插入语、插曲
11. numeral ['nju:mərəl] n. 数码
12. recipe ['resipi] n. 食谱、处方、秘诀
13. oxo ['ɔksəu] n. 氧代
14. phosphite ['fɔsfait] n. 亚磷酸盐
15. borate ['bɔureit] n. 硼酸盐; vt. 使与硼酸混合

Phrases

1. trivial name 俗名,惯用名
   common name 常用名
2. slaked lime 熟石灰,消石灰
3. blue stone 胆矾
4. grain alcohol 酒精
5. washing soda 洗涤碱,晶碱
6. binary (ternary) compound 二元(三元)化合物
7. International Union of Pure and Applied Chemistry (IUPAC) 国际纯粹与应用化学联合会
8. oxy-acid 含氧酸,羟基酸
9. acid hydrogen 酸式氢

10. coordination compound 配位化合物
11. central atom 中心原子
12. neutral ligand 中性配(位)体
13. central ion 中心离子
14. formal charge 形式电荷
15. *fac* = facial 面式的
    mer = meridional 经式的

5.2 NOMENCLATURE OF ORGANIC COMPOUNDS

5.2.1 The Need for Classification of Organic Compounds

The field of organic chemistry is vast, for it includes not only the composition of all living organisms but also of a great many other materials that we use daily. Examples of organic materials are foodstuffs (fats, proteins, carbohydrates), fuels of all kinds, fabrics (cotton, wool, rayon, and nylon), wood and paper products, paints and varnishes, plastics, dyes, soaps and detergents, cosmetics, medicinals, vitamins, rubber products, and explosives.

An immense and ever-growing number—estimated to be on the order of two million at present—of organic chemical compounds have been prepared and are described in the chemical literature.

It is physically impossible for one to study the properties of each of the hundreds of thousands of known organic compounds. Hence organic compounds with similar structural features are grouped into series or classes. Some of the classes of organic compounds are hydrocarbons, alcohols, aldehydes, ketones, ethers, carboxylic acids, esters, carbohydrates, and proteins. Each of these classes of compounds is identified by certain characteristic structural features.

Beginning with a meeting in Geneva in 1892, an international system for naming compounds was developed. In its present form, the system recommended by the International Union of Pure and Applied Chemistry is systematic, generally unambiguous, and internationally accepted. It is called the IUPAC System. Despite the existence of the official IUPAC System, a great many well-established common or trivial names, and even abbreviations such as TNT and DDT, are used and, because of their brevity and convenience, will doubtless continue to be used. So it is necessary to have a knowledge of both the IUPAC System and of many non-IUPAC common names.

Hydrocarbons are composed entirely of carbon and hydrogen atoms bonded to each other by covalent bonds. Several series of hydrocarbons are known. These include the alkanes, alkenes, alkynes, and aromatic hydrocarbons.

Fossil fuels—natural gas, petroleum, and coal—are the principal sources of hydrocarbons. Natural gas is primarily methane. Petroleum is a complex mixture of hydrocarbons. Gasoline, kerosene, fuel oil, lubricating oil, paraffin wax, and petrolatum—all of which are simply mixtures of hydrocarbons—are separated from petroleum. Coal tar, a volatile product driven off in the process of making coke for use in the steel industry, is the source of many valuable chemicals including the aromatic hydrocarbons, benzene, toluene, and naphthalene.

5.2.2 The alkanes

The alkanes, also known as "paraffins" or "saturated hydrocarbons," are straight or branched-chain hydrocarbons having only single covalent bonds between the carbon atoms. It is necessary to learn the names of the first 10 members of the alkane series as these names are used, with slight modifications, for corresponding compounds belonging to other classes.

Methane, CH_4, is the first member of the alkane series. Succeeding members having two, three, and four carbon atoms are ethane, propane, and butane, respectively. The names beginning with the fifth member pentane, are derived from Greek numbers and are relatively easy to recall through memorizing "numerical prefixes". The names of the higher members of this series consist of a numerical term, followed by "-ane", with elision of terminal "a" from the numerical term.

Learning numerical prefixes is like learning to count in organic chemistry. If you take time to memorize only the first 10 prefixes, you will, through application of some rules, learn the names for nearly 100 000 alkenes, 100 000 alkynes, 100 000 alcohols and 100 000 aldehydes.

The numerical prefixes are given in Table 5.8.

Alkyl radicals have the general formula C_nH_{2n+1} (one less hydrogen atom than the corresponding alkane). The name of the radical is formed from the name of the corresponding alkane by simply dropping *ane* and substituting a -yl ending. The names and formulas of selected alkyl radicals are given in Table 5.9. The letter "R" is often used in formulas to mean any of the many possible alkyl radicals.

Table 5.8 Numerical Prefixes

n	prefixes	n	prefixes
1	mono-	17	heptadeca-
2	di-	18	octadeca-
3	tri-	19	nonadeca-
4	tetra-	20	eicosa-
5	penta-	21	heneicosa-
6	hexa-	22	docosa-
7	hepta-	23	tricosa-
8	octa-	30	triaconta-
9	nona-	31	hentriaconta-
10	deca-	40	tetraconta-
11	undeca-	50	pentaconta-
12	dodeca-	60	hexaconta-
13	trideca-	70	heptaconta-
14	tetradeca-	80	octaconta-
15	pentadeca-	90	nonaconta-
16	hexadeca-	100	hecta-

The following relatively simple rules are all that are needed to name a great many alkanes according to the IUPAC System. In later sections these rules will be extended to cover other classes of compounds.

IUPAC rules for naming alkanes:

① Select the longest continuous chain of carbon atoms as the parent compound, and consider all alkyl groups attached to it as branch chains that have replaced hydrogen atoms of the parent hydrocarbon. The name of the alkane consists of the name of the parent compound prefixed by the names of the branch chain alkyl groups attached to it.

② Number the carbon atoms in the parent carbon chain from the end that gives the lowest numbers to the branch chains.

③ Name each branch chain alkyl group and designate its position on the parent carbon chain by a number.

④ When the same alkyl group branch chain occurs more than once, indicate this by a prefix, di-, tri-, tetra-, and so forth, written in front of the alkyl group name. The numbers indicating the positions of these alkyl groups are placed in front of the name.

⑤ When several different alkyl groups are attached to the parent compound, name them in order of increasing size, such as methyl before ethyl (e.g., 4-methyl, 3-ethyloctane).

Table 5.9   Names and Formulas of Selected Alkyl Radicals

Formula	Name	Formula	Name
CH_3-	Methyl	CH_3CH- with CH_3 branch	Isopropyl
CH_3CH_2-	Ethyl	CH_3CHCH_2- with CH_3 branch	Isobutyl
$CH_3CH_2CH_2-$	n-Propyl	CH_3CH_2CH- with CH_3 branch	sec-Butyl (secondary butyl)
		CH_3-C- with CH_3 and CH_3 branches	tert-Butyl (tertiary butyl)
$CH_3CH_2CH_2CH_2-$	n-Butyl		
$CH_3(CH_2)_3CH_2-$	n-Pentyl		
$CH_3(CH_2)_4CH_2-$	n-Hexyl		
$CH_3(CH_2)_5CH_2-$	n-Heptyl		
$CH_3(CH_2)_6CH_2-$	n-Octyl		
$CH_3(CH_2)_7CH_2-$	n-Nonyl		
$CH_3(CH_2)_8CH_2-$	n-Decyl		

The lowercase n in n-Propyl, n-Butyl etc., means that the hydrocarbon chains are normal, that is, the radical represents an unbranched hydrocarbon with a hydrogen atom missing from carbon number 1.

5.2.3 Alkenes and Alkynes

Alkenes (also known as olefins) contain at least one double bond between carbon atoms. Alkynes (also known as acetylenes) contain at least one triple bond between carbon atoms. The simplest alkene is ethene (or ethylene), and the simplest alkyne is ethyne (or acetylene).

The names of alkenes and alkynes are derived from the names of the corresponding alkanes. To name an alkene (or alkyne) by the IUPAC System:

① Select the longest carbon-carbon chain that contains the double or triple bond.

② Name this parent compound as you would an alkane but change the -*ane* ending to -*ene* for an alkene or to -*yne* for an alkyne.

③ Number the carbon chain of the parent compound so that the double or triple bond carries the lowest possible numbers. Use the smaller of the two numbers on the double- or triple-bonded carbon atoms to indicate the position of the double or triple bond. Place this number in front of the alkene or alkyne name.

④ Side chains and other groups are treated as in naming alkanes.

Study the following examples of named alkenes and alkynes.

$$\overset{4}{C}H_3\overset{3}{C}H_2\overset{2}{C}H=\overset{1}{C} \qquad \overset{1}{C}H_3\overset{2}{C}H=\overset{3}{C}$$

1-Butene　　　　　2-Butene

$$\underset{CH_3CHCH=CH_2}{\overset{CH_3}{|}} \quad CH_3CH_2C\equiv CH \quad \underset{\underset{CH_3}{|}}{\overset{\overset{CH_3}{|}}{\overset{1}{C}H_3-\overset{2}{C}\equiv\overset{3}{C}-\overset{4}{C}H-\overset{5}{C}H-\overset{6}{C}H_3}}$$

3-Methyl-1-Butene   1-Butyne   4,5-Dimethyl-2-hexyne

5.2.4 Naming Aromatic Compounds

As is the case with aliphatic compounds, the naming of aromatic hydrocarbons and their derivatives is not completely systematic. Compounds containing the benzene nucleus are named systematically as benzene derivatives. But they are also known by common popular or trivial names.

The systematic name of a monosubstituted benzene is obtained by giving the name of the substituting group followed by the word benzene.

Using the IUPAC System to designate the positions of groups on a benzene ring, the positions are located by numbering the carbon atoms. Numbering is started at one of the groups and may go clockwise or counterclockwise. However, numbering must be done so that the attached groups have the smallest possible numbers.

In the other and older system the positions on the ring relative to a given group, X, are designated ortho, meta and para-abbreviated o-, m-, and p-.

Other aromatic ring systems, called condensed or fused rings, are known. Naphthalene, anthracene, and phenanthrene are examples.

Naphthalene $C_{10}H_8$ Anthracene $C_{14}H_{10}$ Phenanthrene $C_{14}H_{10}$

Aromatic hydrocarbons such as benzene, toluene, xylene, naphthalene, and anthracene, were first obtained in quantity from coal tar. The demand for aromatic hydrocarbons used in the production of a vast number of materials such as drugs, dyes, detergents, explosives, insecticides, plastics, and synthetic rubber, became too great to satisfy from coal tar alone. Currently about one-third of our benzene supply, and the greater portion of our toluene and xylene supplies, are obtained from petroleum.

5.2.5 Naming Alcohols and Ethers

Both common and IUPAC names are used for alcohols. The common name is usually formed from the name of the organic radical, which is attached to the -OH group, followd by the word "alcohol." See examples in Table 5.10. To name an alcohol by the IUPAC System:

① Select the longest continuous chain of carbon atoms containing the hydroxyl group.

② Number the carbon atoms in this chain so that the one bearing the —OH group has the lowest possible number.

③ Form the alcohol name by dropping the final -e from the corresponding alkane name and adding -ol. Locate the —OH group by putting the number (hyphenated) of the carbon atom to which it is attached immediately before the alcohol name.

④ Name each alkyl side chain (or other group) and designate its position by number.

Study the application of this method of naming alcohols in the examples in Table 5.10.

Table 5.10 Naming and Classification of Alcohols

Class	Formula	IUPAC Name	Common Name
Primary	CH_3OH	Methanol	Methyl alcohol
Primary	CH_3CH_2OH	Ethanol	Ethyl alcohol
Primary	$CH_3(CH_2)_3CH_2OH$	1-Pentanol	n-Amyl or n-pentyl alcohol

Primary	$CH_3(CH_2)_6CH_2OH$	1-Octanol	n-Octyl alcohol
Primary	CH_3CHCH_2OH $\|$ CH_3	2-Methyl-1-propanol	Isobutyl alcohol
Secondary	CH_3CHCH_3 $\|$ OH	2-Propanol	Isopropyl alcohol
Secondary	$CH_3CH_2CHCH_3$ $\|$ OH	2-Butanol	sec-Butyl alcohol
Tertiary	CH_3 $\|$ CH_3-C-OH $\|$ CH_3	2-Methyl-2-propanol	t-Butyl alcohol

Alcohols may be oxidized to aldehydes, ketones, and carboxylic acids. Primary alcohols yield aldehydes and carboxylic acids; secondary alcohols are oxidized to ketones; tertiary alcohols resist oxidation.

Alkyl halides can be made from alcohols by reacting them with phosphorous trihalides, PCl_3 or PBr_3, PI_3, and thionyl chloride, $SOCl_2$.

Alkenes and ethers are formed by the dehydration of alcohols. This is an reaction, in that the alkene is formed at one temperature and the ether at another temperature, both using the same reagents.

Alcohols (ROH) and ethers (ROR) are isomeric, having the same molecular formula, but different structural formulas.

The names of ethers consist of the names of the two radical groups attached to the oxygen followed by the word "ether."

R—O—R CH_3—O—CH_3 CH_3—O—CH_2CH_3
                         ↑   ↑   ↑   ↑   ↑   ↑
                   Methyl Ether Methyl   Methyl Ether Ethyl

Ether   Methyl ether or Dimethyl ether   Methyl ethyl ether

5.2.6 Aldehydes and Ketones

Two classes of organic compounds that are similar in structure are aldehydes and ketones.

Aldehydes are named by dropping the *e* in the parent hydrocarbon name and adding the letters al. The first member of the series is methanal (one carbon aldehyde). Its common name is formaldehyde. Several aldehydes are listed below:

Formula	Common Name	IUPAC Name
HC(H)=O	Formaldehyde	Methanal
CH$_3$C(H)=O	Acetaldehyde	Ethanal
CH$_3$CH$_2$CH$_2$C(H)=O	Butyraldehyde	Butanal
C$_6$H$_5$C(H)=O	Benzaldehyde	

In the IUPAC system, ketones are named by dropping the *e* in the parent hydrocarbon name and adding the letters one. The first member of the homologous series has three carbon atoms and is called propanone. Its common names are acetone and dimethyl ketone. When the carbon chain contains five or more carbon atoms, the longest chain containing the ketone group is numbered so that the ketone group will have the smallest number possible.

An alternate method commonly used for naming simple ketones is shown below. The name includes the name of the two organic radicals attached to the ketone group, followed by the word ketone. Several ketones in the series are listed below:

Formula	Common Name	IUPAC Name
$CH_3\underset{O}{\overset{\parallel}{C}}CH_3$	Acetone	Propanone
$CH_3\underset{O}{\overset{\parallel}{C}}CH_2CH_3$	Methyl ethyl ketone	Butanone
$CH_3CH_2\underset{O}{\overset{\parallel}{C}}CH_2CH_3$	Diethyl ketone	3-Pentanone
$CH_3CH_2CH_2\underset{O}{\overset{\parallel}{C}}CH_3$	Methyl-n-propyl ketone	2-Pentanone

The simplest aromatic ketone is

$C_6H_5\underset{O}{\overset{\parallel}{C}}CH_3$ Acetophenone or Methyl phenyl ketone

5.2.7 Acids and Esters

Organic acids, known as carboxylic acids, are characterized by the functional group called a carboxyl group.

Carboxylic acids are named by dropping the e of the parent hydrocarbon name and adding the letters oic followed by the word acid. Thus the one carbon acid, HCOOH, is called methanoic acid. Many carboxylic acids are found in nature and their common names often reflect the natural source. Animal and vegetable fats and oils are an important source of organic acids that range up to eighteen carbon atoms (for example, stearic acid, $CH_3(CH_2)_{16}COOH$). Several acids in the series are listed below:

Formula	Common Name	IUPAC Name
HCOOH	Formic acid	Methanoic acid
CH_3COOH	Acetic acid	Ethanoic acid
CH_3CH_2COOH	Propionic acid	Propanoic acid
$CH_3CH_2CH_2COOH$	Butyric acid	Butanoic acid
$CH_3(CH_2)_{16}COOH$	Stearic acid	Octadecanoic acid

The simplest aromatic acid is benzoic acid. Ortho-hydroxy-benzoic acid is known as salicylic acid, the basis for many salicylate drugs such as aspirin.

Benzoic acid   Salicylic acid   Acetylsalicylic acid (Aspirin)

Esters are named like salts, giving them an *ate* ending. The alcohol is named first, followed by the name of the acid modified to end in ate. The *ic* of the acid name is dropped and the letters *ate* are added. Thus, acetic becomes acetate and ethanoic becomes ethanoate. The ester derived from methyl alcohol and acetic acid is called methyl acetate or methyl ethanoate (CH_3COOCH_3).

CH_3C——O——CH_3 Methyl acetate or Methyl ethanoate

Acetic acid   Methyl alcohol

Table 5.11 summarizes the structure of the common classes of organic compounds. The R groups used in the table can be either aliphatic or aromatic groups. Only aliphatic examples are shown in the table.

Table 5.11 Classes of Organic Compounds

Class	Structure of Functional Group	Structural Formula	Name Common	Name IUPAC
Alkane	R—H	CH_4	Methane	Methane
		CH_3CH_3	Ethane	Ethane
Alkene	>C=C<	$CH_2=CH_2$	Ethylene	Ethene
		$CH_3CH=CH_2$	Propylene	Propene
Alkyne	—C≡C—	$CH≡CH$	Acetylene	Ethyne
		$CH_3C≡CH$	Methyl acetylene	Propyne
Alkyl halide	—X, X = F, Cl, Br, I	CH_3Cl	Methyl chloride	Chloromethane
		CH_3CH_2Cl	Ethyl chloride	Chloroethane
Alcohol	—OH	CH_3OH	Methyl alcohol	Methanol
		CH_3CH_2OH	Ethyl alcohol	Ethanol
Aldehyde	H–C=O	$H-\underset{\parallel}{\overset{H}{C}}=O$	Formaldehyde	Methanal
		$CH_3-\underset{\parallel}{\overset{H}{C}}=O$	Acetaldehyde	Ethanal
Ketone	R—C—R′, ‖O	$CH_3-\underset{\parallel O}{C}-CH_3$	Acetone	Propanone
		$CH_3-\underset{\parallel O}{C}-CH_2CH_3$	Methyl ethyl ketone	2-Butanone

Table 5.11 Classes of Organic Compounds (Continued)

Class	Structure of Functional Group	Structural Formula	Name Common	Name IUPAC
Carboxylic acid	−C(=O)−OH	HCOOH CH$_3$COOH	Formic acid Acetic acid	Methanoic acid Ethanoic acid
Ester	−C(=O)−OR	HCOOCH$_3$ CH$_3$COOCH$_3$	Methyl formate Methyl acetate	Methyl methanoate Methyl ethanoate
Amide	−C(=O)−NH$_2$	HC(=O)−NH$_2$ CH$_3$C(=O)−NH$_2$	Formamide Acetamide	Methanamide Ethanamide
Ether	R−O−R′	CH$_3$−O−CH$_3$ CH$_3$CH$_2$−O−CH$_2$CH$_3$	Dimethyl ether Diethyl ether	Methoxymethane Ethoxyethane
Amine	−NH$_2$	CH$_3$NH$_2$ CH$_3$CH$_2$NH$_2$	Methylamine Ethylamine	Methanamine Ethanamine
Nitrile	−C≡N	CH$_3$C≡N CH$_3$CH$_2$C≡N	Acetonitrile propiononitrile	Ethanonitrile Propanonitrile
Amino acid	−CH(NH$_2$)−COOH	H$_2$NCH$_2$COOH CH$_3$CH(NH$_2$)−COOH	Glycine Alanine	2-Aminoethanoic acid 2-Aminopropanoic acid

Vocabulary

1. cosmetic [kɔz'metik] n. 化妆品；a. 化妆用的
2. brevity ['breviti] n. 短暂、简短
3. paraffin ['pærəfin] n. 石蜡
4. alkyl ['ælkil] n. 烷基、烃基
5. alkene ['ælki:n] n. 烯烃
6. alkyne ['ælkain] n. 炔、炔烃
7. aliphatic [æli'fætik] a. 脂肪族的、脂肪质的
8. anthracene ['ænθrəsi:n] n. 蒽
9. insecticide [in'sektisaid] n. 杀虫剂
10. dehydration [di:hai'dreiʃən] n. 脱水
11. aldehyde ['ældihaid] n. 乙醛
12. tertiary ['tɔʃiəri] a. 第三的、第三位的、第三世纪的
13. halide ['hælaid] n. 卤化物；a. 卤化物的
14. isometric [aisou'metrik] a. 同容积的、同大的
15. formaldehyde [fɔ:'mældəhaid] n. 甲醛、蚁醛
16. homologous [hou'mɔləgəs] a. 相应的、对应的、一致的
17. benzoic [ben'zouik] a. 安息香的
18. salicylic [sæli'silik] a. 得自水杨酸的、水杨酸的

Phrases

1. aromatic hydrocarbon   芳(香)烃
2. fossil fuel   化石燃料,矿物燃料
3. paraffin wax   石蜡
4. saturated hydrocarbon   饱和烃
5. straight-chain hydrocarbon   直链烃
   branched-chain hydrocarbon   支链烃
6. single bond   单键
7. parent compound   母体化合物
8. carbon chain   碳链

9. double bond 双键
   triple bond 三键
10. side chain 侧链,支链
11. aliphatic compound 脂肪族化合物
12. substituting group 取代基
13. condensed ring = fused ring 稠环
14. in quantity = in large quantities 大量
15. thionyl chloride 亚硫酰(二)氯
16. homologous series 同系列

6

Reading Materials

6.1 THE GENESIS OF POLYMER

6.1.1 What are Polymers

What are polymers? For one thing, they are complex and giant molecules and are different from low molecular weight compounds like, say, common salt. To contrast the difference, the molecular weight of common salt is only 58.5, while that of a polymer can be as high as several hundred thousands. These big molecules or 'macro-molecules' are made up of much smaller molecules. The small molecules, which combine to form a big molecule, can be of one or more chemical compounds. To illustrate, imagine a set of rings of the same size and made of the same material. When these rings are interlinked, the chain formed can be considered as representing a polymer from molecules of the same compound. Alternatively, individual rings could be of different sizes and materials, and interlinked to represent a polymer from molecules of different compounds. The two situations are shown in Figure 6.1.

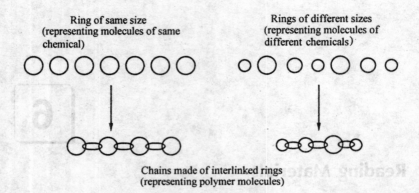

Figure 6.1　Conceptual representation of formation of polymer molecule by linking together of molecules of simple chemicals

This interlinking of many units has given the polymer its name, poly meaning 'many' and mer meaning 'part' (in Greek). As an example, a gaseous compound called butadiene, with a molecular weight of 54, combines nearly 4 000 times and gives a polymer known as polybutadiene (a synthetic rubber) with about 200 000 molecular weight. The picture is simply as follows:

$$\text{butadiene} + \text{butadiene} + \cdots + \text{butadiene} \rightarrow \text{polybutadiene}$$
$$(4\ 000\ \text{times})$$

One can thus see how a substance with as small a molecular weight as 54 grows to become a giant molecule of $(54 \times 4\ 000) \approx 200\ 000$ molecular weight. It is essentially the 'giantness' of the size of the polymer molecule that makes its behaviour different from that of a commonly known chemical compound such as benzene. Solid benzene, for instance, melts to become liquid benzene at 5.5℃ and, on further heating, boils into gaseous benzene. As against this well-defined behaviour of a simple chemical compound, a polymer like polyethylene does not melt sharply at one particular temperature into a clean liquid. Instead, it becomes increasingly softer and, ultimately, turns into a very viscous, tacky molten mass. Further heating of this hot, viscous, molten polymer does convert it into various gases but they are no longer polyethylene(Figure 6.2)[1].

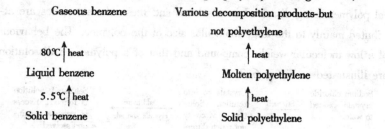

Figure 6.2   Difference in behaviour on heating of a low molecular weight compound(benzene) and a polymer(polyethylene)

Another striking difference with respect to the behaviour of a polymer and that of a low molecular weight compound concerns their solubility pattern. Let us take, for example, sodium chloride and add it slowly to a fixed quantity of water. The salt, which represents a low molecular weight compound, dissolves in water up to a point (called saturation point) but, thereafter, any further quantity added does not go into solution but settles at the bottom and just remains there as solid. The viscosity of the saturated salt solution is not very much different from that of water. But if we take a polymer instead, say, polyvinyl alcohol and add it to a fixed quantity of water, the polymer does not go into solution immediately. The globules of polyvinyl alcohol first absorb water, swell and get distorted in shape and after a long time go into solution. Also, we can add a very large quantity of the polymer to the same quantity of water without the saturation point ever being reached. As more and more quantity of the polymer is added to water, the time taken for the dissolution of the polymer obviously increases and the mix ultimately assumes a soft, dough-like consistency. Another peculiarity is that, in water, polyvinyl alcohol never retains its original powdery nature as the excess sodium chloride does in a saturated salt solution. In conclusion, we can say that ① the long time taken by polyvinyl alcohol for dissolution, ② the absence of a saturation point, and ③ the increase in the viscosity are all characteristics of a typical polymer being dissolved in a solvent and these characteristics are attributed mainly to the large molecular size of the polymer. The behaviour of a low molecular weight compound and that of a polymer on dissolution are illustrated in Figure 6.3.

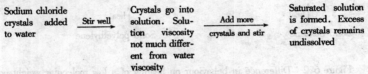

Figure 6.3 Difference in the solubility behaviour of a low molecular weight compound (sodium chloride) and a polymer (polyvinyl alcohol)

6.1.2 How are Polymers Made

We have seen that a polymer is made up of many small molecules which have combined to form a single long or large molecule. The individual small molecules from which the polymer is formed are known as *monomers* (meaning, single part) and the process by which the monomer molecules are linked to form a big polymer molecule is called 'polymerisation'.

We now give a single analogy to illustrate the polymerisation phenomenon. Some children are relaxing on a playground – some in groups and some alone. Then appears a drill master and orders them to form a line, which the children very grudgingly do. With another order, he directs each child to hold the hand of the child on either side of him or her so as to form a long row. Now there is a problem. The very mischievous ones, who are interspersed in the row, decide not to extend their left hands. What would then happen to the hand-to-hand row of the lot? Naturally, the line will extend only up to a particularly naughty boy; he having decided not to extend his one hand to the neighbouring child, the chain breaks. Well, the two hands of the child, essential for the row to grow on either side, represent the two reactive sites of the monomer, which are essential if the monomer molecule is to combine with another to grow in size and form a giant molecule of the polymer. It should be noted that the two reactive sites—or what is termed bifunctionality—of the monomer are prerequisites, if the molecular build-up process is to continue, just as the extension of both the hands of every child is essential for their hand-to-hand line to grow.

We have stated earlier that polymerisation is possible with molecules of the same or of different monomeric compounds. When molecules just add on to form the polymer, the process is called 'addition' polymerisation. The monomer species in this case retains its structural identity when it gets converted into a polymer. For example, the molecules of ethylene monomer can add on to form polyethylene, in which, the structural identity of ethylene is retained.

When, however, molecules do not just add on but also undergo some reaction in forming the polymer, the process is called 'condensation' polymerisation. Here, the two molecules (of the same or different monomers), condense to form a polymer. The condensation takes place between two reactive functional groups, like the carboxyl group (of an acid) and the hydroxyl group (of an alcohol). While forming the polymer, however, water molecules also get eliminated. We can, therefore, see that while, in 'addition' polymerisation, the molecular weight of the polymer is roughly equal to that of all the molecules which combine to form the polymer, in 'condensation' polymerisation, the molecular weight of the polymer is lesser by the weight of the simple molecules eliminated during the condensation process.

In both 'addition' and 'condensation' polymerisations, the polymer molecule formed contains a structural identity, repeating itself several times. These repeating entities are called the repeat units or the monomeric units of the polymer molecule. The size of the polymer molecule is decided by the number of repeat units present in it. This number denotes the degree of polymerisation.

For example, 5 molecules of ethylene monomer can add on to each other to form a single molecule of polyethylene, as represented in Figure 6.4 (top portion). Here, the repeat unit is —CH_2—CH_2— and the polymer

5 CH_2=CH_2 (Ethylene monomer)
↓ Polymerisation
$\{CH_2-CH_2\}_5$ (Polyethylene molecule containing 5
$\boxed{DP = 5}$ repeat units of —CH_2—CH_2—)

4HO—R—COOH (Hydroxy acid monomer)
↓ Polymerisation
HO $\{R-COO\}_4$ H $\boxed{DP = 4}$ (Polyester molecule containing 4 repeat units of —R—COO—) + 3 H_2O (3 molecules of water eliminated)

Figure 6.4 Illustrations showing the degree of polymerisation (DP) of a polymer molecule

molecule contains 5 such repeat units. Hence, the degree of polymerisation is 5. Similarly, when 4 molecules of a hydroxy acid (HO—R—COOH) condense, a polyester molecule is formed. The repeat unit in this case will be—R—COO—[see Figure 6.4 (bottom portion)]. This polyester molecule has 4 such repeat units and, hence, its degree of polymerisation is 4.

6.1.3 Classification of Polymers

Polymer is a generic name given to a vast number of materials of high molecular weight. These materials exist in countless forms and numbers because of a very large number and types of atoms present in their molecules. Polymers can have different chemical structures, physical properties, mechanical behaviour, thermal characteristics, etc., and can be classified in different ways, as now discussed.

(1) *Natural and Synthetic Polymers*. Depending on their origin, polymers can be grouped as natural or synthetic. Those isolated from natural materials are called natural polymers. Typical examples are: cotton, silk, wool and rubber. Cellophane, cellulose rayon, leather and so on are, in fact, chemical modifications of natural polymers.

Polymers synthesised from low molecular weight compounds are called synthetic polymers. Typical examples are: polyethylene, PVC, nylon and terylene.

(2) *Organic and Inorganic Polymers*. A polymer whose backbone chain is essentially made of carbon atoms is termed an organic polymer. The atoms attached to the side valencies of the backbone carbon atoms are, however, usually those of hydrogen, oxygen, nitrogen, etc.. The majority of synthetic polymers is organic and they are very extensively studied. In fact, the number and variety of organic polymers are so large that when we refer to 'polymers', we normally mean organic polymers. The molecules of inorganic polymers, on the other hand, generally contain no carbon atom in their chain backbone. Glass and silicone rubber are examples of inorganic polymers.

(3) *Thermoplastic and Thermosetting Polymers*. Some polymers soften on heating and can be converted into any shape that they can retain on cooling. The process of heating, reshaping and retaining the same on cooling can be repeated several times. Such polymers, which soften on heating

and stiffen on cooling, are termed 'thermoplastics'. Polyethylene, PVC, nylon and sealing wax are examples of thermoplastic polymers. Some polymers, on the other hand, undergo some chemical change on heating and convert themselves into an infusible mass. They are like the yolk of the egg, which on heating sets into a mass, and, once set, cannot be reshaped. Such polymers, which become an infusible and insoluble mass on heating, are called 'thermosetting' polymers.

(4) Plastics, Elastomers, Fibres and Liquid Resins. Depending on its ultimate form and use, a polymer can be classified as plastic, elastomer, fibre or liquid resin. When, for instance, a polymer is shaped into hard and tough utility articles by the application of heat and pressure, it is used as a 'plastic'. Typical examples are polystyrene, PVC and polymethyl methacrylate. When vulcanised into rubbery products exhibiting good strength and elongation, polymers are used as 'elastomers'. Typical examples are natural rubber, synthetic rubber, silicone rubber, if drawn into long filament-like materials, whose length is at least 100 times its diameter, polymers are said to have been converted into 'fibres'. Typical examples are nylon and terylene. Polymers used as adhesives, potting compounds, sealants, etc., in a liquid form are described as liquid resins. Commercially available epoxy adhesives and polysulphide sealants are typical examples.

Vocabulary

1. interlink [ˌintə'liŋk] vt. 连接; n. 连环
2. butadiene [ˌbjuːtə'daiiːn] n. 丁二烯
3. polymerization [ˌpɔliməraiˈzeiʃən] n. 聚合
4. conceptual [kən'septjuəl] n. 概念性描述
5. viscous ['viskəs] a. 粘滞的
6. solubility [sɔljuˈbiliti] n. 溶解性
7. dissolution [disəˈluːʃən] n. 分解、解散
8. peculiarity [pikjuːliˈæriti] n. 特性

9. analogy [ə'nælədʒi] n. 类似、相似、类推、类比
10. grudgingly ['grʌdʒiŋli] ad. 不愿意地、不情愿地
11. mischievous ['mistʃivəs] a. 淘气的、有害的、恶作剧的、胡搅的
12. intersperse [ˌintə'spə:s] vt. 散布、点缀
13. prerequisite [pri:'rekwizit] n. 先决条件
14. condensation [kɔnden'seiʃən] n. 浓缩
15. entity ['entiti] n. 实体
16. isolate ['aisəleit] v. 离析
17. cellophane ['seləfein] n. 玻璃纸
18. cellulose ['seljuləus] n. 纤维素
19. rayon ['reiɔn] n. 人造丝
20. terylene ['terəli:n] n. 多元脂纤维的一种、涤纶
21. backbone [ˌbækboun] n. 脊椎、志气、骨干、支柱
22. thermoplastic [ˌθə:mou'plæstik] a. 热塑的
n. 热塑性物质、热塑塑料
23. infusible [in'fju:zəbl] a. 不熔化的、不熔化性的
24. elastometry [iˈlæsˈtɔmitri] n. 弹力测定法、弹性测定法
25. polystyrene [ˌpɔli'stairi:n] n. 聚苯乙烯
26. vulcanise ['vʌlkənaiz] v. 用高温和硫磺处理、硫化
27. polysulphide [ˌpɔlisʌlfaid] n. 多硫化物
28. adhesive [əd'hi:siv] a. 粘着的、有粘性的；
n. 粘剂
29. sealant ['si:lənt] n. 密封剂、封闭层

Phrases

1. tacky molten mass   粘性熔铸块
2. fixed quantity of ...   定量的(固定量的)
3. add on to (do)   (另外)加上去(做)
4. infusible and insoluble   不熔不溶
5. set into   开始成为

Note

1. Further heating of this hot,... no longer polyethylene. 句中 does 起强调作用。

6.2 MECHANISM OF DEVELOPMENT OF LARGE STRAINS

6.2.1 Physical States

The chain structure of their macromolecules and the presence of a fluctuation network underlie the most fundamental feature of the mechanical behaviour of polymers—their viscoelasticity.

When an external deforming stress is applied to a polymer, a strain develops depending on the duration of action of the stress. At the beginning of its action, the segments outside of the network linkage points migrate. This is confirmed by the above values of the settled lifetime of such segments(small fractions of a second). The motion of these "free" segments causes the coiled shape of the macromolecules typical of the initial equilibrium state to become distorted, and the coils unfold in the direction of action of the stress. The settled lifetime of the "bound"segments,i.e. those contained in the network linkage points, is longer. This signifies that they do not break up initially, and the intactness of the fluctuation network structure is retained. If the stress is removed, the segments return to their initial position. Hence, a strain arising when a stress acts for a short time is reversible. This is elastic strain(deformation).

If the deforming stress is not removed, then after a definite time the network linkage points begin to break up and the bound segments begin to migrate. The migration of a considerable number of free and bound segments will in the long run cause the macromolecules to become displaced relative to one another. As in a low-molecular liquid, the motion of the molecules relative to one another ensures irreversible deformation-flow. Consequently, when a stress acts in a polymer for a long time, an irreversible or,as it is often called, viscous strain or deformation accumulates in a polymer.

The duration of the settled lifetime of segments is characterized by a time interval[1]. This is of fundamental importance and is due to the circumstance that the network linkage points vary in their structure. The intensity of segment interaction in them varies, therefore both the segments in linkage points and the free segments interact with one another differently. The broad range of intensities of interaction between the free and bound segments results in the absence of a clear time boundary between the instant of completion of elastic and the beginning of viscous strain. Migration of the more firmly bonded free segments is attended by breaking up of the weak linkage points of the fluctuation network. Hence, the deformation of a polymer is both elastic and viscous at the same time. First the segments, and then the macromolecules migrate in a viscous medium such a strain is called viscoelastic.

Viscoelasticity is characteristic of all bodies in principle. Ice deforms elastically, which is sometimes easy to note even with the naked eye. But it is also known that glaciers flow, i.e. during a long time one can note viscous deformation of elastic ice. Water unlike ice is a viscous liquid, but whoever had dived into water noted its elasticity.

Two important circumstances distinguish the viscoelasticity of polymers from that of low-molecular bodies. The first is the scale of time. To detect the elasticity of water, a very high speed of action of the force is needed, while to detect the viscous deformation of ice, a very long time is needed. Polymers exhibit viscoelastic strain at the usual durations of action of a stress. The second circumstance is the scale of elastic strain. The elastic strain up to failure in low-molecular bodies is fractions of a per cent or a few per cent. In polymers, the elastic strain may reach scores, hundreds, and even thousands of per cent. We shall consider the conditions for the manifestation of such a large elastic strain below.

The presence of a barrier to rotation about a simple bond in the backbone of a macromolecule and the existence of fluctuation network linkage points in the bulk of a polymer presume a number of special features in the nature of the temperature dependence of the mechanical properties of a polymer. These features are determined by the fact that a change in the temperature causes a change in the relation between the magnitude of the rotational barrier or the strength of a bond in the fluctuation network linkage points and the magnitude of the heat energy fluctuations. At a low value of these fluctuations (a low temperature), the rotational barrier may be unsurmountable, and a macromolecule loses its ability of being strained. This is also facilitated by the greater strength of the fluctuation network linkage points at a lower temperature. Studying of the temperature dependence of the mechanical properties or, in other words, the obtaining of a thermomechanical relation or curve not only tells us about the properties of the relevant polymer at a given temperature, but also enables us to analyse the structure of the polymer and the flexibility of its macromolecules. When the latter are more flexible, the interaction at the linkage points of the fluctuation network is smaller and the polymer remains elastic up to a lower temperature.

It is convenient to obtain a thermomechanical curve in the form of a temperature dependence of the deformability of a polymer. Let us cut out a cube from a polymer, measure it, put it into a thermostat at a definite temperature T_1 and apply a load P to it for, say, 10 s. We then measure the height of the cube under the load and calculate the strain of the polymer ϵ_1 at the temperature T_1.

Let us heat the specimen to the temperature T_2 and again load it with the load P for 10 s. We then measure the strain ϵ_2 at the higher temperature T_2. We continue this procedure and obtain a number of strain values $\epsilon_1, \epsilon_2, \epsilon_3, \epsilon_4$, etc. for the relevant temperatures T_1, T_2, T_3, T_4, etc. We use these experimental data to plot a thermomechanical curve (Figure 6.5).

At a low temperature, the strain is small and increases only a little with the temperature. An amorphous polymer behaves like glass at low temperatures. The polymer is said to be in the glassy state.

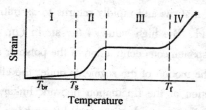

Figure 6.5 Thermomechanical curve of an amorphous polymer
I—region of glassy state; II—transition region; III—region of rubberlike state; IV—region of viscous—flow state
T_{br}—brittle point; T_g—glass transition temperature; T_f—flow temperature; the asterisk indicates the point where thermal destruction begins

If the load when determining a thermomechanical curve is not large and does not exceed 0.1 MPa, the strain is only fractions of a per cent of the initial height of the specimen. Such a small strain is also typical of many low-molecular solids. For polymers, it is a reliable indication that under the applied stress the macromolecule segments do not move and, consequently, the macromolecules do not change the shape of their random coils.

Beginning from a temperature called the glass transition temperature T_g, the strain begins to grow and finally reaches several scores, or if the curve is being obtained under conditions of tension, several hundreds of per cent. Upon further heating, the strain again depends only slightly on the temperature, which is depicted on the curve by the presence of an extended plateau. The intermediate region of temperatures between the glass transition temperature and the temperature of passing out onto the plateau is called the transition region.

The polymer deforms "sluggishly"—it has stopped being a glassy rigid body and has not yet become elastic like good rubber. Its mechanical behaviour reminds one of the behaviour of a strip of natural leather or linoleum after being bent, when it slowly returns to its initial position.

At the temperature of passing out onto the plateau, the polymer transforms into a developed highly elastic or rubberlike state. It readily deforms under the action of a stress and rapidly returns to its initial position after removal of the load. The high value of the strain (up to hundreds of per cent) and the spontaneous contraction of the polymer after removal of the load point to the motion of the segments under the effect of a stress, and also to destruction of the fluctuation network linkage points and the appearance of new, less stressed points at a different site. Such a strain is necessarily associated with a change in the shape of the macromolecules' random coils and is therefore called long-range or highly elastic strain. The polymer within the entire range of temperatures confined by the plateau on the thermomechanical curve is in the rubberlike state.

The highly elastic strain, whose magnitude is determined by the change in the shape of the macromolecular coils, depends only slightly on the temperature. For this reason, the growth in the strain under the effect of the same stress and during the same time interval, as determined above, can be due only to the development of a new type of strainviscous-flow strain. The latter is the result of considerable slipping of the segments relative to their equilibrium positions, which leads to motion of the molecular coils relative to one another. The temperature at which the viscous-flow strain begins to grow sharply in a polymer and causes an inflection to appear on the thermomechanical curve is called the flow temperature. Above this temperature, a polymer is in the viscous-flow state.

If no thermal destruction of a polymer occurs when obtaining a thermomechanical curve, slow cooling enables one to reproduce the curve: the specimen first passes into the rubberlike and then into the glassy state (it becomes vitrified). It is of major importance that at a temperature below the glass transition one a polymer generally retains a set of properties typical only of polymers. We say that a polymer has become vitrified, but it has not become brittle like ordinary silicate(window)glass. A sheet of organic glass [poly(methyl methacrylate), plexiglas] can be thrown onto the floor and it will not shatter into fragments. And nevertheless a glassy polymer can be cooled to a temperature when it will easily be broken by a blow. This temperature is known as the brittle point T_{br}. It does not appear as a typical point on a thermomechanical curve. The methods of determining the brittle point are always associated in some way or other with destruction of the specimen.

The most important destination of a thermomechanical curve is not to characterize the level of the mechanical properties of a polymer, but to determine the temperature limits of existence of the glassy, rubberlike, and viscous-flow states. These three states are called physical or relaxation states.

Each physical state of a polymer has a definite significance for its processing or use. If the region $T_g - T_{br}$ is long enough and includes room temperature, such a polymer will most likely find use as a plastic. If room temperature coincides with the region $T_f - T_g$, such a polymer will most likely find use as an elastomer. This approach to appraising the applicability of polymers as plastics or elastomers(rubber) is very conditional. Of importance here is the fact that there are no differences of principle between plastics and elastomers, except that the glass transition temperature is higher than room temperature for plastics and lower than it for elastomers.

The viscous-flow state is important when processing polymers. when T_f is lower, it is easier to process a polymer because the temperature of processing is in the region of strongly developed flow. If the processing temperature is close to T_f and the polymer retains its elasticity when being shaped, it is difficult to shape it. For example, a sphere made from an elastic polymer can be pressed into a cubical space of a press-mould and cooled to a temperature below T_g. When the mould is opened, the polymer retains the shape of a cube, but this shape is not stable – when the cube is heated above T_g it again acquires the shape of a sphere.

Above, we treated amorphous polymers. If a polymer consists of macromolecules with a regular structure, then short-range order in the arrangement of the segments may at a definite temperature (the crystallization temperature) and during a definite time transform into long-range order. A crystalline structure appears. We shall study the features of polymer crystallization in detail on following pages. We shall note here that a polymer cannot crystallize completely, like low-molecular substances do. Owing to the considerable entanglement of the macromolecule coils, part of the segments cannot participate in the building up of a crystal for purely steric reasons. The degree of crystallinity of polymers therefore ranges within broad limits from 30% to 80%. In very regular polymers, the content of the crystalline part may reach 90% ~ 95%.

Because of the entanglement of the macromolecular coils, all the crystalline microregions are connected to one another by tie macromolecules. Such a structure is quite rigid and deforms only a little even when the degree of crystallinity of a polymer is only 30% ~ 35%. Here a polymer exhibits properties of a typical plastic although its glass transition temperature is below room temperature. An example is polyethylene – a typical plastic. Its glass transition temperature reaches – 70 ℃, while its crystals melt at $T_m = +110 - 135$ ℃. After melting, crystalline polymers can pass either into the viscous-flow or rubberlike state.

If a polymer is in the glassy or rubberlike state, the force of gravity will not cause it to change its shape, and we can therefore say that it is in the solid state of aggregation. If a polymer is in the viscous-flow state, it will not retain its shape under the action of the force of gravity (it slowly spreads out), which corresponds to the liquid state of aggregation. The gaseous state is unknown for polymers because of the large length of their macromolecules. The heat of "vaporization" of macromolecules, i.e. the energy of interaction between them, is so great that it exceeds the energy of breaking of carbon-carbon bonds in the backbone of the molecules. It is easier to destroy a polymer thermally than to convert it into a gas.

Since a polymer cannot evaporate, instead of the term "heat of vaporization" the interaction between polymer macromolecules is characterized by the cohesive energy $E(\text{MJ} \cdot \text{m}^{-3})$. For a number of reasons, it is more convenient to use the solubility parameter $\delta = \sqrt{E}$. Both the cohesive energy and the solubility parameter are a measure of the polarity of a polymer—their values are higher when a polymer is more polar.

6.2.2 Summary

The large length of chain macromolecules causes them to become flexible. The flexibility is limited by the interaction of the atoms and atomic groups joined to the main chain. This interaction restricts the freedom of rotation about the carbon-carbon bonds in a macromolecule. The greater the interaction, the higher is the rotational barrier and the lower is the flexibility of a macromolecule. The flexibility of macromolecules manifests itself in a temperature dependence of the properties typical of polymers and underlies the existence of three physical states of a polymer and the features of its crystalline structure. The presence of two basic structural elements, namely, macromolecules and their segments, underlies the features of the supermolecular structure and , in particular, the existence of a fluctuation network. All this taken together makes a viscoelastic strain the most typical of a polymer instead of a purely elastic or purely viscous(irreversible) one[2].

Vocabulary

1. viscoelasticity ['viskou,elæs'tisiti] n. 粘弹性
2. signify ['signifai] vt. 象征、预示; vi. 要紧
3. intact [in'tækt] a. 原封不动的、完整的

4. fluctuation [flʌktjuˈeiʃən] n. 变动、上下、动摇
5. duration [djuˈreiʃən] n. 持续时间
6. segment [segment] n. 片断、部分、切片、段、节；v. 分割
7. unsurmountable [ʌnsəːˈmauntəbl] a. 不可战胜的、不可克服的、不可凌驾的
8. flexibility [ˈfleksibiliti] n. 弹性、挠性
9. thermostat [ˈθəməstæt] n. 恒温器
10. specimen [ˈspesimən] n. 样品、标本、试料
11. amorphous [əˈmɔːfəs] a. 无定形的、无组织的
12. asterisk [ˈæstərisk] n. 星号（*）
13. plateau [plæˈtou] n. 高地、高原、(上升后的)稳定状态
14. sluggishly [ˈslʌgiʃli] a. 粘滞地
15. linoleum [liˈnouljəm] n. 油布、油毯
16. vitrified [ˈvitrifaid] a. 玻璃化的
17. plexiglas [ˈpleksiglɑːs] n. 塑胶玻璃
18. cubical [ˈkjuːbikəl] a. 立方体的、立方的
19. mould [məuld] n. 模子、模型； v. 铸造、造型、塑造
20. entanglement [inˈtæŋglmənt] n. 纠缠、卷入、缠绕物
21. aggregation [ægriˈgeiʃən] n. 集合、聚合
22. steric [ˈsterik] a. 空间的、立体的
23. cohesive [kouˈhiːsiv] a. 粘着的

Phrases

1. settled lifetime   固有寿命
2. in principle   原则上,大体上
3. in the bulk of   大批,大块
4. coincide with   与……一致,符合
5. short-range order   短程有序

long-range order　　长程有序
6. manifest oneself in　　表明,证明,出现,露出

Notes

1. The duration of the settled lifetime ... by a time interval. 全句可译为:用时间段来表征链段固有寿命的长短"。
2. All this taken ... viscous (irreversible) one. 全句可译为:所有这些因素综合在一起使粘弹性成为聚合物最具典型的特点而不是单纯的弹性或单纯的粘弹性。

6.3　ELASTOMER

Polymers that are rubbery at their use temperature are often called elastomers. These are a very special group of polymers with specific molecular characteristics that result in their unique mechanical behavior. For this reason they are treated separately in this text.

To be considered (ideally) elastomeric, a polymer should possess the following mechanical behavior:

(1) It must be rapidly stretchable to very high extensions (on the order of 500%).

(2) It must possess high tensile strength when fully stretched.

(3) It must snap back when the stress is released.

(4) It must retract completely with no permanent set.

The last three requirements are equivalent to stating that the strain energy is stored elastically in the material (i.e., none is converted into heat).

When this behavior is observed, certain molecular and environmental conditions usually exist:

(1) The material is a polymer.

(2) The material is amorphous.

(3) The temperature is above T_g.

(4) The material is lightly crosslinked.

The reasons for these conditions become evident when one considers the large extension ratios that elastomers must exhibit. In the unstressed state the elastomer consists of coiled, relatively flexible molecules entangled with and lightly crosslinked to one another.

When stressed (and, in particular, when stressed to high elongations), these molecules are stretched into lower entropy conformations:

until the crosslinks prevent further significant elongation. When the stress is released, the stored strain energy restores the molecules to their original conformations.

Unless all the conditions listed previously exist, the material usually does not exhibit this type of behavior. For example, short chains would

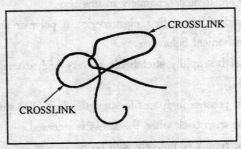

have little opportunity to coil and uncoil, and a material containing them would have a low extension to break. If crystals were present (or the temperature was below T_g), the molecules would have restricted mobility and the bulk material would be inflexible. Finally, without crosslinks some permanent flow would occur under stress and the molecules would not return to their original state.

6.3.1 Properties of Important Commercial Elastomers

Because of the variation that exists in the structures of repeating units, many types of elastomers are commercially available. Many of the common elastomers are copolymers, as opposed to homopolymers. There are several reasons for this. In butyl rubber, for example, a relatively small amount of isoprene comonomer is utilized. The purpose of this comonomer is to provide sites for crosslinking reactions, since the butyl repeating unit is not receptive to either sulfur or peroxide crosslinking. This is also done with acrylate elastomers (using 2-chloroethyl vinyl ether). Each of these systems involves a large proportion of one comonomer and a small proportion of another. Some copolymer elastomers contain large proportions of more than one comonomer. In ethylene-propylene rubber, for example, the propylene content can vary from approximately 30 to 60 mole percent. The purpose of this is to create an amorphous material. Polyethylene and polypropylene homopolymers are both crystalline, but the random copolymer has little tendency to crystallize when substantial quantities of each comonomer are present. The copolymer structure is irregular because the propylene units are positioned at random in the molecules. This discourages crystal formation because of the dissimilar nature of neighboring molecules.

Since most commercial elastomers contain varying amounts of additives, fillers, and crosslinking agents, which can drastically alter their properties, the values in some handbooks are intended only for comparative purposes. Some of the properties listed that are important in elastomeric applications have not been considered to any significant extent in previous discussions. For example, a lower use temperature (as well as an upper use temperature) is important in elastomeric applications. Unlike other amorphous polymers, elastomers are utilized above their glass transition temperatures. If an application requires exposure to very low temperatures, then an elastomer with a low T_g is required. Otherwise, as the ambient temperature approached the glass transition temperature, the material would begin to become glassy and lose its desirable elastomeric properties.

Another property that can be important in elastomeric applications is damping capacity. This is a measure of the ability of a material to absorb vibrational energy as opposed to transmitting it. It is a viscoelastic phenomenon involving stress-strain hysteresis (Figure 6.6). Elastomers differ in their ability to do this , and their applications often depend on this ability (or lack of it).

Since vibration problems occur in many ways in automobiles, automobiles are an important example of a product whose design involves material damping. For example, a primary function of the engine mounts is to reduce the vibration of the engine that would otherwise be transmitted to the frame of the car. The elastomer in the engine mount internally converts much of the vibrational strain energy into heat. This represents energy that cannot be transmitted to the frame as vibration. Butyl rubber is a good choice for this application. On the other hand, using the same elastomer in tires would be disastrous. Because heat is one of the greatest enemies of a tire, a low-damping elastomer is preferred over a high-damping one. In truck and heavy equipment tires, for example, natural rubber is a preferred material because of its relatively low damping qualities.

6.3.2 Thermoplastic Elastomers

Thermoplastic elastomers are an exception to the crosslink requirement in elastomers. An example of these is a styrene butadiene block copolymer in which each molecule consists of two blocks of styrene repeating units separated by a block of butadiene repeating units (The actual mechanisms are complex).

This is in direct contrast to styrene butadiene rubber (SBR), which is a chemically crosslinked, elastomeric random copolymer of styrene and butadiene units. It is because of the block nature of the styrene butadiene thermoplastic elastomer that the crosslinking requirement can be circum-

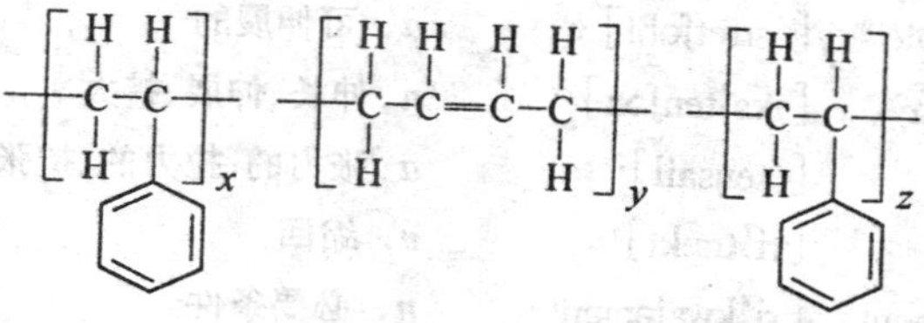

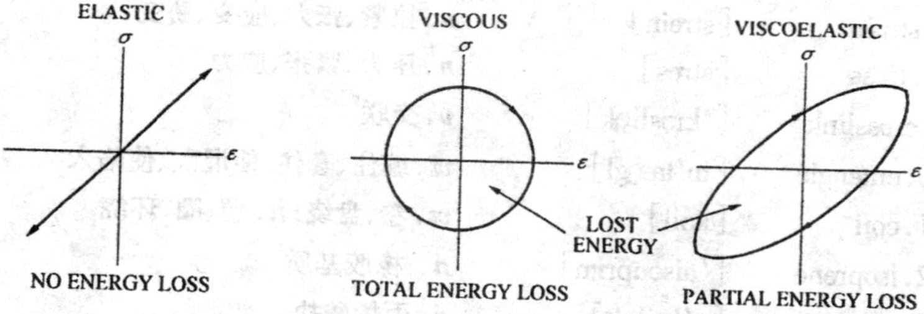

Figure 6.6 Damping in Elastic, Viscous, and Viscoelastic Materials

vented. The styrene segments are not compatible with the butadiene segments, so they tend to segregate in the bulk material. The result is a blend of glassy, brittle, polybutadienelike regions separated by (yet covalently bonded to) soft, rubbery, polybutadienelike regions. The glassy regions act as crosslinks for the butadiene segments.

This material has found applications in a variety of products, including shoe soles, sealants, adhesives, and coatings. A major advantage, of course, is that it can be softened, shaped, and cooled repeatedly, in contrast to other elastomers, which are thermosets. The softening process required to make a thermoplastic elastomer flow involves heating the material above the glass transition temperature of the polystyrene regions. Solidification occurs when the material cools and the glassy, polystyrenelike regions form. Since no degradation occurs in doing this, the scrap is recyclable.

Vocabulary

1. elastomer   [iˈlæstəmə]   n. 弹性体、弹料

2. stretchable [ˈstretʃəbl] a. 可伸展的
3. extension [iksˈtenʃən] n. 伸长、伸展、扩大
4. tensile [ˈtensail] a. 张力的、拉力的、抗张的
5. retract [riˈtrækt] v. 缩回
6. requirement [riˈkwaiəmənt] n. 必要条件
7. strain [strein] n. 拉紧、张力、应变、胁变
8. stress [stres] n. 压力、紧张、应力
9. crosslink [ˈkrɔsliŋk] v. 交联
10. entangle [inˈtæŋgl] vt. 缠住、套住、使混乱、使陷入
11. coil [kɔil] vt. 卷、盘绕；n. 卷、圈、环绕
12. isoprene [ˈaisəupriːn] n. 橡胶基质
13. acrylate [əˈkrileit] n. 丙烯酸盐
14. damping [ˈdæmpiŋ] n. 阻尼、减幅、衰减
15. vibrational [vaiˈbreiʃənəl] a. 振动的、震动的
16. hysteresis [ˌhistəˈriːsis] n. 滞后(现象)、滞后作用
17. circumvent [ˌsəːkəmˈvent] vt. 围绕、包围、智胜、防止……发生
18. mount [maunt] n. 衬纸
19. segregate [ˈsegrigeit] n. (使)分离、(使)分开、(使)隔离
20. sole [soul] n. 底面、基底
21. thermoset [ˈθəːməset] a. 热固性的；n. 热固性

Phrases

1. snap back   很快恢复,迅速恢复
2. permanent set   永久性变(定)形
3. extension ratio   延伸率
4. restore ... to ...   使……恢复到
5. opposed to sb(sth)   与…不同,对立面的
6. ambient temperature   环境温度

6.4 PLASTIC MATRIX MATERIALS

Reinforced with various kinds of fibers, plastics are widely used in composites as matrix materials. They are introduced in this section, since their structure and properties are quite different from those of other matrix materials such as metals. Plastic resins used as matrix materials for fiber-reinforced products are classified as thermosetting and thermoplastic. The properties of a thermosetting plastic are developed by chemical reactions which link together small monomer molecules to form long interlinked polymer molecules. Catalysts or curing agents and the application of energy (heat, microwave, etc.) are usually required to accomplish this reactioin. As a consequence of the formation of the three dimensional network of covalent bonds, the thermosetting plastics are fairly rigid. They will not soften upon further heating since the polymerization reaction is irreversible. At high temperatures, however, the covalent bonds may break. This irreversible destruction of the network causes the resin to lose its rigidity and strength at elevated temperatures. Thermoset polymers such as polyester, epoxy, phenolic, and silicone are widely used as matrix materials of fiber-reinforced composites. Thermoplastic materials, on the other hand, do not possess a rigid network structure. The linear molecular chains are formed prior to the development of the matrix and consequently are linked together mainly through the weak van der waals bonds. A thermoplastic material can be softened upon heating to a temperature known as the glass-transition temperature and the viscosity of the plastic decreases as the temperature rises. Thermoplastic materials softened at temperature somewhat above the glass-transition temperature regain their strength upon cooling. At elevated temperatures the covalent bonds of the polymer are destroyed and the resin decomposes. Since the glass-transition temperature is considerably lower than the temperature required to destroy covalent bonds, thermoplastic resins are limited to lower temperature applications than thermosetting resins. Thermoplastic materials such as polyethylene, polystyrene, and nylon have been used in composites as moulding compounds. Resins can be obtained from suppliers in various forms such as solutions in organic solvents, powders, and granules.

Composite materials with resin matrices and in particular, thermoplastic matrices, are attractive because of their relatively low fabrication cost. The low density, low electrical and thermal conductivity, corrosion resistance and translucence of the resin matrices further add to their attractions. The percentage of fiber composites based upon thermoplastics is much lower than that based upon thermosetting plastics. However, diligent efforts are being made in the development of fiber-reinforced thermoplastics in order to take advantage of their unusual formability. The structure and properties of some common plastic matrix materials are discussed in more detail in the following sections.

6.4.1 Polyester resin

The principles of polymerization of unsaturated organic compounds were discussed. The general reaction in ester polymerization involves the combination of di-basic and di-acidic monomers and can be expressed as $n(HO-R-OH) + n(H-P-H) \rightarrow [R-P]_n + 2nH_2O$, di-basic and di-acidic monomers such as ethylene glyco $HO-CH_2-CH_2-OH$ and fumaric acid $HOOC-CH=CH-COOH$ can condense by splitting out water to form the following unsaturated polyester: $[CH_2-CH_2-OOC-CH=CHCOO]_n$

The double bonds between adjacent carbon atoms indicate that the polyester molecules are unsaturated. These un-saturation sites provide cross-links to be established between a molecular chain and an unsaturated material such as a styrene monomer. A strong three dimensional network thus can be formed. The reaction to form the cross-linked structure is known as curing, and the curing agent then becomes part of the network structure. There is a great variety of choices of reactants, which all tend to affect the resin molecular weight and the distance between the sites of unsaturation along the chain. For example, if the unsaturated polyester is reacted with an unsaturated monomer such as styrene, the cross-linking reaction forms a network as illustrated below:

$$-M-CH-CH-M-CH-CH-M-$$
$$\left(\begin{array}{c} CH_2 \\ | \\ CHC_5H_6 \end{array}\right)_n \quad \left(\begin{array}{c} CH_2 \\ | \\ CHC_6H_5 \end{array}\right)_n \quad \text{where } M=\{COOCH_2CH_2COO\}$$
$$-M-CH-CH-M-CH-CH-M-$$

These various choices give rise to a family of polyester resins with each member of the family possessing different sets of properties. Since the strength as well as the thermal and chemical stability depends on the degree of cross-linking, it is important to choose the appropriate combination of reactants and monomers for the desired end product. The degree and the rate of curing is controlled by catalysts. The function of a catalyst is to act as an initiator for the polymerisation process through the formation of free radicals. The rate of their formation and, hence, the cure time is affected by the temperature. By varying the catalyst, curing can be completed in times ranging from several days to less than a minute at temperatures from 70°F to 300°F (21.1 ~ 148.9°C). The molecular weight and viscosity of the resulting polyester resin can be controlled by the duration of reaction in this temperature range. In view of the catalyst effect, it is desirable to store thermosetting polymer resins at room temperature and away from sunlight. The high viscosity of polyester resin as received from suppliers can be modified by dissolving it in styrene to obtain the desired viscosity. Most polyesters can be used satisfactorily at temperatures up to 482°F (250°C). The strength of polyester resins deteriorates with increase in service temperature. A major disadvantage of polyester resins is that they shrink when cured. The shrinkage of 4 to 8 per cent by volume is due to the shrinking of the monomer at curing. Consequently, it is difficult to obtain smooth surfaces in the final products.

Polyester resins can be incorporated with reinforcement materials very easily. It is often desirable to add mineral filler to polyester resins. In addition to lowering the cost of resins, filler materials also improve the surface appearance, resistance to water, and reduce shrinkage.

6.4.2 Epoxy resin

The structure of epoxy resin is characterised by the epoxy group, also known as epoxide. Epoxy resins of several families are now available ranging from viscous liquids to high-melting solids. Among them, the conventional epoxy resins manufactured from epichlorohydrin and bisphenol A remain the major type used.

Commercially available resins may contain modifiers, or they may be a mixture of resin types. There are a variety of hardeners or curing agents generally used for epoxy resins. The amine type compounds are often used in structural applications. The hardening effect is achieved through the formation of cross-links between the resin polymer chain and the hardener, or by direct linkage among the epoxy groups. The characteristics of the resin-hardener system can be varied through the addition of modifiers. A typical type of modifier is a diluent. Diluents, as the term implies, tend to lower the viscosity of epoxy resins. Some diluents can retard the degradation of resin properties by reacting with the epoxy resins. Epoxy resins are also modified by plasticiser.

Plasticisers are low-molecular-weight additives and serve to separate molecular chains from one another. The addition of plasticisers thus yields a non-crystalline solid with improved impact and low-temperature properties. Just as for polyester resins, filler materials can be added in epoxy resins as modifiers.

Composite materials using epoxy resins show far better resistance to chemicals and water than those using polyester resins. The shrinkage of epoxy is less than 2% and there is no water or volatile by-product generated during curing. As a result there is less chance for the epoxy matrix to pull away from the reinforcement and the subsequent exposure of fibers at the crack interfaces. Epoxy resins can be cured in the temperature range of 5 to 180℃ (41 to 356°F). The cured epoxy has demonstrated thermal stability up to 250℃ (482°F). The most desirable characteristic of epoxy resins lies in the fact that the properties of these resin systems can be 'tailor made' to fit a wide variety of applications through proper choice of resins, curing agents, and modifiers. The cost of epoxy resins is higher than polyesters and phenolics. Care shoud be exercised in handling certain hazardous resins and hardeners. The properties of different epoxy types are characterised by their viscosity, softening point, and epoxide value.

The versatility in from and property of epoxy resins has provided a wide spectrum of applications ranging from building and construction to electrical application. Their uses in filament-wound structures are extremely broad. These include pipes and tanks, electrical conduits, transformer and switchgear components, and pressure vessels, as well as aerospace and hydrospace vehicles.

6.4.3 High-temperature-resistant polymers

① Phenolics. The phenolic resins are derived from the condensation of phenols (C_6H_5OH) and aldehydes such as formaldehyde (HCHO). During curing, phenolics undergo three distinct stages. In the initial stage, the condensation stage, water is formed as a by-product. Upon heating, phenolics transform from a soluble, thermoplastic material, to a hard, insoluble, and infusible thermosetting polymer. The strength of phenolics is attributed to the cross-link formed by the formaldehyde in the polymerized materials. Phenolics can be used at high temperature up to 300℃ (572°F). The cost of phenolic resins is the lowest among thermosetting plastic materials. Some forms of phenolic resins may also generate volatile by-products during curing. Hence, it may by necessary to apply pressure when laminates are made of fibers impregnated with phenolic resin.

② Silicone. Silicone resin is characterised by the structure in which the polymer chains consist of alternating oxygen and silicon atoms. By-products are also generated during condensation. Silicone resins are known for their thermal and oxidative resistance. They have been used in the temperature range of 500 to 1 000°F (260 to 537.8℃). Although the price of silicone resin is relatively high, the excellence in both mechanical and electrical properties at elevated temperature has made fiber-reinforeced silicone resins ideal for radomes and supersonic vehicle components.

③ High performance polymers. Because of the advantages of polymeric resins discussed previously, considerable effort has been expended to increase their service temperature range. Several polymeric materials have been developed which can be used in the temperature range 600 to 1 000 °F (315.6 to 537.8℃). The properties of these high-performance polymers are attributed to the rigidity of the polymer chain achieved by the incorporation of ring structures. Of the various high performance polymers, the polyimides are currently the most widely used. A typical polyimide repeat unit is sketched below.

$$\left[N\begin{matrix}\overset{O}{\overset{\|}{C}}\\ \underset{\|}{\underset{O}{C}}\end{matrix}\bigcirc\begin{matrix}\overset{O}{\overset{\|}{C}}\\ \underset{\|}{\underset{O}{C}}\end{matrix}N-\bigcirc-O-\bigcirc\right]_n$$

Although the increase in rigidity has enhanced the strength and modulus of the polymer, the brittle material offers less resistance to crack propagation. These polymers are also more susceptible to cure and thermal shrinkage.

Vocabulary

1. matrix ['mætriks] n. 基体
2. catalyst ['kætəlist] n. 催化剂
3. rigidity [ri'dʒidəti] n. 刚性
4. decompose [di:kəm'pəuz] v. 分解、(使)腐烂
5. granule ['grænju:l] n. 小粒、细粒
6. matrices pl. (matrix 的复数形式) 基体
7. translucence [træz'lju:səns] n. 半透明
8. formability [,fɔmə'biliti] n. 可成型性
9. ethylene glycol [glaikɔl] n. 乙烯基乙二醇
10. fumaric [fju'mærik] a. 反丁烯二酸的
11. deteriorate [di'tiəriəreit] v. 恶化
12. epoxide [ep'ɔksaid] n. 环氧化物
13. epichlorohydrin [,epiklɔ:rə'haidrin] n. 环氧氯丙烷
14. bisphenol ['bisfenə] n. 双酚(A)
15. amine ['eimi:n] n. 胺
16. diluent ['diljuənt] n. 稀释剂
17. retard [ri'ta:d] v. 阻滞
18. plasticiser ['plæstisaizə] n. 增塑剂
19. versatility [,və:sə'tiləti] n. 多功能性
20. conduit [kəndit] n. 管道
21. phenolics [fi'nɔliks] n. 酚醛树脂
22. phenol ['fi:nl] n. 苯酚
23. impregnate ['impregneit] v. 浸渍
24. radome ['reidəum] n. (雷达)天线屏蔽器
25. polyimide [pɔli'maid] n. 聚酰亚胺

Phrases

1. be classified as 被分类成(为)……

2. curing agent   固化剂
3. as a consequence of   由于,作为……的结果
4. glass (-) transition temperature   玻璃化转变温度
5. moulding compound   浇铸体
6. by-product   副产物
7. be incorporated with   与……混合
8. the degree of crosslinking = the degree of curing   固化度
9. end product = resulting product   终产物
10. a major disadvantage of ... is that ...   ……的主要缺点是……
11. as the term implies   顾名思义
12. pull away from   脱离
13. epoxide value   环氧值
14. the hardening effect   固化效果(焠火)
15. be susceptible to   易受……影响的

6.5 FABRICATION METHODS OF RESIN MATRIX COMPOSITES

1. Hand lay-up. Fabrication by hand lay-up is the simplest method of making fiber reinforced plastics. In this method, a layer of random fiber mat or woven fabric is first laid on a form. Thermosetting plastics are then brushed on to the reinforcing materials. This process is repeated until the desired thickness of the composite is reached. Air trapped in the resin is squeezed out by a roller. The resin which is catalyzed is then allowed to cure at room temperature. Polyester and epoxy are the two most common resins used in hand lay-up fabrications. Very often, organic or inorganic fillers such as wood-flour, sawdust, clay, or sandstone are added to the resin for the purposes of reducing inflammability, providing extra weight, and decoration. The content of reinforcement achieved by this method is relatively low, about 30 per cent by weight.

The moulds used in hand lay-up may be made of a variety of materials such as wax, clay, wood, metal, paper, and plastic sheets. Mould releases, for example, polyvinyl alcohol, silicone, and mineral oils are often used for facilitating release of the final product from the mould. The tooling-up cost for hand lay-up is relatively low. This method is commonly used in making models and prototypes where only limited production is required. It also has been used to fabricate complex products which are not practical to make with matched dies, as well as for extremely large parts such as radomes and boat hulls.

2. Spray-up. In the method of spray-up, fibers are first fed through a chopper and cut into desired lengths. A mixture of fiber, resin, and catalyst is sprayed on to a mould. A roller is then used to smooth the surface and remove entrapped air.

3. Bag moulding. Bag-moulding is used to improve the quality of hand lay up products by further removing the entrapped air. The three basic bag-moulding methods are vacuum bag, autoclave, and pressure bag. In applying the vacuum-bag method, the lay up of resin and reinforcing materials is first covered with a perforated parting film and a layer of jute bleeder material. The combination allows the bleeding of air and excess resin. Then the lay-up is covered with a flexible film diaphragm, such as cellophane or nylon, which is sealed to the mould, the vacuum is then drawn upon the whole system with a pressure about 12 psi (8.28×10^4 N/m^2). The bagging process should immediately follow the lay-up in order to avoid the hardening of resins. The entire bagged system can be cured either in an oven or an autoclave system.

In the case of autoclave curing, a large, metal pressure vessel is used and is pressurized with a gas—typically nitrogen. The pressure applied is in the range from 50 to 100 psi (2.45×10^5 to 6.89×10^5 N/m^2). The autoclave system is heated and the hot gas is circulated to provide a uniform temperature within the vessel. Sophisticated autoclave systems provide electronic controls which produce programmed temperature/pressure-time cycles. After the laminate has been bagged and subjected to vacuum, it is placed inside the autoclave for cure. The vacuum system continues to function during the cure cycle in order to remove additional air and volatiles emitted during polymerisation of the matrix system. Next, the temperature/pressure-time cycle is initiated and carried out. The laminate is then removed from the autoclave for debagging. The pressure bag concept provides an economical alternative to the autoclave system when a heated platen press or comparable equipment is available. The pressure bag may be pressurized with air and is confined by the platens of the press. The pressure bag concept does not, however, possess the general flexibility of the autoclave system.

4. Filament winding. The method of filament winding employs continuous filaments which are wound onto a mandrel in predetermined orientations. Since it is possible to align the reinforcements along the direction of high stress, the strength of filaments can be utilized in an efficient manner. Structural applications of filament winding have been used in the aerospace industry to fabricate rocket-motor cases and radomes, as well as in commercial applications such as storage tanks, pipes, and pressure vessels. The single, most important, reinforcement used in filament winding is fiberglass. The weight percentage of glass fibers attainable in filament winding is the highest among all fabrication methods. At the present time, fibers of boron and graphite have not been used in large quantity for this purpose because of their relatively high cost. Depending upon the temperature range of application, plastic resins such as polyesters, epoxies, silicones and phenolics are used for filament winding. Two types of fabrication, wet-winding and dry-winding, are often employed. In wet-winding, the fibers are impregnated with resin just before winding. The filaments are pre-impregnated in dry-winding. Continuous fiberglass rovings or strands are usually wound over mandrels in two different patterns. In the planar winding, the mandrel is stationary. A layer of fiber is

wound on to the mandrel when the fiber feed arm rotates about a longitudinal axis for one complete revolution. The fibers in each layer lie adjacent to but not crossing over one another. The second basic winding type is called helical winding. Unlike planar winding the mandrel rotates while the feed carriage shuttles back and forth. As a result, the filaments are wound in a helical form. Helical winding is characterized by the filament crossovers on the mandrel. The product of winding is then allowed to cure with or without the applications of heat and pressure. For objects with open-ended shape such as cylindrical pipes, the mandrel can be forced out easily. For closed pressure vessels, the mandrels cannot be removed intact. In these cases, the mandrels are usually made of water-soluble hard salt or low melting-point metals. They are removed by solution or melting.

The tension exerted in the filaments can be controlled during winding. This tension can affect the void content, resin content, and, hence, the thickness of the laminate. The laminated composites resulting from filament winding generally have the filaments oriented in the following manners: (1) an angle-ply consisting of two monolayers oriented at ± 0; (2) a cross-ply consisting of two mono-layers at right angles to each other; and (3) an angle-ply and a mono-layer at 90 degrees. For example, in a closed pressure vessel, the orientation required as well as the amount of filament used can be determined from given hoop and longitudinal stresses.

5. *Continuous production techniques.* Continuous production techniques are the truly automated methods used in the fiber composite industry. One of the continuous pultrusion methods uses continuous glass fibers which are first impregnated with a thermoset resin. These fibers are then drawn through a die to obtain the desired shape of the end product. The curing is achieved by using an oven, or through external heating of the die. Glass content as high as 60 to 80 per cent by weight can be

achieved.

Another typical continuous production method is to be found in the production of continuous laminated sheet. In this method chopped glass fibers, impregnated with polyester resin, are sandwiched between two cellophane webs. The resulting products have been widely used as translucent building panels in the forms of corrugated and flat sheets. Still other continuous fabrication methods have been developed in the production of filament-wound pipes, in facing plywood panels with a reinforced plastic skin, and in manufacturing rigid laminated strips for sporting goods.

6. Closed moulding methods. Matched-die moulding, pre-mix moulding, and injection moulding all belong to this category. They are employed when detail is important and when a two-sided finish is desired. Two-piece male and female moulds are used in these methods. In contrast to closed moulding methods, those discussed previously are known as open-moulding methods, which are suitable for the fabrication of large objects and complex shapes.

Matched-die moulding is usually employed in the fabrication of structural parts where the contour is complex and the tolerance is close. Reinforcing materials in the form of glass mats, chopped glass preforms, or fabrics are used. Plastic resins are then applied onto the reinforcements, and curing is achieved through heating and pressing the moulds at pressures which are sometimes higher than 1 000 psi (6.89×10^6 N/m^2). Chopped glass preforms are used when moulding articles which have considerable contours. The preform is manufactured by first depositing chopped glass fibers onto a screen which has the desired contour. A resin binder is sprayed onto the reinforcements. The dried preform when removed from the screen is ready for matched-die moulding. The content of reinforcement material achieved by this method is higher than that of bag-moulding.

Pre-mix moulding differs from matched-die moulding in that it employs a ready for use moulding compound. This compound is a mixture of chopped glass fibers, resin, catalyst, and filler. The main attraction of this method is the comparatively low cost and the ease with which it can be moulded. It is particularly suitable for moulding objects with variable wall thicknesses and sharp contour changes. Because the reinforced moulding compounds have good electrical properties and resistance to corrosion, heat and flame, they have been used in a variety of applications to replace wood, metal, and ceramics. Common applications can be found in circuit panels for telephones, air-conditioning partitions, shower floors, etc..

The injection moulding process also utilizes a moulding compound which is injected into the cavity of the mould. However, unlike pre-mix moulding, the mouling compound used in injection moulding consists of thermoplastics. Typical resins used include nylon, polyethylene, polypropylene, polycarbonate, polystyrene, and polyvinyl chloride. Glass fibers are the most commonly used reinforcement material. Moulding compounds received from suppliers are usually in the form of pellets which contain fibers, thermoplastics, pigments, and lubricants. Pellets are produced by chopping impregnated glass rovings or by blending chopped glass strands and thermoplastics and feeding them to a compounding extruder. Thermoplastics reinforced with glass have shown marked increases in strength and rigidity as well as decreases in thermal expansion. The commercial applications of reinforced thermoplastics are developing rapidly in recent years in areas of appliances, electronics and household goods.

Vocabulary

1. mat   ['mæt]   n. 席子、毡
2. trap   [træp]   vt. 使受限制
3. squeeze   [skwiːz]   vt. 压榨
4. roller   ['rəulə]   n. 滚筒、辊子
5. sawdust   ['sɔːdʌst]   n. 锯屑
6. clay   [klei]   n. 陶土

7. sandstone ['sændstəun] n. 沙岩
8. decoration [dekə'reiʃən] n. 装饰
9. prototype ['prəutəutaip] n. 原型
10. jute [dʒuːt] n. 黄麻纤维
11. bleeder ['bliːdə] n. 放出管
12. diaphragm ['daiəfræm] n. 横膈膜
13. oven ['ʌvən] n. 烤炉
14. platen ['pleitən] n. 滚筒
15. mandrel ['mændril] n. 心轴
16. orientation [ɔːriə'teiʃən] n. 方位
17. helical ['helikəl] a. 螺旋的
18. shuttle ['ʃʌtl] v. 穿梭、往返
19. crossover ['krɔsəuvə] n. 交叉
20. hoop [huːp] n. 箍
21. corrugated ['kɔrəgeitid] a. 使起波状的
22. panel ['pænəl] n. 层合板
23. contour ['kɔntuə] n. 轮廓
24. tolerence ['tɔlərəns] n. 公差
25. polycarbonate [ˌpɔli'kaːbənit] n. 聚碳酸酯
26. pellet ['pelit] n. 小球
27. lubricant ['luːbrikənt] n. 润滑剂
28. extruder [ek'struːd] v. 挤出机

Phrases

1. hand lay-up 手糊成型
   spray-up (moulding) 喷射成型
   bag-moulding 袋模成型
2. lay on 安装
3. draw upon 利用
4. filament winding 纤维缠绕
5. rocket-motor case 火箭发动机壳体

6. wet-winding 湿法缠绕成型
　 dry-winding 干法缠绕成型

6.6 METAL EXTRACTION AND REFINING

6.6.1 Electrolytic Extraction (Electro-Winning)

The principal reasons for using an electrochemical method for the extraction of a metal from its ores are that ① the reduction with carbon is thermodynamically unfavourable at conveniently attainable temperatures; or ② low grade ores are to be worked, initially by reaction with an aqueous reagent, giving a dilute solution of a salt of the metal. The first reason applies to sodium, magnesium, calcium, aluminium, etc., (see Table 6.1); and as these active metals cannot be deposited by electrolysis from any aqueous solution, electrolysis of molten salts is used (though research on electrodeposition from organometallic compounds in non-aqueous solution appears promising, especially for aluminium).

Table 6.1　Metals Which are Extracted or Refined by Electrolysis

Electrolyte state	Electro-winning	Refining
Molten	Al, Mg, Na, Li, K, Ca, Sr, Ba, Be, B, Th, U, Ce, Ti, Zr, Mo, Ta, Nb…	Al
Aqueous	Cu, Zn, Co, Ni, Fe, Cr, Mn, Cd, Sb, Pb, Sn, In, Ag	Cu, Ni, Co, Sn, Pb, Hg, Ag, Sb, In

General features of fused-salt electrolyses include ① the use of mixtures of salts to reduce the melting-points; ② the use of the heating effect of the electric current to maintain the temperature of the melt; ③ the use of graphite anodes, at which the usual product is chlorine (aluminium electro-winning is an exception); ④ some means of protecting the molten metal from air and from the anode; ⑤ applied potential differences are usually 6 to 8 V, and currents may be as high as 10^5 A.

It is noteworthy, at a time when world resources of many metals are getting alarmingly low, that there will never be a shortage of aluminium so long as there is energy to extract it, since on a molar basis it is the fourth most abundant element in the earth's crust (after O , Si and H) at 4.8 mol%[1]. However, supplies of high-grade bauxite are becoming exhausted, and the cost of obtaining aluminium oxide for the electrolytic cells will increase. Recycling of scrap aluminium will become even more important than it has been hitherto.

The extraction of metals through the electrolysis of aqueous solutions is less widely practised. Low-grade copper ores are leached by sulphuric acid, giving copper(II) sulphate solution which is electrolysed with a copper cathode (which gains mass) and a lead anode (at which oxygen is evolved, and sulphuric acid reformed). Some zinc is produced in a similar way, the low-grade ore being roasted to the oxide, then treated with sulphuric acid.

6.6.2 Electrolytic Refining

Electrolysis of aqueous solutions is used on a large scale to refine metals produced by carbon reduction. Copper, nickel, silver, tin and lead are among the metals commonly treated in this way. Large anodes of the crude metal and thin sheets of the pure metal are set in cells containing a suitable electrolyte ($CuSO_4 + H_2SO_4$, $NiSO_4 + NiCl_2$, $PbSiF_6 + H_2SiF_6$), usually acidified to increase conductivity. As the anode dissolves, impurities with a more negative oxidation potential remain undissolved (e.g. Bi, Sb and Ag in lead), and are collected in bags hung around the anode, and later refined. Metals with a more positive reduction potential dissolve along with the principal metal, of course, but they fail to deposit on the cathode. Current efficiencies of 90% ~ 95% are obtainable, and the energy consumption may be as low as 60 $kJ \cdot mol^{-1}$ (e.g. 0.2 $kW \cdot h \cdot kg^{-1}$

for Cu). This is only one-tenth of the energy required for metal extraction from aqueous solution.

Vocabulary

1. electrodeposition [i,lektrəudi'pɔzit] n. 电沉积
2. winning ['winiŋ] n. 获得、[矿]开采
3. bauxite ['bɔ:ksait] n. [矿]铝土矿、矾土
4. scrap [skræp] n. 废料
5. hitherto [hiðə'tu:] ad. 迄今、至今
6. leach [li:tʃ] v. 过滤
7. refine [ri'fain] vt. 精炼、提纯、精制

Phrases

1. electro-winning 电解沉积
2. fused-salt 熔融盐
3. reduction potential 还原电势
   oxidation potential 氧化电势

Note

It is noteworthy, at a time ... at 4.8 mol%. 句中"so long as"意为"只要";"on a molar basis"指"根据物质的量算法"。

6.7 ELECTROPLATING

Electroplating, the cathodic deposition of a thin layer of metal on the surface of an object, is most often carried out for the purpose of enhancing the appearance, or protecting against corrosion or both. The metals commonly used for plating are copper, silver, gold, tin, zinc, nickel and chromium. Electroplate can be thinner than, and yet because of its uniformity just as effective as, metal plated mechanically, e.g. by dipping[1]. Indeed the rising price of tin would have made tin-plate (on steel) uneconomic had it not been possible to change to the electrolytic method of manufacture: and now tin-plating is, after aluminium and caustic soda/chlorine, the most important electro-technology[2].

In most electroplating cells the anode is a piece of the metal being deposited, since this will dissolve at the same rate as the deposition at the cathode, and the electrolyte solution is not depleted of its metal content. The plating bath is almost invariably a mixture of a current carrying but otherwise inert electrolyte and a compound of the metal, in many cases a complex ion[3]. It is essential that no local displacement of the dissolved metal by the cathode metal should occur, since the metal coating produced in this way is usually powdery and poorly adhesive. For example, silver nitrate solution would be useless as a plating bath for a copper object, since copper can displace silver by local action: the complex dicyanoargentate(I) (argentocyanide) is used instead.

A process which is very similar to electroplating is electroforming, which consists of the deposition of a very thick layer of metal(usually copper or nickel) which, when stripped away from the original article or 'former', reproduces in negative relief all the detail of the surface[4]. The electroformed object then serves as a mould, for example for plastic injection moulding. Gramophone records are reproduced in this way, as are blocks for printing(electrotype)[5].

Objects to be plated need not be of metal, since the surface can be made conducting by graphite, painted on as a suspension known as aquadag, or by silver deposited by chemical reaction (a 'silver mirror'). This procedure permits the eletroplating of plaster, plastic, wax, wood etc..

The best constitution of the plating bath, its temperature, pH, and other factors are often determined by experiment, and electroplating was for a long time more a craft than a science. Thorough removal of grease, dirt and oxide films from the surface is always essential. A high current is desirable to save time but too high a current density may produce a loose spongy deposit, or even a metal powder. The appearance of the plate may be enhanced by 'brighteners', and other addition agents, usually colloids such as gelatin or glue, improve the adhesion. These agents encourage the formation of the metal as small crystals, but the way in which they act is still imperfectly understood.

A most important characteristic of a successful plating bath is that it should have good 'throwing power', that is, that the plating should be of uniform thickness at points near to and far from the anode, and that plating should occur in crevices etc. where diffusion may be restricted[6].

Vocabulary

1. electroplate [iˈlektroupleit] vt. 电镀; n. 电镀物品
2. chromium [ˈkroumjəm] n. 铬
3. caustic [ˈkɔstik] a. 腐蚀性的、苛性的
4. deplete [diˈpliːt] v. 耗尽
5. dicyanoargentate [ˌdaiˈsaiənəuˌaːdʒənteit] n. 二氰化银
6. argentocyanide [ˈaːdʒəntəuˌsaiənaid] n. 银氰化物
7. electroform [iˈlektroufɔːm] vt. 电铸
8. gramophone [ˈgræməfəun] n. 留声机
9. aquadag [ˈækwədæg] n. 胶体石墨、导电敷层
10. grease [griːs] n. 油脂
11. spongy [ˈspʌndʒi] a. 松软的、多孔的
12. crevice [ˈkrevis] n. 裂缝

Notes

1. Electroplate can be thinner than ... e.g. by dipping. 句中"metal plated mechanically"是指"用机械方法镀金属";"dipping"此处指"浸镀"。
2. Indeed the rising price of tin would have made ... important electro-technology. 句中"had it not been possible to ... manufacture"是倒装形式,表示虚拟语气,正常句式为"if it had not been ...",表示假设。
3. The plating bath is almost ... a complex ion. 全句可译为:镀液几乎总为导电

混合物或另外的惰性电极和金属化合物,大多情况下是配离子。
4. A process which is ... the detail of the surface. 句中"in negative relief"意为"以(用)负模的形式"。
5. Gramophone records are reproduced ... for printing. 句中"as are blocks for printing"为倒装语序。
6. A most important characteristic of ... be restricted. 句中"throwing power"指"分散能力"。

6.8 CORROSION OF IRON

One of the most widespread electrochemical processes is also a most unwelcome one: in Britain alone the corrosion of iron costs an estimated £ 10^9 p.a. (3.5% of the gross national product), of which about one-third could be saved by improved protective measures. In the power industries an estimated £ 25 m p.a. is saved by cathodic protection. Rusting has been known throughout history, but it was not until 1902 that a satisfactory explanation (proposed by W.R. Whitney) was available. It is now generally accepted that the principle cause of the corrosion is the establishment of a short-circuited galvanic cell, thus:

$$Fe(s) \rightarrow Fe^{2+}(aq) + 2e$$

$$\frac{1}{2}O_2(aq) + H_2O(l) + 2e \longrightarrow 2OH^-(aq)$$

followed by two subsequent non-electrochemical steps leading to hydrated iron (III) oxide.

Rusting requires liquid water, not merely water vapour, because there must be an electrolyte solution in contact with the metal if the above cell is to be formed. Pure water is not sufficient, but it is normally impossible to exclude some electrolyte, if only dissolved CO_2, salt from sea-spray or highway ice-clearing is particularly deleterious.

It is found that if oxygen is totally excluded, or destroyed chemically (e.g. by hydrazine) rusting is prevented. The possible alternative reduction half-reaction, the production of hydrogen, does not generally occur unless the solution is very acidic. Even so, the presence of water, electrolyte, oxygen and iron together does not necessarily lead to corrosion. If conditions are uniform at all points on the metal surface rusting is extremely slow. It appears that the oxidation and reduction half-reactions must occur at different sites. One situation which permits this separation of half-reactions is when the iron is in contact with a more noble metal (e.g. copper), for then that metal acts as the cathodic site (for the reduction of O_2), and the neighbouring iron becomes an anodic site, and is attacked.

However, even in the absence of a dissimilar metal corrosion can occur at points which are partly hidden from the air, such as under bolts or rivets, under specks of insoluble material, or in a corner, crack or hole. It appears that rusting, the oxidation of iron occurs at places furthest from the source of dissolved oxygen, and it was to explain this strange behaviour that the theory of differential aeration was devised.

Consider the conditions within a drop of impure water held in a crevice on an iron surface. Near the water-air interface there will be a plentiful supply of oxygen, but in the remoter parts of the drop there may be a much lower concentration. At the oxygen-rich site, the reduction of oxygen to hydroxide ions will tend to occur, consuming electrons there, and making the metal positive relative to the solution. This polarity hinders the oxygen reduction, particularly at sites where oxygen concentration is low, but encourages the anodic oxidation of the iron metal. Any hole or crevice tends to deepen.

Of the various protective measures against rusting, one is very directly related to the electrochemical mechanism: namely cathodic protection, in which the iron structure is made cathodic (negative with respect to the electrolyte). There are three ways in which this can be done: ① plating with zinc ('galvanized iron'); ② sacrificial anodes; and ③ direct current from a generator.

In the case of galvanized iron, the zinc is constantly under attack (but fortunately the zinc oxide produced affords a degree of physical protection), but the iron underneath is thereby made cathodic.

For iron pipes or other underground structures, protection is often given by nearby buried blocks of magnesium alloy, each electrically connected to the iron. Steel ships are also protected by sacrificial anodes (following the method devised in 1824 to protect the copper bottoms of the ships of the Royal Navy). A sufficient number of these anodes must be provided, because protection ceases where the length of the current path through the electrolyte becomes too great. (Why? The iron, a good conductor, will be at almost the same potential all along its length.) In places where a generator and cable may be positioned, it is sometimes cheaper to replace the magnesium by scrap iron or carbon, and to maintain them at a small negative potential by an external source.

Vocabulary

1. corrosion [kə'rouʒən] n. 腐蚀、腐蚀状态、减退
2. gross [grəus] n. 毛重
3. galvanic [gæl'vænik] a. 流电的、抽搐的
4. spray [sprei] n. 水花、喷雾、喷雾器;
   vt. 喷雾、扫射、喷射;
   vi. 喷、溅散
5. deleterious [ˌdeli'tiəriəs] a. 有害于、有毒的
6. aeration ['eiəreiʃən] n. 通风、充气、分解
7. sacrificial [ˌsækri'fiʃəl] a. 牺牲的、献祭的、具有牺牲性的
8. galvanize ['gælvənaiz] v. 电镀、镀于

Phrase

1. short-circuited cell 短路电池

6.9 FUEL CELLS

In an earlier section some design features of good cells were mentioned: high ratio of electrical capacity to mass, low internal resistance, a steady potential difference which does not fall excessively when a current is drawn, and a long 'shelf life'. Three categories were introduced: primary cells, secondary (rechargeable) cells, and fuel cells, but these categories are not mutually exclusive, and refer to the function and design of the cell as much as to the electrochemical reactions which occur. Primary cells are portable sources of small quantities (say 1 kJ) of electrical energy, and the price is justified by the convenience rather than by the value of the electricity. Secondary cells, also, are relatively small, self-contained units, designed so that the reactants in the current producing reactions may be regenerated in position when a direct current is passed through the cell in the reverse direction. The energy which can be stored in a lead-acid cell of mass 1 kg is about 70 kJ, which is about the same as in an equal mass of torch batteries, and less than a tenth of that in some recently developed rechargeable batteries such as the sodium-sulphur battery. However, the lead-acid cell is particularly suitable for use for the starter motor of a petrol-driven car, because it can provide heavy currents (200 to 400 A) for short periods without suffering damage. As the source of power for driving (rather than starting) the car, lead-acid batteries are less satisfactory, for they can only provide about 100 $W \cdot kg^{-1}$ on continuous discharge, which means that the economic top speed is equal to a brisk walking pace (e.g. milk floats, fork-lift trucks). For electrically driven vehicles to be viable, the mass and the cost per unit of power must be reduced to about one-hundredth of those of the present car battery. It now seems quite possible that the fuel cell will achieve this.

The principles of the fuel cell may be shown by considering one of the simplest systems—the hydrogen-oxygen cell. Hydrogen is supplied to one of the porous catalytic electrodes, and oxygen or air to the other. The electrolyte is a concentrated solution of an acid (e.g. phosphoric(V)acid) or, more commonly, an alkali (usually potassium hydroxide). The overall reaction, $H_2 + \frac{1}{2}O_2 \longrightarrow H_2O(1)$, has $\Delta G^\ominus = -237$ kJ·mol^{-1} and hence $E^\ominus_{cell} = +1.23$ V at 298 K. The enthalpy change for the combustion of H_2 to $H_2O(g)$ is -242 kJ·mol^{-1}. Therefore if the fuel cell could be operated at its maximum (reversible) cell potential, an efficiency of 98% chemical energy conversion would be achieved. This may be compared with the theoretical maximum efficiency of 40% ~ 50% for any 'heat engine' type of energy convertor, with its Carnot limitation of $(T_2 - T_1)/T_2$, where T_1 and T_2 are the temperatures of the heat sink and source respectively. This apparent doubling of conversion efficiency was like a mirage which deluded and eventually disappointed early researchers into fuel cells.

As was pointed out previously, the working cell potential is invariably lower than the reversible (or 'no-load') potential, by an amount comprising the overpotentials at the two electrodes, together with any concentration polarization and the 'IR drop' due to internal resistance[1]. There can be no current without overpotential, although the size of the overpotential depends profoundly upon the catalytic properties of the electrode surface. For this reason research into cheaper and more effective catalytic electrodes is one of the principle directions that the fuel cell development programme is taking. There is little hope of finding the 'perfect catalyst' (cf. philosophers' stone), and the goal of anything like 100% efficiency has now been abandoned − in fact fuel cell technologists might well be content with the same 40% limit that applies to heat engines, in return for electrodes that were cheap, durable and easy to maintain. In this respect it appears that fuel cells will have the important advantage that they can achieve 40% energy conversion even on such a small scale as a 25 kW unit (whereas the modern gas turbine generator does not approach this unless it is capable of 10^5 kW) and so their first use may be for small domestic or vehicle units.

If the overpotential of the hydrogen-oxygen cell presents a problem at normal temperatures, that of a hydrocarbon fuel cell (e.g. CH_4-O_2) is quite impossible[2]. The rate of reaction is negligible below about 200℃, yet hydrocarbon fuels, either natural gas or 'cracked' petroleum oil, are so much cheaper (and easier to store) than hydrogen that a considerable part of the $ 50m fuel cell effort has gone towards attempts to utilize them. One way is to convert the hydrocarbon to hydrogen immediately before use, by catalytic reaction with steam

e.g.
$$CH_4 + H_2O \longrightarrow CO + 3H_2 \quad (ca.\,900℃)$$
$$CO + H_2O \rightleftharpoons CO_2 + H_2 \quad (ca.\,300℃)$$

The technical difficulties of the conversion and purification are serious but not insuperable.

The other solution to the unreactivity of hydrocarbons is to raise the temperature of the cell, to 650℃ or so, by using a molten salt electrolyte such as mixed lithium and sodium carbonates. At these high temperatures the overpotentials are low, and no expensive catalytic electrodes are necessary. For this reason, high temperature hydrogen-oxygen cells have also been designed, either using molten carbonates, or a very concentrated potassium hydroxide solution under pressure (as in the first practical fuel cell, built by Bacon and Frost at Cambridge in 1959). In all such cells, corrosion is one of the most serious obstacles, and most of the construction materials of the most advanced Pratt and Whitney cell have been developed specially for the purpose in the last 7 years.

A heavy duty fuel cell which consumes carbon monoxide at 1 000℃ is being designed by Westinghouse and uses a solid state electrolyte of ZrO_2 containing a little Y_2O_3 (which gives it an O^{2-} deficiency). To avoid corrosion troubles at this high temperature, the air electrode is made of a solid state conductor (SnO_2) rather than a metal.

It must not be thought that low temperature aqueous cells are being neglected, however. Esso in the U.K. and Exxon-Alsthom (U.S.A./France) are developing methanol fuel cells which will require electrodes of high specific catalytic power and large surface to volume ratio. That a low temperature fuel cell is possible was proved in the General Electric Company's hydrogen-oxygen cell which provided the electric power in the Gemini spacecraft; but this, using an ion-exchange resin electrolyte and platinum-coated PTFE electrodes, was far too expensive for terrestrial use. Fuel cell technology is part science, part economics.

Vocabulary

1. brisk   [brisk]   a. 敏锐的、凛冽的、活泼的;
                                   vt. 使活泼; vi. 活跃起来
2. viable   ['vaiəbl]   a. 能养活的、能生育的
3. portable   ['pɔːtəbl]   a. 手提的
4. torch   [tɔːtʃ]   n. 喷灯
5. convertor   [kən'vəːtə]   n. 变换器
6. mirage   ['mirɑːʒ]   n. 雾气、幻想
7. delude   [di'luːd]   vt. 迷惑、盅惑
8. insuperable [in'sjuːpərəbl]   a. 不能制胜的、不能克服的
9. lithium   ['liθiəm]   n. 锂
10. deficiency [di'fiʃənsi]   n. 缺乏、不足
11. terrestrial [tə'restriəl]   n. 地球上的人; a. 地球的、
                                   地上的、陆地的、人间的

Phrase

1. shelf life   (产品的)货架寿命

Notes

1. As was pointed out ... due to internal resistance. 句中"concentration polarization"是指"浓差极化";"internal resistance"指"(系统的)内阻"。
2. If the overpotential of the hydrogen-oxygen ... is quite impossible. 句中"that"指

代"the overpotential (超电势)"。

6.10 SECONDARY BATTERIES—NICKEL-CADMIUM BATTERY

6.10.1 Introduction

In nickel-cadmium batteries, the energy is stored as the reaction enthalpy of the couple Cd and NiOOH. During current generation, i.e., during the discharging phase, the following overall chemical reaction takes place:

$$Cd + 2NiOOH + 2H_2O \longrightarrow Cd(OH)_2 + 2Ni(OH)_2$$

In the idealized case, the total "free enthalpy of reaction" is released as electrical energy.

The given reaction can be reversed by passing electrical energy into the system. This occurs during the charging phase.

The active components of individual nickel-cadmium cells are outlined in Figure 6.7. The negative and the positive electrodes contain cadmium and nickel(Ⅲ)-oxyhydroxide, respectively, as the active masses. The electrodes are interconnected over aqueous, alkaline electrolyte.

Nickel-cadmium cells are classified as alkaline accumulators, which were developed considerably later than the lead-acid battery. The motive of the efforts of Edison and Jungner starting at about 1890 was the search for sturdy storage cells which would be suitable for electrical vehicles[1]. Thus Edison invented the Fe-Ni system, while Jungner came up with the Ni-Cd system. There were great hopes at that time that the electric motor could even compete with the internal combustion engine as a propulsion unit for motor vehicles.

Though the idea of electrical propulsion has been revived today and extensive developments have been made in several countries, the Ni-Cd system is not seriously discussed in this context. Availability and costs of

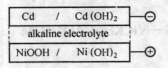

Figurt 6.7 Set-up of the nickel-cadmium cell.

materials rule out such a broad application.

However, the Ni-Cd system is invariably preferred where such qualities as mechanical rigidity, good low-temperature behavior, high-rate discharge, and simple handling are desired. A special impact was made by the sealed version, which was based on the initial work of Neumann in Germany.

In this contribution, a treatment of thermodynamics and kinetics will be followed by a description of the preparation of materials and the construction of different types of electrodes and cells, as well as a discussion of their technical performance and application.

6.10.2 Electrodes and Cells

(1) Electrode Types. A high specific interface between the active material and the electrolyte phase is desired to obtain a high current loading during charging and discharging. At the same time, car has to be taken to maintain good electronic contact with the current-collecting structure in the electrodes. A series of practical solutions has emerged from the efforts to match the structural specifications of electrodes. Tests with nickel foam as structural material and with organic binders are under progress. All attempts are concentrated in realizing a long-life, high-load, low-cost nickel electrode.

① Pressed Electrodes. Powdery active material is pressed into tablets or cylinders. A conducting material such as graphite can be admixed to improve the electrical contact. The pressed materials are often enclosed between nickel screens.

② Tubular Electrodes. Edison used this sturdy type of electrode in which the active material is filled in tubes made of perforated, nickel-plated steel ribbons. The individual tubes have a diameter of 7 ~ 8 mm and are welded together to form larger electrode areas. In the case of poorly conducting nickel hydroxide, nickel flakes are added that are galvanically produced and have a thickness of about 1 μm. The supporting structure is subjected to heavy mechanical stresses created by the great volume changes occurring during the exchange of charge[2]. The stability can be ensured , however, by mounting steel rings around the tubes.

③ Pocket Electrodes. Electrodes as sturdy as the tubular type can also be prepared by incorporating the active mass in the hollow space between two perforated sheets. This type is found both on the cadmium side and on the nickel side. In this case, very coarse and open structures are adequate as spacers between the electrodes: the resistance in the electrolyte space is only slightly increased by them.

④ Sintered Electrodes. The shielding of active material by means of the encasing foil used in tubular and pocket electrodes proves to be particularly disadvantageous at a high current loading. This can be considerably improved by incorporating the reactive materials in a sintered body of carbonyl nickel. The volume porosity is adjusted to approximately 80%. Thin supporting bodies can be stabilized by sintered, perforated metal ribbons.

In order to load the supporting body, it is soaked with nickel or cadmium nitrate solution, dried , and then immersed in potassium hydroxide solution to precipitate the hydroxide. The last stage of preparation can also be replaced by polarizing the electrodes cathodically; owing to the diffusion limited mass transfer between the external electrolyte and the porous system, the pH value increases until the hydroxide is precipitated.

(2) Cell Types. In the past few years, development of Ni-Cd cells has been directed toward a large number of specific applications, leading to a correspondingly colorful palette of cell types.

The following types can be distinguished according to the constructive design:

Plate Cells. They are constructed on the lines of the lead-acid battery. Coarse plastic screens serve adequately as spacers where tubular and pocket electrodes are used. Thin sintered electrodes are separated by intermediate layers of nonwoven fabrics (fleeces).

Cylindrical Cells. These cells are formed from rod-shaped, pressed electrodes or coiled sintered electrodes. This design is often selected for the sealed type.

Button Cells. Button cells are constructed from pressed or sintered electrodes. They can be easily stacked to form a battery.

Attempts were made quite early to make secondary cells as maintenance free as primary cells. Thus T. A. Edison had a patent granted which covered the catalytic combustion of gases that are liberated inside the cell upon charging.

The concealed objective behind the search over the decades was to develop a sealed accumulator that could be utilized in any position without spilling the aggressive electrolyte, without leakage of vapor and without any maintenance, especially the refilling of distilled water.

After the plate sintered from nickel powder had lead to a boom in production of the nickel-cadmium accumulator around 1940, a few years later a German research group had succeeded in sealing this system according to the initial patents of Dassler and co-workers Peters and Neumann.

The task imposed was to prevent the formation of an explosive gas mixture (oxyhydrogen gas) under all working conditions, because otherwise this gas would burst the sealed cell. This was achieved by an excess of the uncharged negative material $Cd(OH)_2$. During the charging process, oxygen will be evolved at the positive electrode before the negative material is completely reduced at the counterelectrode. If the electrolyte matrix is provided in such a manner that it not only contains the electrolyte-filled fine pores but also free coarse pores, oxygen can find its way through these pores to the cadmium electrode where it will be reduced. Thus under overloading there will be no net change in the chemical composition inside the cell: the oxygen evolved at the positive electrode will be consumed at the negative electrode.

The prevention of a pressure rise during overcharging is not the only concern. There is also a risk of excessive gas formation during deep discharge by pole changing. Owing to manufacturing deviations, not all cells have exactly the same storage capacity. In the cells connected in series, explosive oxyhydrogen gas can appear when the weakest cell undergoes a pole change after a total discharge of its capacity. Such an event will be accompanied by decomposition of the electrolyte. This difficulty was surmounted by introducing an "antipolar mass." In the types common today, $Cd(OH)_2$ is added to the nickel electrode to such an extent that unreduced $Cd(OH)_2$ still remains in the positive electrode even when cadmium in the negative electrode is completely oxidized. H_2 evolution is thus prevented. When the oxygen pressure exceeds a limiting value it can react again electrochemically at the Cd electrode, but this time at the cadmium of the antipolar mass.

Several thousand cycles are possible at a 25% discharge, and still several hundred at a 75% discharge.

A cell which is maintenance free and independent of position is designed according to the button, cylindrical, or rectangular cell type. In larger cells, electrodes with a sintered nickel skeleton in which the active mass is precipitated are prevailing more and more.

The capacities of sealed cells fabricated since 1953 range at present from 4 mAh up to about 50 Ah. The upper limit is given by the heat evolution during the O_2 reduction which leads to an additional pressure rise. Since the heatradiating surface does not increase as rapidly as the capacity, which is almost proportional to the volume, the charging and discharging current must be brought into the correct relation to the electrode area and the capacity.

6.10.3 Application

In view of the variety of cell types with different dimensions that have been designed and developed, the practical application of Ni-Cd cells covers a wide range. Thus they are used for stationary emergency power supplies. The sequential computer, for which relatively short current failures of only a few minutes have to be bridged, is a special case. The ability of sintered plaque cells with small internal resistance to produce the required high discharge current densities commend their choice in this case. These cells are thus also suitable to serve as batteries for starting emergency diesel-generating sets.

In comparison to other storage batteries, nickel-cadmium cells show excellent resistance to mechanical shocks, accelerations, and vibrations. These qualities have made them desirable for use in train lighting and for vehicle and rocket technology.

In view of their high specific energy content and cycle life, tubular batteries have prevailed as traction batteries for electrically driven vehicles, for instance, electric boats, mine locomotives, electric carts for factories, trucks, and tractors.

The low maintenance and the independency of position of sealed nickel-cadmium accumulators have increasingly commended their use in past years as power sources in satellite technology and for portable equipments such as hand-held calculators, measuring instruments, radio receivers, hearing aids, electronic watches, cameras, security systems, tools, toys and tape recorders. They can also be mounted rigidly in printed circuits and in solid state memories.

It is expected that the market for alkaline nickel-cadmium batteries will strongly expand in the future. At present the world turnover is estimated at about $150 million per year.

Vocabulary

1. oxyhydroxide   n. 氢氧化物
2. hexagonal [hek'sægənl] a. 六角形的、六边的、底面为六角的
3. passivation [ˌpæsi'veiʃən] n. 钝化
4. binder ['baində] n. 缚者、用以绑缚之物、夹器
5. perforate [ˌpəːfəreit] a. 多孔的
6. flake [fleik] n. 薄片
7. galvanical [gæl'vænikəl] a. 通电的
8. sinter ['sintə] v. 使烧结
9. encase [in'keis] vt. 装入、包住、围
10. porosity [pɔː'rɔsəti] n. 多孔性
11. palette ['pælit] n. 调色板、颜料
12. antipolar ['æntipəulə] a. 相反极性的
13. cylindrical [si'lindrikəl] a. 圆筒、圆柱状
14. rectangular [rek'tæŋgjulə] a. 矩形的、成直角的
15. plaque [plaːk] n. 匾、饰板、名牌、血小板
16. diesel ['diːzəl] n. 柴油机

Notes

1. The motive of the efforts of ... for electrical vehicles. 句子的主干结构为"The motive was the research."
2. The supporting structure is subjected to ... the exchange of charge. 句中"be subject to"是指"经受,遭受,受……支配"。

6.11 FUNDAMENTAL PRINCIPLES OF ELECTROCHEMISTRY

6.11.1 General

Electroplating is the process of depositing a coating having a desirable form by means of electrolysis. Its purpose is generally to alter the characteristics of a surface so as to provide improved apearance, ability to withstand corrosive agents, resistance to abrasion, or other desired properties or a combination of them, although occasionally it is used simply to alter dimensions. Electrolysis is carried out in a bath which may consist of fused salts or of solutions of various kinds; in commercial practice it is almost invariably a water solution.

The laws of electrolysis, formulated by Faraday in 1833, afforded the first quantitative demonstration of the electricil nature of matter, and have long defined the unit quantity of electricity. Today, unmodified by newer discoveries, they remain among the foundations of physical science. For the electroplater, Faraday's laws specify the current-time product required to produce a definite weight of deposit. One Faraday (96,490 ± 2.4 amp-sec) of electricity yields 1 gram-equivalent of substance. The gram-aquivalent is obtained by dividing the gram-atomic weight of the deposited metal by the number of electrons required per atom.

If more than one substance is deposited, as in the codeposition of hydrogen or in alloy plating, then Faraday's laws apply to the total number of equivalents of all the substances produced but do not specify proportions. No exceptions have been found, but electrode processes do not invariably lead to deposition of metal or gas; Fe^{3+} ion, for example, may be reduced to Fe^{2+} ion at the cathode, or the reverse process may proceed at the anode, accounting for corresponding amounts of current.

To complete an electric circuit through the bath, two electrodes are required, an anode and a cathode, and each may consist of several pieces; Faraday's laws apply separately to each electrode. When a soluble anode is used, the same number of equivalents of metal dissolve from the anode as deposit on the cathode, if no other reactions occur. In this way, the bath would be maintained at its original concentration, if side reactions did not interfere and no losses of solution occurred. In practice, however, solution is carried out of the bath on plated work-pieces as they are removed ("drag-out"); consequently replacements of chemicals and water are required. Furthermore, since secondary reactions often consume certain constituents, replenishment is generally not as simple as restoring a fresh portion of the original bath; it is discussed in following chapters for individual baths.

A simple example of a plating cell comprises two copper electrodes immersed in a water solution of copper sulfate and connected externally to a source of direct current, such as a battery, generator, or rectifier. It is usually considered that copper sulfate is completely ionized, so that only copper ions and sulfate ions are found in the bath (together with minute quantities of hydrogen and hydroxide ions resulting from the ionization of water; in commercial copper sulfate baths, however, sulfuric acid is also added). Electrons are supplied to one electrode, imparting a negative charge, and removed from the other, at which a positive charge therefore remains. A copper ion in the bath, being positively charged, is attracted by and attaches itself to the negative electrode or cathode, accepting two of the surplus electrons; that is, it is discharged and thus becomes a copper atom. At the opposite electrode, or anode, a copper atom gives up two electrons in response to the positive charge and so becomes an ion, passing into the solution.

The anode process is, accordingly, oxidation of copper metal to the ion, and, conversely, the cathode process is reduction of the ion to metal. The sulfate ion plays no direct part in the electrode processes. Since equivalent changes must occur simultaneously at anode and cathode in order to maintain net electrical neutrality in the solution, another copper ion enters the bath whenever a copper atom is deposited. Therefore the system undergoes no total change, except for the disappearance of a copper atom from the anode and the appearance of another at the cathode.

6.11.2 The plating process

According to modern concepts, a metal crystal or grain consists of metal ions located in regular lattice positions, together with a cloud of moving valence electrons, equal in number to the charges of the metal ions. The valence electrons are more or less free to distribute themselves among the ions according to certain distribution laws. Each ionic charge is neutralized by the electronic charges when averaged over a period of time or over a number of ions; but individual valence electrons are not confined to any particular ion. According to Pauling, this arrangement gives rise to resonating bond forces responsible for cohesion and the resulting physical properties of the crystal. On immersion in water a few metal ions leave their lattice sites, become hydrated, and diffuse out into solution as dissolved cations. The corresponding electrons, however, remain as part of the electron "atmosphere" of the solid metal, and this surplus of electrons constitutes a negative charge in the metal. The dissolved cations are attracted by this charge, and some re-enter the metal, occupying vacant lattice sites. As the charge in the metal increases, the rate of return of ions to the metal is accordingly increased, and soon becomes equal to the rate of dissolving, which is diminished by the accumulating negative charge on the metal. This balanced condition in rates constitutes a dynamic equilibrium, with the metal negatively charged relative to the solution. The negative potential so produced is related to the conventional electrode potential but is not numerically identical with it ; a positive reversible electrode potential means only that the true potential is less negative than that of the hydrogen electrode.

The number of ions released in building up this potential is usually too small to be detected chemically. Many are held in the interface between the metal and the solution by electrical attraction. The negatively charged metal is thus enveloped by a film of solution containing mostly positive ions. This arrangement constitutes an electrical double layer, the capacitance of which can readily be measured. The flow of ions to and from the metal at equilibrium (zero net current) is equivalent to a current termed the exchange current. Methods of measuring this quantity have recently come into use.

The magnitude of the electrode potential reflects the algebraic sum of two quantities of energy: lattice energy, required to remove the ion from its crystal lattice, and hydration energy available from its interaction with solvent or complexing agents in the solution. Lattice energy depends partly on orientation of the crystal face from which the ion is removed; in the randomly oriented polycrystalline metal usually dealt with in technical electroplating, slightly differing potentials from several faces are mixed. Energy of hydration or coordination depends in part on the concentrations, or more correctly the activities, of the reacting species, which are related to potential by the Nernst equation.

The magnitude of the exchange current density is determined by the potential difference and the activation needed to permit the dissolution and deposition reactions to proceed. This last quantity, which may be considered a potential barrier to be surmounted, depends on the nature of the mechanisms by which the ions react. In turn, these apparently depend in part on the number, location, and nature of growth sites in the metal lattice. Finally, both electrode potential and exchange current may be profoundly influenced by the presence of adsorbed material at the interface, either intentional (addition agents) or unintentional (contaminants).

In the plating cell used earlier as an example, the copper electrodes, being identical, have equal potentials and the net potential is accordingly zero; hence no net current flows if the electrodes are connected (short-circuited). If an external potential is applied between the electrodes, added negative potential at the cathode increases the rate of transfer of ions across the double layer to the electrode, where they are deposited. At the same time, excess negative potential diminishes the rate of metal dissolution (at the cathode). As a result, deposition of metal prevails over dissolution, and plating takes place. At potentials used in technical processes, dissolution is completely negligible. At the anode, the situation is reversed in most but not all aspects.

Vocabulary(略)

6.12 ATMOSPHERIC CLEANSING PROCESSES

The atmosphere, like a stream or a river, has natural, built-in self-cleansing processes without which the troposphere would quickly become unlivable for humans. Most of the air-pollution-control devices discussed in this chapter, for both stationary and mobile sources, make use of some of the principles involved in the natural atmospheric cleansing processes.

Dispersion, gravitational settling, flocculation, absorption (involving washout and scavenging), rainout and adsorption are some of the most significant natural removal mechanisms at work in the atmosphere. Though not literally a removal mechanism, dispersion of pollutants by wind currents lessens the concentrations of pollutants in any one place. Gravitational settling is one of the most important natural mechanisms for removing particulates from the atmosphere, especially particles larger than 20 μm.

Gravitational settling also plays a key role in several of the other natural atmospheric cleansing processes. For example, through flocculation, particles smaller than 0.1 μm can be settled out. In this phenomenon, larger particles act as receptors for smaller ones. Two particles bump together to form a unit, and the process is repeated until a small floc particle is formed that will be large enough and heavy enough to settle out.

In the natural absorption process, particulates or gaseous pollutants are collected in rain or mist, then settle out with that moisture. This phenomenon, known as washout or scavenging takes place below cloud level.

The potential for scavenging gases and particulates depends on many factors, including the intensity of rainfall and the nature of the contaminants being scavenged. Under ordinary circumstances, only a fraction of the particles in the path of descent of a raindrop will be collected, with most small particles remaining in the air that flows around the falling drop. Recent research indicates that washout may be negligible for particles less than 1 μm in diameter.

Gases may be dissolved without being chemically changed, or they may, in some instances, enter into chemical reactions with the rainwater. For example, SO_2 gas, which is simply dissolved into rain, falls with the droplets as SO_2. However, SO_2 may also react with rainwater to form H_2SO_3 (sulfurous acid) or H_2SO_4 (sulfuric acid) mists, mists known as "acid rains" and potentially far more harmful than the original SO_2.

Rainfall through uncontaminated air has a pH of 5.6. However, in the western hemisphere, pH values as low as 2 have been recorded for rainfall from a single precipitation event. This low pH of rainfall can have far-reaching effects. As noted earlier, acid rain runoff can cause extensive erosion of some surfaces (notably limestone) and can change the pH in streams and rivers, thereby influencing the species of algae which predominate in those streams.

Rainout is another natural atmospheric cleansing process involving precipitation. Whereas washout occurs below cloud level when falling raindrops absorb pollutants, rainout occurs within clouds when submicron particulates serve as condensation nuclei around which drops of water may form. The phenomenon has resulted in increased rainfall and fog formation in urban areas.

Adsorption occurs primarily in the friction layer of the atmosphere, the layer closest to the earth's surface. In this phenomenon, gaseous, liquid, or solid contaminants are attracted (generally electrostatically) to a surface, where they are concentrated and retained. Natural surfaces such as soil, rocks, leaves, and blades of grass can adsorb and retain pollutants. Particles may be brought into contact with an adsorption surface by gravitational setting or by inertial impaction, a process by which particulates or gaseous pollutants are transported to surfaces by wind currents[1]. Impaction is particularly effective for particles in the 10 to 15 μm range, and numerous small surfaces—such as blades of grass and leaves of trees —are more effective than larger surfaces in removing particulates in this size range.

When the various and sundry natural atmospheric cleansing mechanisms are overwhelmed by gaseous and particulate emissions, the effects of air pollution become increasingly more evident. Clothing is soiled, particles are deposited on buildings and other surfaces, plants are damaged, visibility is reduced, and human respiratory problems are increased. To prevent these and other evidences of air pollution, it is necessary to establish control procedures or to install control devices. But even with the application of the best available technology, low-level emissions will still inevitably be made into the atmosphere, and these emissions must ultimately be removed by natural atmospheric cleansing mechanisms.

Vocabulary

1. troposphere [ˌtrɔpəsˈfiə] n. 对流层
2. gravitational [ˌɡrævɪˈteɪʃənəl] a. 引力的、地球引力的、重力的
3. floc [flɔk] n. 絮状物、絮凝物
4. flocculating n. 絮凝化
5. scavenging [ˈskævɪndʒɪŋ] n. 清理垃圾
6. descent [dɪˈsent] n. 降下、降落
7. erosion [ɪˈrouʒən] n. 腐蚀、冲蚀、侵蚀
8. algae [ˈældʒiː] n. 藻类、海藻
9. submicron n. 亚微米
10. friction [ˈfrɪkʃən] n. 摩擦、摩擦力、不和、冲突

11. inertial	[i'nɔːʃəl]	a. 不活泼的、惰性的
12. sundry	['sʌndri]	a. 各式各样的
13. overwhelm	[əuvə'welm]	v. 淹没

Phrase

1. condensation nuclei 凝结核

Note

1. Particles may be brought into … by wind currents. 句中从"a process"至句尾部分作"inertial impaction"的同位语。

6.13 DEGREES OF WASTEWATER TREATMENT AND WATER QUALITY STANDARDS

The degree of treatment required for a wastewater depends mainly on discharge requirements for the effluent. Table 6.2 presents a conventional classification for wastewater treatment processes. Primary treatment is employed for removal of suspended solids and floating materials, and also conditioning the wastewater for either discharge to a receiving body of water or to a secondary treatment facility through neutralization and /or equalization[1]. Secondary treatment comprise conventional biological treatment processes. Tertiary treatment is intended primarily for elimination of pollutants not removed by conventional biological treatment.

Table 6.2　Types of Wastewater Treatment

Primary treatment
　Screening
　Sedimentation
　Flotation
　Oil separation
　Equalization

Neutralization
Secondary treatment
Activated sludge process
Extended aeration (or total oxidation) process
Contact stabilization
Other modifications of the conventional activated sludge process: tapered aeration, step aeration and complete mix activated sludge processes
Aerated lagoons
Wastewater stabilization ponds
Trickling filters
Anaerobic treatment
Tertiary treatment (or "advanced treatment")
Microscreening
Precipitation and coagulation
Adsorption (activated carbon)
Ion exchange
Reverse osmosis
Electrodialysis
Nutrient removal processes
Chlorination and ozonation
Sonozone process

These treatment processes are studied in following chapters. The approach utilized is based on the concepts of unit processes and operations. The final objective is development of design principles of general applicability to any wastewater treatment problem, leading to a proper selection of process and the design of required equipment. Consequently, description of wastewater treatment sequences for specific industries, e. g., petroleum refineries, steel mills, metalplating plants, pulp and paper industries, breweries, and tanneries, is not included in this book.

Water quality standards are usually based on one of two criteria: stream standards or effluent standards, stream standards refer to quality of receiving water downstream from the origin of sewage discharge, whereas effluent standards pertain to quality of the discharged wastewater streams themselves.

A disadvantage of effluent standards is that it provides no control over total amount of contaminants discharged in the receiving water. A large industry, for example, although providing the same degree of wastewater treatment as a small one, might cause considerably greater pollution of the receiving water. Effluent standards are easier to monitor than stream standards, which require detailed stream analysis. Advocates of effluent standards argue that a large industry, due to its economic value to the community, should be allowed a larger share of the assimilative capacity of the receiving water.

Quality standards selected depend on intended use of the water, some of these standards include: Concentration of dissolved oxygen (DO, mg/liter), pH, color, turbidity, hardness (mg/liter), total dissolved solids (TDS, mg/liter), suspended solids (SS, mg/liter), concentration of toxic (or otherwise objectionable) materials (mg/liter), odor, and temperature.

Vocabulary

1. effluent   ['efluənt]   a. 流出的、射出的; n. 流出物
2. equalization [ˌi:kwəlai'zeiʃən]   n. 使均等、同等化、平等化
3. tertiary   ['tə:ʃiəri]   a. 第三的、第三位的、第三世纪的
4. assimilative [ə'similətiv]   a. 同化的、同化力的
5. screening   ['skri:niŋ]   n. 遮蔽、屏蔽、隔离
6. sludge   [slʌdʒ]   n. 泥渣、污泥
7. taper   [teipə(r)]   n. 锥形、逐渐(变)细
8. brewery   ['bru:əri]   n. 啤酒厂

9. tannery ['tænəri] n. 制革厂
10. criteria [kraitiərə] n. 准则、要求
11. pertain [pə'tein] v. 适合、相称、从属于、关于

Phrases

1. trickling filters 滴滤池
2. microscreening 微滤筛
3. sonozone 超声臭氧(化)

Note

1. Primary treatment is employed … and/or equalization. 句中"a receiving body of water"是"受纳水体"或"承受……的水体"之意。

6.14 SOLID WASTE

Solid Wastes are all the wastes arising from human and animal activities that are normally solid and that are discarded as useless or unwanted. The term as used in this chapter is all-inclusive, and it encompasses the heterogeneous mass of throwaways from residences and commercial activities as well as the more homogeneous accumulations of a single industrial activity.

6.14.1 Types of Solid Wastes

The types and sources of solid wastes and the physical and chemical composition of solid wastes are considered in this section. The term solid wastes is all-inclusive and encompasses all sources, types of classifications, composition, and properties. As a basis for subsequent discussions, it will be helpful to define the various types of solid wastes that are generated. Three general categories are considered: (1) municipal wastes, (2) industrial wastes, and (3) hazardous wastes.

(1) Municipal Wastes. It is important to note that the definitions of terms and the classifications used to describe the components of solid

waste vary greatly in practice and in the literature. Consequently the use of published data requires considerable care, judgment, and common sense[1]. The definitions presented in Table 6.3 are intended to serve as a guide for municipal solid wastes.

(2) Industrial Wastes. Industrial wastes are those wastes arising from industrial activities and typically include rubbish, ashes, demolition and construction wastes, special wastes, and hazardous wastes.

Table 6.3 Classification of Materials Comprising Municipal Solid Waste

Component	Description
Food wastes	The animal, fruit, or vegetable residues (also called garbage) resulting from the handling, preparation, cooking, and eating of foods. Because food wastes are putrescible, they will decompose rapidly, especially in warm weather
Rubbish	Combustible and noncombustible solid wastes, excluding food wastes or other putrescible materials. Typically, combustible rubbish consists of materials such as paper, cardboard, plastics, textiles, rubber, leather, wood, furniture, and garden trimmings. Noncombustible rubbish consists of items such as glass, crockery, tin cans, aluminum cans, ferrous and nonferrous metals, dirt, and construction wastes
Ashes and residues	Materials remaining from the burning of wood, coal, coke, and other combustible wastes. Residues from power plants normally are not included in this category. Ashes and residues are normally composed of fine, powdery materials, cinders, clinkers, and small amounts of burned and partially burned materials
Demolition and construction wastes	Wastes from razed buildings and other structures are classified as demolition wastes. Wastes from the construction, remodeling, and repairing of residential, commercial, and industrial buildings and similar structures are classified as construction wastes. These wastes may include dirt, stones, concrete, bricks, plaster, lumber, shingles, and plumbing, heating, and electrical parts

Special wastes	Wastes such as street sweepings, roadside litter, catch-basin debris, dead animals, and abandoned vehicle are classified as special wastes
Treatment-plant wastes	The solid and semisolid wastes from water, wastewater, and industrial waste treatment facilities are included in this classification

(3) Hazardous Wastes. Wastes that pose a substantial danger immediately or over a period of time to human, plant, or animal life are classified as hazardous wastes. A waste is classified as hazardous if it exhibits any of the following characteristics. ① ignitability, ② corrosivity, ③ reactivity, or ④ toxicity. A detailed definition of these terms was first published in the Federal Register on May 19, 1980 (p.33.121,122).

In the past, hazardous wastes were often grouped into the following categories: ① radioactive substances, ② chemicals, ③ biological wastes, ④ flammable wastes, and ⑤ explosives. The chemical category includes wastes that are corrosive, reactive or toxic. The principal sources of hazardous biological wastes are the hospitals and biological research facilities.

6.14.2 Solid Waste Management—An overview

Recognizing that our world is finite and that the continued pollution of our environment will, if uncontrolled, be difficult to rectify in the future, the subject of solid-waste management is both timely and important. The overall objective of solid-waste management is to minimize the adverse environmental effects caused by the indiscriminate disposal of solid wastes, especially of hazardous wastes. To assess the management possibilities it is important to consider ① materials flow in society, ② reduction in raw materials usage, ③ reduction in solid-waste quantities, ④ reuse of materials, ⑤ materials recovery, ⑥ energy recovery, and ⑦ day-to-day solid-waste management[2].

(1) Materials Flow in Society. An indication of how and where solid wastes are generate in a technological society is shown in the simplified materials-flow diagram presented in Figure 6.8. Solid wastes (debris) are generated at the start of the process, beginning with the mining of raw material. Thereafter, solid wastes are generated at every step in the process as raw materials are converted to goods for consumption. It is apparent from Figure 6.8 that one of the best ways to reduce the amount of solid wastes to be disposed is to reduce the consumption of raw materials and to increase the rate of recovery and reuse of waste materials. Although the concept is simple, effecting this change in a modern technological society has proved extremely difficult.

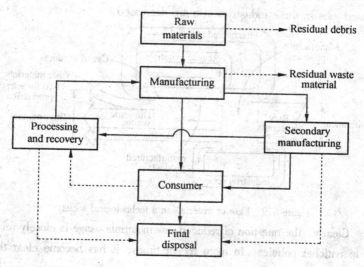

—Raw materials, products, and recovered materials ··· waste materials

Figure 6.8　Generalized flow of materials and the generation of solid wastes in society

(2) Reduction in Raw Materials Usage. The general relationships shown in Figure 6.8 can be quantified relatively, as shown in Figure 6.9.

To satisfy the principle of conservation of mass the input must equal the output. Clearly, if a reduction in the usage of raw materials is to occur either the input or output must be reduced. Raw materials usage can be reduced most effectively by reducing the quantity of municipal and industrial wastes. For example, to meet EPA mileage restrictions American cars are now (1984) on the average, 20 percent smaller than they were in the late 1950s and early 1960s[3]. This reduction in size has also reduced the demand for steel by about 20 percent. The reduced demand for steel has in turn resulted in less mining for the iron ore used to make steel. While most people would agree that it is desirable to reduce the usage of raw materials, others would argue that as the usage of raw materials is decreased jobs in those industries also are decreased.

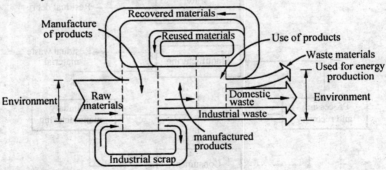

Figure 6.9 Flow of materials in a technological society

Clearly, the question of reduced raw materials usage is closely related to national policies. In more recent times, it has become clear that such usage is also related to the world economy. For example, the increase in oil prices has led to more usage of wood as an alternative source of energy.

Vocabulary

1. encompass　　[in'kʌmpəs]　　vt. 围绕、包围、包含
2. hazardous　　['hæzədəs]　　a. 碰运气的、危险的

3. demolition	[ˌdeməˈliʃən]	n. 破坏、毁坏、遗迹、拆除
4. pose	[pouz]	n. 姿势、伪装；vi. 摆姿势、假装；vt. 使摆好、提出、造成
5. putrescible	[pju:ˈtresəbl]	a. 易腐烂的
6. trimming	[ˈtrimiŋ]	n. 修饰物
7. crockery	[ˈkrɔkəri]	n. 陶器(罐)
8. cinder	[ˈsində(r)]	n. 煤渣、焦木块
9. clinker	[ˈkliŋkə]	n. 炉渣、渣滓
10. raze	[reiz]	v. 拆毁、夷为平地
11. shingle	[ˈʃiŋ(ə)l]	n. 鹅卵石、木瓦
12. plumbing	[ˈplʌmiŋ]	n. 水管、铅管品制造
13. ignitability	[igˌnaitəˈbiliti]	n. 可燃性
14. flammable	[ˈflæməbl]	a. 易燃的、可燃性的
15. rectify	[ˈrektifai]	vt. 订正、矫正、改正、调整、校正、精馏
16. indiscriminate	[ˌindisˈkriminit]	a. 无差别的、不分皂白的、杂陈的
17. assess	[əˈses]	vt. 评价、评估、评判、评定、征收
18. mileage	[ˈmailidʒ]	n. 英里数(里程)

Notes

1. Consequently the use of ... and common sense. 句中"common sense"意为"常识"。
2. To assess the management ... solid-waste management. 句中"material flow"指"物质流"。
3. For example, to meet EPA mileage ... and early 1960 s. 句中 EPA = Environmental Protection Agency.

6.15 DIRECT AND INDIRECT REUSE OF WASTEWATER

It is generally impossible to reuse a wastewater completely or indefinitely. The reuse of treated effluent by direct or indirect means is a method of disposal that complements the other disposal methods. The amount of effluent that can be reused is affected by the availability and cost of fresh water, transportation and treatment costs, water quality standards, and the reclamation potential of the wastewater. A complete discussion and analysis of wastewater reuse is beyond the scope of this text, but some of the possible reuses for treated water are briefly reviewed in the following discussion.

Water reuse may be classified according to use as ① municipal, ② industrial, ③ agricultural, ④ recreational, and ⑤ groundwater recharge. Direct and indirect reuse applications for these use are shown in Table 6.4.

Table 6.4 Potential Uses of Renovated Water

Use	Direct	Indirect
Municipal	Park or golf course watering; lawn watering with separate distribution system; potential source for municipal water supply	Groundwater recharge to reduce aquifer overdrafts
Industrial	Cooling tower water; boiler feed water; process water	Replenish groundwater supply for industrial use
Agricultural	Irrigation of certain agricultural lands, crops, orchards, pastures, and forests; leaching of soils	Replenish groundwater supply for agricultural overdrafts
Recreational	Forming artificial lakes for boating, swimming, etc.; swimming pools	Develop fish and waterfowl areas
Other	Groundwater recharge to control saltwater intrusion; salt balance control in groundwater; wetting agent—solid waste compaction	Groundwater recharge to control land subsidence problems; oilwell repressurizing; soil compaction

Municipal Reuse

Direct reuse of treated wastewater as drinking water, after dilution in natural waters to the maximum possible extent and after coagulation, filtration, and heavy chlorination for disinfection, is practicable on an emergency basis. This practice varies only in degree from the situation existing on many rivers that are used for both water supply and waste disposal[1]. One example is the Merrimack River in Massachusetts. Advanced methods of wastewater and water treatment, such as demineralization and desalination, are capable of almost complete removal of impurities, and water treated by such methods, after chlorination, is safe to drink. These methods are very expensive and, where they are found to be necessary due to inadequate water supplies, may be economically feasible only if a dual supply system is adopted. In such cases, adequately treated and disinfected wastewater effluents could be reused for flushing toilets, yard watering, and other direct applications.

Industrial Reuse

Industry is probably the single greatest user of water in the world, and the largest of the industrial water demands is for process cooling water. Waters with a high mineral content and those that do not meet bacterial drinking water standards have been used by industry in some cases. Public health dangers and aesthetic concerns are generally eliminated because of the use of closed cycle processes.

Agricultural Reuse

The types of crops that can be irrigated with reclaimed wastewater depend on the quality of the effluent, the amount of effluent used, and the health regulations concerning the use of treated and untreated waste water on crops. Health considerations in this country have dictated against the use of untreated wastewater. Furthermore, field crops that are normally consumed in a raw state cannot be irrigated with wastewater of any kind. Preliminary treated or undisinfected wastewater effluent is usually allowed for field crops, such as cotton, sugar beets, and vegetables for seed production.

Recreational Reuse

Golf-course and park watering, establishment of ponds for boating and recreation, and maintenance of fish or wildlife ponds are methods for the recreational reuse of water. Today's technology allows the production of an excellent effluent that is well suited for the purposes described. The use of treated effluent for park watering has been practiced for many years in this country. The Santee Project outside San Diego is an example of recreational reuse of wastewater in forming a series of lakes suitable for boating, fishing, and other recreational purposes.

Groundwater Recharge

Groundwater recharge is one of the most common methods for combining water reuse and effluent disposal. Recharge has been used to replenish groundwater supplies in many areas. The effluent from the Whittier Narrows Plant operated by the Los Angeles county Samitation Districts, is used for replenishment of the groundwater in the Rio Hondo River Basin. In New York, California, and other coastal areas, rapid development of industry and increase in population have caused a lowering of the groundwater table, resulting in saltwater intrusion into the freshwater aquifers[2]. Treated effluent has been used to replenish the groundwater and to stop this intrusion. Another possible effluent use is in the recharging of oil-bearing strata. Oil companies have conducted much research on flooding techniques to increase the yield of oil-bearing strata.

Vocabulary

1. reclaim   [ri'kleim]   *vt*. 开垦、改造、感化、纠正、回收
                                                    vi. 改造、感化
2. dilution   [dai'lju:ʃən]   *n*. 稀释、稀释度、稀薄物品
3. coagulation   [ˌkouægju'leiʃən]   *n*. 凝结、凝结物
4. filtration   [fil'treiʃən]   *n*. 过滤、筛选

5. disinfection   [dis'infekʃən]   n. 消毒
6. demineralization   n. 矿化作用
7. desalination   [diˌsæli'neiʃən]   n. 除去盐分
8. aesthetic   [es'θetik]   a. 美学的、审美的、有美感
9. irrigate   ['irigeit]   v. 浇、冲洗
10. dictate   [dik'teit]   vt. 口述、口授、使听写；
    vi. 口述、口授、听写、命令、支配；
    n. 命令、支配
11. replenish   [ri'pleniʃ]   vt. 重新补足、再装满
    vi. 再装满、充满
12. intrusion   [in'truːʒən]   n. 闯入、侵扰
13. aquifer   n. 地下含水层、含水地带
14. strata   ['streitə]   n. (stratum 的复数)地层、阶层、薄片

Notes

1. This practice varies only ... and waste disposal. 其中"varies in ... from ..."意思是"在……方面与……有所不同"。全句可译为：这一实际(状况)与提供给水和接收排放废水双重作用的许多河流所存在的状况只是在程度上有所不同。
2. In New York, California, and other ... into the water aquifers. 句中"resulting in"至句末部分可译为：其结果导致海水浸入淡水含水层。

6.16 WASTEWATER TREATMENT PROCESSES

The main objectives of conventional wastewater treatment processes are reduction of biochemical oxygen demand, suspended solids, and pathogenic organisms. In addition, it may be necessary to remove nutrients, toxic components, nonbiodegradable compounds, and dissolved solids. Since most contaminants are present in low concentrations, the treatment processes must be able to function effectively with dilute streams.

Many operations are used to purify water before discharge to the environment. A partial listing of these operations and their purposes is given in Table 6.6. These operations will be discussed briefly to show where they fit into an overall treatment plant.

Classification of Processes

Conventional wastewater treatment processes are often classified as pretreatment, primary treatment, secondary treatment, tertiary treatment, and sludge disposal.

Table 6.6 Wastewater Treatment Processes and Major Purposes

Operation	Purpose of operation
Bar screens and racks	Coarse solids removal
Comminutor	Grinding up of screenings(筛余物粉碎)
Grit chamber	Grit and sand removal
Skimmer and grease trap	Floating liquid and solid removal
Equalization tank	Smoothing out flow and concentration
Neutralization	Neutralizing acids and bases
Sedimentation and flotation	Suspended solids removal
Activated sludge reactor, trickling filter, aerated lagoon	Biological removal of soluble organics
Activated carbon adsorber	Soluble nonbiodegradable organics removal
Chemical coagulation	Precipitation of phosphates
Nitrification-denitrification	Biological removal of nitrates
Air stripping	Ammonia removal
Ion exchange	Charged species removal
Bed filtration	Fine solids removal
Reverse osmosis and electrodialysis	Dissolved solids removal
Chlorination and ozonation	Pathogenic organism destruction

Since many different combinations of these operations are possible, each situation must be evaluated to select the best combination.

Pre-and Primary Treatment

Pretreatment processes are used to screen out coarse solids, to reduce the size of solids, to separate floating oils and to equalize fluctuations in flow or concentration through short-term storage. Primary treatment usually refers to the removal of suspended solids by settling or floating.

Sedimentation is currently the most widely used primary treatment operation. In a sedimentation unit, solid particles are allowed to settle to the bottom of a tank under quiescent conditions. Chemicals may be added in primary treatment to neutralize the stream or to improve the removal of small suspended solid particles. Primary reduction of solids reduces oxygen requirements in a subsequent biological step and also reduces the solids loading to the secondary sedimentation tank.

Secondary Treatment

Secondary treatment generally involves a biological process to remove organic matter through biochemical oxidation. The particular biological process selected depends upon such factors as quantity of wastewater, biodegradability of waste, and availability of land. Activated sludge reactors and trickling filters are the most commonly used biological processes.

In the activated sludge process, wastewater is fed to an aerated tank where microorganisms consume organic wastes for maintenance and for generation of new cells. The resulting microbial floc (activated sludge) is settled in a sedimentation vessel called a clarifier or thickener. A portion of the thickened biomass is usually recycled to the reactor to improve performance through higher cell concentrations. Trickling filters are beds packed with rocks, plastic structures, or other media. Microbial films grow on the surface of the packing and remove soluble organics from the wastewater flowing over the packing. Excess biological growth washes off the packing and is removed in a clarifier.

A typical flowsheet of an activated sludge treatment plant is shown in Figure 6.10. The plant includes primary sedimentation for removal of solids and chlorination to reduce the pathogen content of the effluent water.

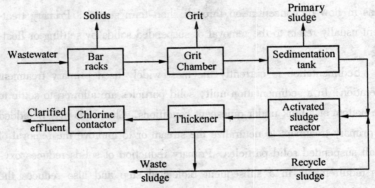

Figure 6.10 Typical flowsheet of an activated sludge treatment plant.

Tertiary Treatment

Many effluent standards require tertiary or advanced wastewater treatment to remove particular contaminants or to prepare the water for reuse. Some common tertiary operations are removal of phosphorus compounds by coagulation with chemicals, removal of nitrogen compounds by ammonia stripping with air or by nitrification-denitrification in biological reactors, removal of residual organic and color compounds by adsorption on activated carbon and removal of dissolved solids by membrane processes (reverse osmosis and electrodialysis). The effluent water is often treated with chlorine or ozone to destroy pathogenic organisms before discharge into the receiving waters.

A prominent example of a wastewater treatment plant employing tertiary processes is the facility at Lake Tahoe, California. The entering wastewater is first treated in conventional primary and secondary processes to remove most of the settleable solids and soluble organics. The tertiary treatment steps are phosphate removal by granular bed filtration, residual organic removal by activated carbon removal and disinfection by chlorine. The sewage sludges are incinerated, the spent granular carbon is regenerated thermally for reuse and the lime sludge is recalcitrated and reused.

Physical-chemical Treatment

Physical-chemical treatment processes are alternatives to the biological processes. In a physical-chemical plant, the main processes are chemical coagulation, carbon adsorption, and filtration. Suspended solids and phosphates are precipitated together in a sedimentation vessel after addition of suitable chemicals, such as alum, ferric chloride, or lime. Adsorption on granular activated carbon extracts the remaining soluble organics and filtration is used to remove residual suspended solids. The granular carbon column may serve the dual function of adsorbing organics and filtering out solids. Physical-chemical treatment is usually applied to wastes containing toxic or nonbiodegradable compounds that are not amenable to biological processes.

Sludge Disposal

Wastewater treatment processes generate significant quantities of sludge from suspended solids in the feed, biomass generated by biological operations, and precipitates from added chemicals. Some common sludge disposal operations and their functions are listed in Table 6.7. Selection of a treatment sequence for sludges depends upon the nature of the sludge, environmental factors, and ultimate disposal options.

Concentration operations, such as gravity or flotation thickeners, increase the solids concentration and achieve a significant reduction in sludge volume. Stabilization operations, such as anaerobic digestion, convert sludges into a less offensive form in terms of odor, degradability, and pathogen content.

Table 6.7 Sludge Treatment Processes and their Major Purposes

Operation	Purposes of Operation
Thickening	Increase solids concentration and reduce volume
Gravity	
Flotation	
Stabilization	Reduce sludge solids, pathogens and odor
Anaerobic digestion	
Aerobic digestion	
Conditioning	Improve dewatering rate and solids capture
Chemical addition	
Heat treatment	
Dewatering	Reduce volume and form a damp cake
Vacuum filtration	
Centrifugation	
Sand beds	
Drying and Oxidation	Dry or oxidize sludge cake
Incineration	
Heat drying	
Wet air oxidation	
Ultimate disposal	Fertilize or dispose of sludge solids
Landfill	
Spreading on soil	
Lagoons	
Ocean	

In dewatering operations, the water content of sludges is reduced to a level where they can be handled as damp solids. Vacuum filtration, centrifugation, and sand beds are the most common dewatering methods. Thermal processes, such as heat drying and incineration, are used to either dry the sludge or to oxidize its organic content. Residual sludge and ash from sludge treatment processes must be disposed of in the ocean or on land. Some of the options for ultimate disposal on land are landfill, land reclamation, and crop fertilization.

Industrial Wastes

Since industrial wastes have a broader range of characteristics than domestic wastes, they are treated by a wider variety of processing schemes. Industrial wastes are more likely to contain toxic and non-biodegradable compounds that require physical-chemical instead of biological treatment. In some cases, industrial wastes are discharged to a municipal plant directly or after limited pretreatment. In other cases, they are treated in a separate plant designed for the specific wastes. The wastewater load in an industrial plant can often be reduced by recirculating slightly contaminated water, segregating low and high strength wastes or separate treatment, substituting less polluting chemicals or processes, and recovering selected contaminants as by products or for reuse.

Design of a wastewater treatment process for industrial or domestic wastes depends upon many factors, such as characteristics of the wastewater, required effluent quality, avaliability of land, and options for sludge disposal. In addition to capital and operating costs, stability, reliability, and flexibility are important considerations when selecting a process from the various alternatives.

Vocabulary

1. pathogenic [ˌpæθəˈdʒenik] a. 使生病的、成为病原的、病原性的
2. comminutor [ˌkɔmiˈnjuːtə] n. 粉碎器
3. grit [grit] n. 粗砂；v. 研磨
4. skimmer [skimə] n. 撇乳器
5. equalize [ˈiːkwəlaiz] vt. 使相等、补偿；vi. 相等
6. sedimentation [ˌsedimenˈteiʃən] n. 沉淀、沉降
7. quiescent [kwaiˈesnt] a. 静止的、不活动的、寂静的
8. trickling [ˈtrikliŋ] n. v. 移动
9. microbial [maiˈkroubiəl] a. 微生物的、由细菌引起的
10. clarifier [ˈklærifaiə] n. 澄清器
11. biomass [ˈbaioumæs] n. 生物量
12. pathogenesis [ˌpæθəˈdʒenisis] n. 发生、病因

13. residual	['rezidju:l]	a.	残渣的、剩余的 n. 残渣、剩余、余数
14. recalcitrate	[ri'kælsitreit]	vi.	不服从、执拗、反抗、回踢
15. amenable	[ə'mi:nəbl]	a.	应服从的、会接纳的、肯服从的
16. flotation	[flou'teiʃən]	n.	漂浮、开创、发行、募集
17. anaerobic	[ˌæneiə'roubik]	a.	没有空气而能生活的、厌氧性的
18. incineration	[ˌinsinə'reiʃən]	n.	烧成灰、焚化、灰化
19. upgrade	['ʌp'greid]	n.	提高等级、叠积作用

6.17　TYPES OF WATER SUPPLY AND CLASSIFICATION OF WATER CONTAMINANTS

According to their origin, water supplies are classified into three categories: (1) surface waters, (2) ground waters, and (3) meteorological waters. Surface waters comprise stream waters (e.g., rivers), oceans, lakes, and impoundment waters. Stream waters subject to contamination exhibit a variable quality along the course of the stream. Waters in lakes and impoundments, on the other hand, are of a relatively uniform quality. Ground waters show, in general, less turbidity than surface waters. Meteorological waters (rain) are of greater chemical and physical purity than either surface or ground waters.

Water contaminants are classified into three categories: (1) chemical, (2) physical, and (3) biological contaminants. Chemical contaminants comprise both organic and inorganic chemicals. The main concern resulting from pollution by organic compounds is oxygen depletion resulting from utilization of DO in the process of biological degradation of these compounds. This depletion of DO leads to undesirable disturbances of the environment and the biota. In the case of pollution resulting from the presence of inorganic compounds the main concern is their possible toxic effect, rather than oxygen depletion. There are, however, cases in which inorganic compounds exert an oxygen demand, so contributing to oxygen depletion. Sulfites and nitrites, for example take up oxygen, being oxidized to sulfates and nitrates, respectively (Eqs. (6.1) and (6.2))

$$SO_3^{2-} + \frac{1}{2}O_2 \longrightarrow SO_4^{2-} \qquad (6.1)$$

$$NO_3^- + \frac{1}{2}O_2 \longrightarrow NO_3^- \qquad (6.2)$$

Heavy metal ions which are toxic to humans are important contaminants. They occur in industrial wastewaters from plating plants and paint and pigment industries. These include Hg^{2+}, As^{3+}, Cu^{2+}, Zn^{2+}, Ni^{2+}, Cr^{3+}, Pb^{2+}, and Cd^{2+}. Even their presence in trace quantities (i.e., minimum detectable concentrations) causes serious problem.

Considerable press coverage has been given to contamination of water by mercury. Microorganisms convert the mercury ion to methylmercury (CH_3Hg) or dimethylmercury [$(CH_3)_2Hg$]. The dimethyl compound, being volatile, is eventually lost to the atmosphere. Methylmercury, however, is absorbed by fish tissue and might render it unsuitable for human consumption. Mercury content in fish tissue is tolerable up to a maximum of 15 ~ 20 ppm. Methylmercury present in fish is absorbed by human tissues and eventually concentrates in certain vital organs such as the brain and the liver. In the case of pregnant women it concentrates in the fetus. Recently in Japan, there were several reported cases of deaths from mercury poisoning, due to human consumption of mercury-contaminated fish. Analysis of fish tissue revealed mercury concentrations of approximately 110 ~ 130 ppm. These high mercury concentrations, coupled with the large fish intake in the typical Japanese diet, caused this tragedy.

Contamination by nitrates is also dangerous. Fluorides, on the other hand, seem actually beneficial, their presence in potable waters being responsible for appreciable reduction in the extent of tooth decay. There is, however, considerable controversy concerning fluoridization of potable water.

Some physical contaminants include(1) temperature change (thermal pollution). This is the case of relatively warm water discharged by industrial plants after use in heat exchangers (coolers); (2) color(e.g., cooking liquors discharged by chemical pulping plants); (3) turbidity (caused by discharges containing suspended solids); (4) foams detergents such as alkylbenzene sulfonate (ABS) constitute important cause of foaming; and (5) radioactivity.

Biological contaminants are responsible for transmission of diseases by water supplies. Some of the diseases transmitted by biological contamination of water are cholera, typhoid, paratyphoid, and shistosomiasis.

Vocabulary

1. meteorologic	[ˌmiːtrərə'lɔdʒik]	a.	气象的、气象学的
2. impoundment	[im'paudmənt]	n.	蓄水、扣留
3. depletion	[di'pliːʃən]	n.	竭尽、用尽
4. disturbance	[di'stəːbəns]	n.	骚动、干扰、损伤、破坏
5. biota	[bai'əutə]	n.	生物(区)、生命
6. fetus = foetus	['fiːtəs]	n.	胎儿、形成期
7. potable	['pəutəb(ə)l]	a.	可饮的、适于饮用的
8. detergent	[di'təːdʒənt]	n.	洗涤剂、清洁剂
9. cholera	['kɔlərə]	n.	霍乱
10. typhoid	['taifɔid]	n.	伤寒
11. paratyphoid	[pærə'taifɔid]	n.	副伤寒
12. shistosomiasis		n.	血吸虫病

Phrases

1. water supply   给水
2. oxygen depletion   氧亏
3. DO = Dissolved oxygen   溶解氧

4. press coverage   新闻版面

6.18 BIOLOGICAL EFFECTS OF RADIATION

6.18.1 Ionizing Radiations

Radiation which is capable of producing ionization and which is absorbed can cause injury or other damage. As the radiation penetrates the organism or cell, it comes into contact with atoms and molecules in the living protoplasm, altering the structure or electrical charges with the formation of other molecules and substances which in turn may cause other effects. The radiation may pass through the cell causing no damage; or if damaged the cell may partially repair itself. But if a cell is damaged and not repaired it may reproduce in damaged form, or the cell may be destroyed. The unknown longterm effects of radiation, including possible cancers, birth defects, and hereditary changes that may be passed on to future generations, heighten public concern and individual emotions.

6.18.2 Somatic and Genetic Effects

The biological effects of radiation on all living organisms, including human beings, are termed somatic, meaning effects occuring during the lifetime of the exposed organism, as opposed to genetic, which are the effects on generations yet unborn. Somatic or genetic effects from radiation exposure may not be evident for months, years, or a lifetime. Even relatively large doses usually show no immediate apparent injury. But many cells, tissues, and organs of the body are interdependent, and the destruction of one may eventually result in the poor functioning or death of the other. All cells, with the exception of sperm cells and nerve cells, can apparently replace themselves or recover some extent from radiation exposure, if the dose is not excessive and is administered in small increments over an extended period of time. The sixth report of the Royal Commission on Environmental Pollution noted that at levels of radiation likely to be permitted in relation to possible somatic effects, the genetic effects should be of little concern.

Hence the somatic risks should govern the required standards of radiological protection.

In assessing radiation hazard the sum total of all exposures should be considered, that is, natural exposures, and radiation ingested through the air, water, and food. Exposure of the gonads is necessary to cause genetic effects from ionizing radiations. Everyone is subjected to natural background radiation, as well as heat and trace amounts of chemicals that cause a small number of socalled spontaneous mutations in genes. The genes determine all inherited characteristics. Hence any additional radiation of the reproductive cells should be avoided to keep down further mutations. The genetic harm done by radiation is cumulative up to the end of the reproductive period and depends on the total accumulated gonadal dose. Not all mutant genes are equally harmful. Some may cause serious harm, most cause very little or no apparent damage. Nevertheless, radiation exposure of large population groups should be kept as low as practicable since any mutation is generally considered harmful. It is estimated, for example, that a person receives on the average 125 millirems/yr from natural sources and 50 to 70 millirems/yr (genetically significant dose—GSD) from medical exposures. The actual amounts received in different parts of the world will vary with altitude, medical practices, and other factors. For example, a round trip flight between New York and Los Angeles will produce an exposure of 5 mR. It is estimated that 75 to 230 additional cancer deaths would result in a hypothetical population of one million persons exposed to a one-time radiation dose of 10 rads. The national average is less than one-fifth rad per year, mostly cosmic rays and medical X rays. The death rate due to all types of cancer in the U.S. in 1975 was 1 717 per million, or 343 400 in a population of 200 million.

In another example of the relative health hazard associated with natural radiation, fallout radiation, and similar effects from all other causes in the United States, the Federal Radiation Council gave the following summary estimates for the next 70 years:

	Leukemia	Bone Cancer
Total number of cases from all causes	840 000	140 000
Estimated number caused by natural radiation	0 to 84 000	0 to 14 000
Estimated number of additional cases from all nuclear tests through 1962	0 to 2 000	0 to 700

Knowledge of the long-term effects on human beings of both acute doses (large doses received in a short period of time) and chronic doses (those received frequently or continuously over a long period of time), from either external or internal sources of radiation, is far from complete. Relatively speaking however, more is known about radiation effects than effects of trace amounts of toxic substances.

6.18.3 Factors to Consider

Some of the factors that determine the effect of radiation on the body are the following:

① Total amount and kind of radiation absorbed, external and internal.

② Rate of absorption.

③ Amount of the body exposed.

④ Individual variability.

⑤ Relative sensitivity of cells and tissues; parts of the body exposed.

⑥ Nutrition, oxygen tension, metabolic state.

For example, a dose of 600 R whole-body X radiation in one day would mean almost certain death. On the other hand, 600 R administered in weekly increments over a period of 30 years may result in little or no detectable effect. This would amount to

$$\frac{600 \text{ R}}{30 \text{yr} \times 52 \text{ wk}} = 0.385 \text{ R/wk, or } 385 \text{ mR/wk}$$

large doses of radiation can be applied to local areas, as in therapy, with little danger. A person could expose a finger to 1 000 R and experience a localized injury with subsequent healing and scar formation.

Effects of whole-body radiation are given in another Table. It is known that the bloodforming organs—spleen, lymph nodes, and bone marrow—are the most radiosensitive organs when the whole body is irradiated and that protection of these organs from exposure lessens the whole-body effect.

Vocabulary

1. protoplasm   ['prəutəuplæz(ə)m]   n. 原生质、细胞
2. hereditary   [hi'reditəri]   a. 世袭的、遗传的
3. somatic   [sə'mætik]   a. 身体的、肉体的
4. sperm   [spə:m]   n. 精液、鲸油
5. increment   ['inkrimənt]   n. 增加、增量
6. gonad   ['gəunæd]   n. 生殖器官、性腺
7. mutation   [mju:teiʃ(ə)n]   n. 变异、突变
8. hypothetical   [haipə'θetik(e)l]   a. 假设的、有前提的
9. rad   n. 拉德(辐射剂量单位)
10. leukemia   [lu:'ki:miə]   n. 白血病
11. administer   [əd'ministə(r)]   v. 管理、给予、执行
12. therapy   ['θerəpi]   n. 治疗、疗法
13. spleen   [spli:n]   n. 脾脏、坏脾气、忧郁
14. lymph   [limf]   n. 淋巴(液)管
15. bone marrow   ['mærəu]   n. 骨髓
16. fallout   n. 放射性尘埃、原子尘、附带结果

6.19 WHAT CAN CHEMISTS DO FOR NANOSTRUCTURED MATERIALS

Nanostructured Materials

Nanostructured materials have now been investigated for more than a decade using a rather wide range of experimental methods. The structures and properties of these new materials, which are artificially synthesized from nanometer-sized "building blocks", such as clusters, grains or layers, have been elucidated in a number of important areas and the relationships among these areas are beginning to be understood[1]. Various investigations of their mechanical, chemical, electrical, magnetic, and optical behavior have demonstrated the possibilities to engineer the properties of nanostructured materials through control of the sizes of their constituent clusters, grains or layers and the manner in which these constituents are assembled[2]. There are, however, tremendous opportunities remaining for creative new tailored chemical synthesis and processing methods and for developing an understanding of the important role of surface and interface chemistry in the assembly and resulting properties of these materials. Some aspects of our present understanding of nanostructured materials and their properties are briefly presented here, along with some thoughts regarding a few critical future research needs in various areas of chemistry that would add greatly to the field of nanostructured materials.

Interest in the physics of condensed matter at size scales larger than that of atoms and smaller than that of bulk solids has grown rapidly because of the increasing realization that the properties of these mesoscopic atomic ensembles are different than those of conventional solids. Thus interest in artificially assembling materials from nanometer sized building blocks arose from discoveries that, by controlling their sizes in the range of 1 ~ 100 nm and the assembly of such constituents, one could begin to alter and prescribe the properties of the assembled nanostructures. Nature had apparently already learned well the value of nanostructuring since many examples of naturally created nanostructures, formed chemically under ambient conditions, can be found in biological systems from sea shells to the human body.

Nanostructured materials are modulated over nanometer length scales. They can be assembled with modulation dimensionalities of zero (atom clusters or filaments), one (multilayers), two (ultrafine-grained overlayers or coatings or buried layers), and three (nanophase materials), or with intermediate dimensionalities. Thus, nanocomposite materials containing multiple phases can range from the most conventional case in which a nanoscale phase is embedded in a phase of conventional sizes to the case in which all the constituent phases are of nanoscale dimensions. All nanostructured materials share three features: atomic domains (grains, layers or phases) spatially confined to less than 100 nm in at least one dimension, significant atom fractions associated with interfacial environments, and interactions between their constituent domains.

Multilayered materials have had the longest history among the various artificially synthesized nanostructures, one that has already seen applications in semiconductor devices, strained-layer superlattices, and magnetic multilayers. The technological potential of multilayered quantum heterostructure semiconductor devices was recognized in the 1970s in the rapidly expanding electronics and computer industries and helped to drive advances in this exciting new field. Since the 1980s there has been a rapid expansion of research on isolated atom clusters and an increased understanding of their potential as the constituents of new materials.

A knowledge of the variation in cluster properties, both physical and chemical, with cluster size is important to both our fundamental understanding of condensed matter and our ability to use cluster-assembled materials in a variety of technological applications[3]. The manner in which the structure and properties of collections of condensed atoms in a single cluster vary with cluster size, from atomic or molecular to bulk solid-state behavior, is a fundamental touchstone in the development of realistic theoretical models for condensed matter. Theoretical areas that are impacted by this variation include understanding the forces acting among atoms, the structures of atom collections, electronic effects (quantum size effects) caused by spatial confinement of delocalized valence electrons, and cooperative (many body) atom phenomena such as lattice vibrations or melting.

Synthesis and Properties

A number of methods exist for the synthesis of nanostructured materials. They include synthesis from atomic or molecular precursors (chemical or physical vapor deposition; gas-condensation; chemical precipitation; aerosol reactions; biological templating), from processing of bulk precursors (mechanical attrition; crystallization from the amorphous state; phase separation), and from nature. It is generally preferable to synthesize nanostructured materials from atomic or molecular precursors, in order to gain the greatest control over a variety of microscopic aspects of the condensed ensemble, but other methodologies can often yield very useful results. Novel chemical synthesis and processing methods for creating tailored nanostructures are sorely needed, especially ones that can carefully control the interface chemistry and ones that can be effected at ambient temperatures.

Foremost in importance in nanostructuring is the ability to control the size and size distribution of the constituent phases or structures. The desirable sizes are generally below 100 nm, since it is in this size range (and frequently below 10nm) that various properties begin to change significantly owing to confinement effects. Spatial confinement can in general affect any property when the size of the atomic ensemble becomes comparable to or smaller than a critical length scale for the mechanism that is responsible for that property[4]. Examples can be as diverse as the "blue" (high-frequency, short-wavelength) shifts of the optical absorption in semiconducting clusters when their sizes fall below the Bohr radii (ca. 5 ~ 50 nm) of the excitonic (electron-hole pair) states responsible for absorption and the increased strengthening of normally soft metals when their grain sizes fall below the critical length scales (ca. < 50 nm) for the sources of dislocations (the defect responsible for easy deformation) to easily operate at conventional applied stresses.

Second, the chemical compositions of the constituent phases in a nanostructured material are of fundamental importance, as they are also to the performance of conventional materials. This not only relates to average compositions, but also invariably to chemical gradients within the constituent phases or structures and especially those near surfaces and interfaces. Indeed, the chemistry of the surfaces of the constituents, as well as that of the subsequently formed interfaces on assembly, can often play a crucial role in the ability to use these materials. The third aspect of nanostructured materials that one would like to be able to control in their synthesis is the nature of the interfaces created between constituent phases and, hence, the nature of the interactions across the interfaces. Both the local structure and chemistry of the interfaces are important in this regard.

The properties of nanostructured materials are determined by the interplay among these three features (domain size, composition, and interfaces). In some cases, one or more of these features may dominate the particular property in question. Thus, one wants to be able to synthesize nanostructured materials under controlled condition, but with an eye toward the particular property or properties of interest. The degree of control available, of course, depends upon the particular synthesis method being used to create the given nanostructured material.

Future needs

There are tremendous opportunities for synthesizing nanostructured materials with new architectures at nanometer length scales from atomic or molecular precursors via the assembly of atom layers and clusters and by a myriad of other techniques now becoming available, such as nanoscale lithography and biological templating. Important keys to the future of nanostructured materials will be ① our ability to continue to significantly improve the properties of materials by artificially structuring them on these nanometer length scales and ② developing the methods for producing these materials in commercially viable quantities. There are, thus, very significant opportunities for developing creative new chemical synthesis and processing routes for the tailored nanostructuring of materials with new engineered properties in a manner not unlike that so successfully applied to polymer synthesis and processing[5]. It is also imperative that an understanding of the important role of surface and interface chemistry in the assembly and resulting properties of these materials be developed. Because of the nanometer scales involved, new or improved experimental probes with accessible lateral length scales in the nanometer regime will certainly need to be developed.

Various aspects of our present understanding of nanostructured materials and their properties have been briefly presented here, along with some thoughts regarding a few critical future research needs in various areas of chemistry that would add greatly to the field of nanostructured materials. Based upon what we have learned to date about condensed matter in the nanoscale regime, it appears that the future holds great promise for nanostructured materials and that chemists can continue to make seminal contributions to this important and challenging new area.

Vocabulary

1. cluster   ['klʌstə]   n. 类、族、基、串、群、团
2. grain   [grein]   n. (结晶)粒度、颗粒
3. elucidate   ['iljuːsideit]   vt. 阐明、解释
4. tailored   ['teilərəd]   a. 特别的、干净利索的
5. mesoscopic   [mesəu'skɔpik]   a. 亚微观的

6. ensemble　　　［aːnˈsɑːmbl］　　　n . ［法］全体、全体演出者
7. embed　　　　［imˈbed］　　　　 vt . 栽种(花等)、埋置、放入、嵌进
8. confinement　 ［kɔnˈfainmənt］　　n . 限制、界限、密封
9. delocalize　　 ［diːˈloukəlaiz］　　 vt . 使离开原位、去别的地方
10. aerosol　　　 ［ˈɛərɔsɔl］　　　　n . 气溶胶、烟雾剂
11. template　　　［ˈtemplit］　　　　n . 样板
12. attrition　　　［əˈtriʃən］　　　　n . 摩擦、消耗
13. amorphous　　［əˈmɔːfəs］　　　 a . 非结晶的、乱七八糟的
14. precursor　　　［priˈkəːsə］　　　 n . 先驱、初级粒子、产物母体
15. radii　　　　　［ˈreidiai］　　　　n . (radius［ˈreidjəs］的复数)半径、辐射线
16. excitonic　　　［ˌeksiˈtɔnik］　　　a . 激发电子-空穴对的
17. defect　　　　［diˈfekt］　　　　 n . 缺点、不足、故障
18. myriad　　　　［ˈmiriəd］　　　　n . 无数；a . 数不清的
19. lithography　　［ˈliθəgrɑːfi］　　　n . 石板、石印
20. imperative　　 ［imˈperətiv］　　　a . n . 不可避免的、紧急的
21. regime　　　　［reiˈdʒiːm］　　　n . 政体、社会制度、方式方法、领域

Notes

1. The structures and properties of these ... to be understood. 句中由"and"连接并列句。句子主干结构为"The structures and properties have been elucidated ... and the relationships are beginning to be understood"。全句可译为：这些由束、颗粒和层(片)等纳米尺寸的"结构单元"人工合成的新材料的结构和性质已在许多重要领域中得到阐述；且这些领域之间的关系正开始逐渐被了解。

2. Various investigations ... are assembled. 句中"possibilities"后的"to do sth"结构作宾语补语；宾补结构中"through control of sth"作状语，表示方式；其中的"sizes"和"manners"是并列关系的介词宾语。

3. A knowledge of the variation ... of technological applications. 句子的主干结构为"A knowledge is important (to both ... understanding ... and our ability ...)"。句中"the variation in ... with ..."之意为"随……的变化"。

4. Spatial confinement can ... for that property. 句中"comparable to"及"smaller than"意为"接近"和"小于"；"a critical length scale"意为"临界尺寸范围；be

responsible for"意为"决定,对……负责"。

5. There are, thus, very significant opportunities ... to polymer synthesis and processing. 句中前一个"for"为"对于,针对"的意思;后一个"for"为"用于"的意思,"chemical synthesis and processing"指"化学合成和操作(加工)路线"。

6.20 MODERN SYNTHETIC SURFACTANT SHAMPOOS

Function and Properties

Cleanliness of hair and scalp are among the most important personal grooming considerations today. The emphasis place on cleanliness is a relatively new cultural phenomenon and , as mass markets have developed, shampoos have increased in importance beginning with soap-based products and expanding as a growing number of synthetic surfactants became available[1].

Hair and scalp accumulate a diversity of substances. Sebum (the normal exudate of the sebaceous glands), scales (normal or excessive, sloughing off of the dead epidermal cells), residues from hair sprays, conditioners and the like, and airborne dust and dirt are often included in this soil. Qualitative and quantitative aspects of this mixture may vary from one individual to the next because of different biochemical, environmental and grooming factors. Sooner or later the accumulated debris becomes significant, and may exceed 5% of the total hair mass. Extensively soiled hair takes on a lackluster appearance, becomes oily and gritty to the touch and may even exude an unpleasant odor. Cleansing of hair and scalp requires an effective shampoo that with some degree of hand and finger manipulation, safely and efficiently removes the soil and yet leaves hair and scalp in relatively good condition.

Much has been said about the highly efficient cleansing action of modern synthetic surfactant shampoos. Their propensities to overclean and dry out hair and scalp, to leave the hair in a fly away state with a harsh and raspy feel, have long been grist for the competitive mill[2]. Several authors have advised against complete removal of the natural oils lest the condition of the hair and scalp be impaired. Yet, complete extraction of the hair with an efficient grease solvent apparently produces no adverse conditioning effects. Some highly efficient cleansing shampoos remove as little as 10% of the total solvent extractable material on hair. Furthermore, the wide variations in physical and cosmetic effects produced by absorbed surfactants, various cosmetic treatments, exposure to sunlight and atmospheric conditions, and the normal variation in hair diameter must be considered[3]. Obviously, the path to the clean state is not easily followed nor is it likely to be straight and narrow.

In addition, the formulator must deal with a number of essential and desirable shampoo properties if a satisfactory product is to be evolved. During application, the shampoo must spread easily over the head and into the hair. It should foam quickly and copiously in both soft and hard water and then rinse out thoroughly, leaving no detectable residues. Neither hair nor scalp should feel drawn or dry and they should convey a fresh clean scent. The hair should be left in a sott, lustrous, full bodied, and manageable state. The product must be nontoxic, nonirritating to the hand and scalp, and properly preserved against microbial and fungal contamination. In the event of accidental introduction into the eyes, no permanent damage should result. Finally, the particular blend of product color, clarity, pH, viscosity, and fragrance including the appropriate stability may often affect success or failure of the product.

Physical Forms and Formulations

Physical Forms.

Clear Liquids. Clear liquid shampoos may range from simple aqueous solutions of a single cleansing agent and a minimum of preservative and esthetic components, to more complex clear emulsions employing several cleansing agents and a variety of modifying additives[4]. The clarity imparts the impression of superior cleansing, a quality often attributed to clear shampoos. Because these shampoos are often packaged in clear containers, clarity and color stability are of utmost concern to the formulator. Product clarity can usually be ensured through careful selection of cleansing agents, additives, emulsifiers, and solubilizers. Various salts of the alkyl sulfates and alkyl ether sulfates are commonly employed as cleansing agents because of their excellent solubility and low cloud-point features. Control of viscosity without adversely affecting product clarity is frequently accomplished by adjusting the salt concentration or pH. Fatty acid alkanolamides and cellulose thickeners are also used for this purpose.

Clear Gels. The clear gel shampoo, in a compact flexible tube, allows convenient storage in travel bag or toiletry case. Except for higher levels of cleansing and thickening or gelling agents, these shampoos are very similar in composition to the clear liquids.

Opaque Lotions. The popularity of the opaque lotion or liquid cream shampoo has risen sharply during the past decade. These shampoos are elegant in appearance and offer the consumer a range of aesthetically pleasing rich creamy consistencies. They are generally emulsions or hybrid emulsion-dispersion systems, often incorporating several cleansing agents, opacifiers, grooming oils, medicaments, and other additives (see Emulsions). Because product clarity is not a consideration, there is a wider choice of cleansing agents. Many opacifiers (e.g., glycol stearates) can lend a pearliness to the shampoo. Thickening agents such as fatty alkanolamides, cellulose derivatives, and the less soluble stearate salts may be used to achieve the required viscosities.

Cream Pastes. The popularity of cream shampoos has passed its peak, and today, they are usually presented as convenience items to an existing product line of shampoos. They are closely related to the opaque lotions except that the cream pastes are much thicker in consistency. Viscosity is enhanced by suspending insoluble stearate salts in the lotion matrix, or by electrolyte control, or occasionally by the admixture of synthetic and soap-based surfactants.

Synthetic Surfactants. The rapidly expanding cosmetics and toiletries industry makes widespread use of synthetic surfactants in a large assortment of product applications.

Formulations.

Typical shampoo formulations are given in Table 6.8. The names used are names designated by the Cosmetic, Toiletry, and Fragrance Association (CTFA), and manufacturers of these materials may be found in their Cosmetic Ingredient Dictionary.

Table 6.8 Typical Shampoo Formulations

Ingredient	CAS Registry No.	w/%	Function
clear liquid			
sodium laureth sulfate, 30% active	[9004-82-4]	40.0	cleansing agent
lauramide DEA	[120-40-1]	4.0	foam stabilizer
disodium EDTA	[139-33-3]	0.1	sequestering agent
formaldehyde	[50-00-0]	0.04	preservative
fragrance		0.5	fragrance
FD&C Blue No.1	[2650-18-2]	0.001	color
FD&C Yellow No.5	[1934-21-0]	0.004	color
deionized or distilled water		55.355	
opaque, pearlescent			

Table 6.8 (Continued)

TEA lauryl sulfate, 40% active	[17961-18-1]	20.0	cleanser
sodium lauryl sulfate, 28% active	[151-21-3]	20.0	cleanser
cocoamide DEA		5.0	foam stabilizer
glycol stearate	[111-60-4]	1.0	opacifier, pearlescent agent
disodium EDTA		0.1	sequestering agent
methylparaben	[99-76-3]	0.1	preservative
propylparaben	[94-13-3]	0.01	preservative
fragrance		0.5	fragrance
deionized or distilled water		53.29	

Vocabulary

1. exudate ['eksjuːdeit] n. 渗(流)出物
2. exude [ig'zjuːd] v. (使)渗出、(使)发散
3. sebaceous glands [si'beiʃəs glænds] n. 皮脂腺
4. sebum ['siːbəm] n. 脂肪、皮脂
5. slough(off) [slʌf] n.v. 碎落、脱落、废弃
6. epidermal [ˌepi'dəːmal] a. 表皮的
7. groom [grum] v. 修饰
8. gritty ['griti] a. 砂粒质的、多砂的
9. propensity [prə'pensiti] n. 倾向、习性
10. hash [hæʃ] n. 材料堆成的东西、杂乱、大杂烩
11. grist [grist] n. 谷粉、大量、许多
12. copiously ['coupjəsli] ad. 丰富地、多产地

13. scent	[sent]	n.	香味、嗅觉
14. microbial	[mai'kroubiəl]	a.	微生物的、(因)细菌(而引起)的
15. fungal	['fʌŋgəl]	a.	真菌(状类)的
16. contamination	[ˌkəntæmi'neiʃən]	n.	污染、杂质
17. fragrance	['freigrəns]	n.	芬芳、香味
18. lotion	['louʃən]	n.	洗剂、洗净
19. aesthetically	[iːs'θetikəli]	a.	美的、审美的
20. consistency	[kən'sistənsi]	n.	稠度
21. medicament	[me'dikəmənt]	n.	药物、药剂
22. glycol	['glaikɔl]	n.	乙二醇
23. stearate	['stiəreit]	n.	硬脂酸盐
24. alkanolamide	['ælkeinəl'æmiːd]	n.	链烷醇酰胺
25. admixture	[əd'mikstʃə]	n.	混合物、掺合剂
26. toiletry	['tɔilətri]	n.	化妆用品、梳妆用具
27. sodium laureth sulfate		n.	月桂基硫酸钠
28. lauramide DEA		n.	月桂酰二乙醇胺
29. formaldehyde	[fɔː'mældihaid]	n.	甲(蚁)醛
30. sequestering (agent)	[si'kwestəriŋ]	n.	掩蔽(剂)、整合(剂)
31. FD & C (Food, Drug and Cosmetic)			食品、药物和化妆品
32. deionize	[diː'aiənaiz]	v.	去离子
33. TEA(triethanolamine)		n.	三乙醇胺
34. cocoamide DEA		n.	可可二乙醇胺

Notes

1. The emphasis place on … became available. 全句可译为：对于清洁的重视是一个相对较新的社会现象，且随着市场的发展，以肥皂基产品开始，伴随适用化的表面活性剂数目增加而使品种得以拓展的香波的重要性已有增加。句中"as"均为"由于、伴随"的意思。

2. Their propensities to overclean … for the competitive mill. 句中"have long been grist for competitive mill"意为"一直对有竞争力的生产厂家有利"。

3. Furthermore, the wide variation in physical ... must be considered. 全句可译为：而且，表面活性剂的吸附、不同的化妆处理、暴露于阳光和大气环境所产生的身体的和化妆效果的较大不同，以及通常头发直径的正常变化都是必须加以考虑的。句中"the wide variations"和"the normal variation"作并列主语。

4. Clear liquid shampoos may range from ... of modifying additives. 句中含有"from ... to"结构。全句可译为：透明香波制品的范围可以从单一的清洁剂与很少量起防腐和美学作用的成分所组成的简单水溶液直至几种清洁剂和一系列改性添加剂组成的较复杂的澄清乳液。句中"modifying additives"指"改性添加剂"。

6.21 COSMETIC

Cosmetic Emulsions

Cosmetic lotions and creams are emulsions of water-based and oil-based phases. An emulsion is a two-phase system consisting of two incompletely miscible liquids, the internal, or discontinuous phase dispersed as finite globules in the other. Special designations have been devised for oil and water emulsions to indicate which is the dispersed and which the continuous phase. Oil-in-water(o/w) emulsions have oil as the dispersed phase in water as the continuous phase. In water-in-oil(w/o) emulsions, water is dispersed in oil, which is the external(continuous) phase.

Properties of Emulsions

The properties that are most apparent, and thus are usually most important, are: ease of dilution, viscosity, color, and stability. For a given type of emulsification equipment, these properties depend upon. ① the properties of the continuous phase; ② the ratio of the external to the internal phase; ③ the particle size of the emulsion; ④ the relationship of the continuous phase to the particles(including ionic charges), and ⑤ the properties of the discontinuous phase. In any given emulsion, the properties depend upon which liquid constitutes the external phase, i. e., whether the emulsion is o/w or w/o. The resulting emulsion type is controlled by : ① the emulsifier: type, and amount ; ② the ratio of ingredients, and ③ the order of addition of ingredients during mixing.

The dispersibility (solubility) of an emulsion is determined by the continuous phase; thus if the continuous phase is water-soluble, the emulsion can be diluted with water, conversely, if the continuous phase is oil-soluble, the emulsion can be diluted with oil.

The ease with which an emulsion can be diluted can be increased by decreasing the viscosity of the emulsion. The viscosity of an emulsion when the continuous phase is in excess is essentially the viscosity of the continuous phase. As the proportion of internal phase increases the viscosity of the emulsion increases to the point that the emulsion is no longer fluid[1]. When the volume of the internal phase exceeds the volume of the external phase, the emulsion particles become crowded and the apparent viscosity is partially structural viscosity[2].

An emulsion is stable as long as the particles of the internal phase do not coalesce. The stability of an emulsion depends upon: ① the particle size; ② the difference in density of the two phases; ③ the viscosity of the continuous phase and of the completed emulsion; ④ the charges on the particles; ⑤ the nature, effectiveness, and amount of the emulsifier used; and ⑥ conditions of storage, including temperature variation, agitation and vibration, and dilution or evaporation during storage or use. The stability of an emulsion is affected by almost all factors involved in its formulation and preparation in formulas containing sizable amounts of emulsifier, stability is predominantly a function of the type and concentration of emulsifier[3].

Cosmetic Creams and Deodorants

① Cosmetic Creams. Materials used in creams may be prepared in o/w or in w/o emulsions. The esthetic effect and degree of emolliency depend to a great extent on the emulsion type as well as on the emulsion composition. O/w emulsions produce a cooling effect on application to the skin owing to water evaporation. W/o emulsions do not produce this effect since water evaporation is slowed by the film of the oil in the continuous phase.

The classical example of a cream was the USP Unguentum Aquae Rosae which was prepared from 3.0% beeswax, 11.8% spermaceti, 40.2% sweet almond oil, and 45.0% rose water. In 1890 the formula was changed to 12.1% beeswax, 12.6% spermaceti, 55.4% sweet almond oil, 0.5% borax, and 19.4% rose water.

This was the basic formula for the familiar cold cream that is now made with mineral oil instead of almond oil. Its occlusive action aided in rehydration of the corneum when allowed to remain on the skin for an appreciable length of time[4]. Because the solvent action of mineral oil tends to remove skin surface lipids when the cream is applied for a short period of time, partial replacement with a vegetable oil is needed. These emulsions are w/o, the emulsifier is sodium cerotate formed by reaction of borax and free cerotic acid in the beeswax. If the water content is raised to approximately 45% or more the composition changes to an o/w emulsion.

② Deodorants and Antiperspirants. Active Ingredients. Aluminum chloride, once commonly used in deodorants and antiperspirants, may cause fabric damage and skin irritation because of its low pH. This has led to the development of various basic aluminum compounds. The most widely used compound is aluminum chlorhydroxide (ACH). Other compounds that have been or are now used are: basic aluminum bromide, iodide, and nitrate, and basic aluminum hydroxychloride-zirconyl hydroxy oxychloride, with and without glycine. Zirconium salts, formerly used in antiperspirants, were banned for use in aerosol antiperspirants in 1977.

Deodorant-Antiperspirant Sticks. Sodium stearate, the primary gelling agent for deodorant sticks, constitutes about 7~9 wt% of these products. The grade employed depends on the fatty acid used in making the stearate. Derivatives such as acetylated sucrose distearates also can be used but are more common in antiperspirant sticks to gel cetyl alcohol-based formulations (typical use level, 28%).

There are three main types of stick formulations for antiperspirants: hydroxyethyl stearamide, which produces clear-melting, homogeneous sticks; stearamides and cetyl alcohol dry-powder dispersions of ACH; and stearamides couples with propoxylated alcohol. The stearamide wax content is 26%~29%; the active concentration, 20%~25%.

Vocabulary

1. miscible   ['misibl]   a. 能融合的、易混合的
2. globule   ['glɔbjuːl]   n. 小球、液滴
3. coalesce   [ˌkouə'les]   vi. 结合
4. emolliency   [i'mɔliənsi]   n. 润肤剂、柔软剂
5. unguentum   ['ʌŋgwəntəm]（拉丁语）n. 软膏(剂)
6. aqua   ['ækwə]（拉丁语）   n. 水(剂)
7. spermaceti   [ˌspəːmə'seti]   n. 鲸脑油
8. almond   ['aːmənd]   n. 杏仁
9. borax   ['bɔːræks]   n. 硼砂
10. occlusive   [ɔ'kluːsiv]   a. 闭塞的、堵塞的
11. corneum   ['kɔːniəm]   n. 角膜
12. cerotate   ['siərouteit]   n. 蜡酸脂
13. antiperspirant   [ˌæntipəs'pairənt]   n. 抗汗剂
14. chlorhydroxide   [ˌklɔːhai'drɔksaid]   n. 含氯的氢氧化物
15. hydroxychloride [haiˌdrɔksi'klɔːrid]   n. 羟基氯化物
16. zirconyl   ['zəːkounil]   n. 氧锆基
17. glycine   ['glaisiːn]   n. 甘氨酸
18. deodorant   [diː'odərənt]   n. 清香剂、除嗅剂
19. acetylate   [ə'setileit]   v. 乙酰化；n. 乙酰化物
20. sucrose   ['sjuːkrous]   n. 蔗糖
21. cetyl   ['siːtail]   n. 十六(烷)基

Notes

1. As the proportion of internal phase ... is no longer fluid. 句子可译为：当把增加乳液粘度的内相的比例增加至一定值时，乳液就不再是流体了。
2. When the volume of the internal phases ... structural viscosity. 句中"apparent viscosity"意为"表现粘度"；"structural viscosity"意为"构成粘度"。
3. The stability of an emulsion is ... concentration of emulsifier. 全句可译为：乳液的稳定性受包含在配方中以及含有一定乳化剂的乳液配制过程中几乎所有的因素影响，稳定性最主要是乳化剂类型和浓度的函数。句中"a function of"指"……的函数"。
4. Its occlusive action aided in ... length of time. 句中"aided"作谓语，是"辅助，提供，促进"的意思；"occlusive action"指"闭塞，阻塞作用"。全句可译为：如果让这种冷霜在皮肤上停留较长的时间，这种阻塞作用会加速角膜的再水合作用。

6.22 INORGANIC PIGMENTS

White Pigments

Among the white pigments titanium dioxide is undisputedly foremost; its production surpasses the combined production of leaded and unleaded zinc oxide which, in turn, exceeds that of white lead, first in line only a few years ago. The tonnages of lithopone and zinc sulfide blends are also significant. Carbon black is the principal black pigment. The oxides of iron, both natural and synthetic, are produced in large quantities. Prussian blue and phthalocyanine blue are the prominent blues, with ultramarine used in smaller amounts. Lead chromate (yellow), chrome oxide (green), and other chrome greens are used extensively; however, several of the older natural pigments are also in demand. Use of metallic salts of organic dyes, among them other phthalocyanines, is increasing. Metallic oxides are employed for ceramics; metal powders, particularly of aluminum and copper, are used to provide metallic finishes. Brilliant luminescent pigments are now available in several hues.

Titanium Dioxide. The "hiding power" of a pigment in a given system depends on particle size and differences in refractive indexes between pigment and binder. Both commercial forms of titanium dioxide, rutile and anatase, have higher refractive index values than other white pigments and therefore give excellent covering power per pound of pigment.

Titanium ores are very widespread, the chief ores being ilmenite ($FeTiO_3$) and rutile (TiO_2). Although much is imported, important deposits occur along the eastern seaboard of the United States. Preparation of a high quality white pigment requires that the removal of impurities be carefully controlled; in the case of titanium dioxide this is a fairly expensive process necessitating many separate operations.

The pigment is prepared from the concentrated ore by mixing with sulfuric acid; the reaction which takes place is exothermic and violent. After digestion and solution, the iron is reduced to the ferrous state and much of it removed by crystallization as ferrous sulfate. The liquor is filtered, concentrated in vacuum evaporators, and boiled with sulfuric acid to precipitate the titanium dioxide pigment which is washed, dried, and calcined at a temperature of about 900℃. The calcining operation converts the titanium dioxide from the amorphous to the crystalline state, thereby raising the refractive index. Controlled grinding and bagging follow.

After the purification steps the process varies depending on the grade and type of product desired. Each manufacturer makes several grades, so it is essential to use the proper one to obtain the best results for a particular purpose. A type developed for a chalking house paint would hardly be suitable for an automotive finish.

"Extended" titanium pigments are prepared either by mixing or coprecipitating TiO_2 with cheaper pigments of low hiding power. Titanium calcium may be prepared by two methods: (1) by precipitating hydrated TiO_2 in the presence of $CaSO_4$, the coprecipitate being filtered and washed, calcining and dry grinding; (2) TiO_2 and $CaSO_4$ may be mixed as a wet slurry, filtered, dried, calcined and dry-ground. The composite pigment contains 30 per cent TiO_2 and 70 per cent $CaSO_4$, and has much better hiding power than would be obtained from a simple dry mix of the two components in the same proportion. Titanium-barium and titanium-magnesium pigments, each containing 30 per cent TiO_2, are also available.

Titanium dioxide in the pure state or as a composite pigment is used extensively in the paint, paper rubber, ceramic, floor covering, and cosmetic industries. Its chief value stems from its bright white color and opacity. The tinting strength and opacity of titanium dioxide surpass that of any other white pigment.

Titanium dioxide paints with controlled chalking are extensively used for house paints; they stay white longer because of the gradual erosion of the soiled surface. A large proportion of anatase, sometimes combined with a small amount of an oxide of antimony or aluminum, is used for this purpose.

White Lead, used as a pigment before the time of Christ, is still the only which, when used as the sole pigment in an oil-based paint, will give an exterior finish that has good adhesion and is tough, elastic, and durable. It is a basic carbonate of the general formula $4PbCO_3 \cdot Pb(OH)_2 \cdot PbO$. Several methods have been used for its manufacture, different processes requiring from one day to several months for completion.

The Dutch Process. Lead, in the form of cast buckles, is corroded by the vapors of acetic acid and water in the presence of carbon dioxide emanating from fermenting tanbark. Lead acetate, the first compound formed, is transformed by carbon dioxide and moisture partly into lead carbonate and partly into lead hydroxide. The buckles are placed in small earthehware crocks, 8 inches in diameter and 10 inches high, into which a pint of 28 per cent acetic acid has been poured. The buckles lie over the acid, resting on a small shoulder in the inner wall of the crock. A layer of these crocks is placed in a thick bed of spent tanbark, then covered by boards over which more tanbark is shoveled, then by another tier of crocks. This is repeated until 10 or more tiers have been made. Some ventilation to the center is provided by a stack-like opening in the middle. The room is then closed. Fermentation of the tanbark proceeds and the temperature rises to about 70℃. The pots remain undisturbed for three months; after that they are unpacked and the buckles, now white, are lifted out, crushed free from any uncorroded center, and ground dry or wet.

Other Processes. The Dutch process is slow and requires much hand labor. Extremely pure lead is required, otherwise the corrosion does not proceed far enough. Although the high quality of the product obtained by this process justifies its continued use, several other processes are employed, all of which require less time and turn out an even better product. The Carter process, which uses powdered lead, acetic acid, and carbon dioxide, requires one week. The French process, starting with litharge (PbO) instead of lead, yields a product in two days. Rowley uses atomized lead in water to form a hydroxide suspension through which carbon dioxide is passed. The Sperry process uses a lead anode and iron cathode, with sodium acetate and sodium carbonate as electrolytes, in an electrolytic cell in which the electrode compartments are separated by a membrane.

White lead is often wet ground since the poisonous nature of dry dust constitutes a health hazard. After wet grinding, the suspended solid is thickened by settling, giving "pulp lead." If it is intended for paint purposes, linseed oil may be mixed with the pulp. The oil displaces the water and the resulting paste is ground on roller mills and sold as lead-in-oil. The pulp may also be dried and disintegrated, and sold as dry powder.

Zinc Oxide. Zinc oxide is made in several ways. The original method, still in use, is the French or indirect process: zinc metal (spelter) is heated in stoneware retorts, vaporized, and burned in a combustion chamber placed at the mouth of the retort. An exhauster draws the white dust first to an air chamber where the heavier, less desirable particles are deposited, then to a filter chamber where the fine dust is collected.

In order to obtain the whitest pigment a very pure spelter must be used; if it contains lead, for instance, a frequent impurity in zinc, the lead burns to litharge and gives a yellowish tinge—a serious fault. To render the lead harmless, a modified indirect method has been developed where the burning is done in air mixed with carbon dioxide. The lead is changed to white lead, thus the yellow tinge is avoided; the zinc oxide is unaffected. This modification has been extremely successful and has made possible the use of spelters containing appreciable quantities of lead.

A direct process, that is, one producing the oxide directly from the ore, was developed by Weatherill about 80 years ago for the New Jersey Zinc Company. In the Weatherill furnace, franklinite, an oxide of zinc containing iron and manganese (Zn 18 per cent), is mixed with coal and burned on a grate. The natural oxide is first reduced and then re-formed by the air and carbon dioxide from the fire. Provision is made to admit more air over the fire if unburned zinc vapors should rise. The grate is a casting with tapering holes rather than the usual bars. The residue on the hearth is made into spiegeleisen, a manganese-iron mixture which also contains carbon.

Approximately 15 per cent of the zinc oxide pigment used in 1958 was "leaded" zinc oxide. This pigment is essentially a mechanical mixture of zinc oxide with 5 to 50 per cent of either lead sulfate or basic lead sulfate. The lead content of the pigment results in better through-dry and toughness in the film. This pigment is commonly used in exterior house paints.

Colored Pigments

Litharge. Litharge (lead oxide) may be prepared in several ways. The oldest method, still practiced, is to heat metallic lead in a low-arched reverberatory furnace with the usual bridge wall dividing the fireplace from the hearth. The atmosphere is kept oxidizing by allowing a large quantity of air to enter; the temperature is just above the melting point of lead oxide (PbO). As the oxide forms it floats on the surface and is pushed to one side by iron hoes. When enough oxide has collected it is drawn off by means of the hoes, cooled, ground and levigated. The color is buff. A continuous furnace has been used successfully in which air jets impinge on the surface of the lead, sweeping the molten oxide to the front end of the oval furnace; fresh lead is added in a stream so regulated that the amount of lead in the furnace remains constant. The molten oxide overflows into a conical receiver and is cooled, broken up, ground, and airfloated to remove unoxidized lead particles.

Red Lead. Red lead is made by calcining litharge in a muffle furnace. A current of air is admitted into the muffle; the temperature must be maintained within narrow limits, near 340°C, and the period is usually 48 hours. Whereas the use of white lead has been steadily decreasing, the use of red lead in priming paints is increasing, and the production of red lead now surpasses that of white lead.

Carbon Black. There are several kinds of carbon black: thermal black, produced by the thermal decomposition of natural gas; channel black, produced by the impingement of numerous small regulated flames against a relatively cold steel surface which is constantly scraped free of the deposit; furnace black, produced by the partial combustion of the gas in a furnace, with recovery of the carbon product in cyclones and electrical precipitators; lampblack, used mainly as a tinctorial pigment; and oil black. Thermal black is used as a filler and an extender in rubber. Until recently, channel black has been the important black in rubber compounding. Following the introduction of the synthetic GR-S, and more recently of "cold rubber", it has been found that certain furnace blacks with particle size larger than that of the standard channel black (such as 45 to 100 millimicrons) give superior reinforcement. The great advantage of furnace black is that its yield is higher (as high as 8 pounds per 1 000 cubic feet of gas) than that of channel black (1 to 2 pounds per 1 000 cubic feet of gas).

Carbon black may be pelleted, by either a dry or a wet process, for greater convenience and success in handling. Thus carbon blacks which have been treated with zinc naphthenates, and the "beaded" blacks, from which entrapped air has been removed, can be wetted much more quickly and effectively by the vehicle.

Lampblack is an older pigment; it has been made for thousands of years by the Chinese, Egyptians, and other ancient races for the manufacture of ink. It consists of the free soot or smoke collected in chambers from burning oils of hydrocarbon gases. In many of its earlier uses it has been replaced by carbon black.

Iron Oxide. Iron oxide (Fe_2O_3) is made on a large scale by roasting ferrous sulfate obtained from the vats used for pickling steel. Water and sulfur oxides are driven off and led through a stack to the atmosphere. The snade can be varied by altering the firing time, the temperature, and the atmosphere. It is a relatively cheap pigment and is usually used in red barn paint and metal prime. The use of selected grades for polishing glass and lenses is determined by their resistance to grit and the hardness of the glass; such grades of iron oxide are called rouges.

Prussian Blue. Prussian blue or iron blue is manufactured by reacting sodium ferrocyanide, ferrous sulfate, and ammonium sulfate. The precipitate is then oxidized with sodium chlorate or sodium dichromate.

Phthalocyanine Blue. The phthalocyanine pigments are greens and blues of extremely high tinting strength. Structurally they are quite similar to chlorophyll, but copper is the chelating metal rather than magnesium. Although expensive, they are such effective tinters that they are now used extensively in both oil- and water-based paints.

Ultramarine Blue. A bright blue inorganic pigment of complex strutcure, ultramarine is made by heating a mixture of soda ash, clay, and sulfur with charcoal or pitch. Although used for tinting and for outside paints, its primary use is in colored granules for asphalt shingles.

Chrome Yellows. To prepare chrome yellows, soluble lead salts are precipitated from solution by adding sodium or potassium dichromate. The colors are clean and bright, and have a high tinting strength. The shades may be varied by adjusting the pH of the precipitating solutilon.

Chrome Greens. Chrome greens are manufactured by coprecipitating Prussian blue and chrome yellow. The color is dependent on the proportions of the two pigments used.

Natural Pigments. Among the natural pigments still in use are umber (brown), ochre (a poor yellow), and Sienna (a deeper yellow).

Vocabulary

1. phthalocyanine	[θæləu'saiəni(:)n]	n.	酞菁
2. ultramarine	[ˌʌltrəmə'riːn]	n.	群青
3. luminescent	[ljuːmi'nesnt]	a.	发光的
4. rutile	['ruːtil]	n.	金红石
5. anatase	['ænəteiz]	n.	锐钛石(八面石)
6. ilmenite	['ilminait]	n.	钛铁矿
7. calcine	['kælsain]	v.	焙烧、煅烧
8. amorphous	[ə'mɔːfəs]	a.	无晶形的、无定形的
9. opacity	[əu'pæsiti]	n.	不透明(体)、暗度、浑浊度
10. buckle	['bʌkl]	n.	扣子、带扣
11. emante	['eməneit]	vi.	发源、起源
12. crock	[krɔk]	n.	瓦罐、坛子
13. tier	[tiə]	n.	(一)排、层、等级

14.	litharge	[liθɑːdʒ]	n. 正方铅矿
15.	retort	[riˈtɔːt]	n. 蒸馏瓶、曲颈瓶
16.	reverberatory	[riˈvəːbərətəri]	a. 反射的、反射炉的、回响的
17.	levigate	[ˈlevigeit]	vt. 磨光、水磨、使成糊
18.	buff	[bʌf]	n. 浅黄色、米色
19.	impinge	[imˈpindʒ]	v. 冲击、撞击(on upon, against)
20.	conical	[kɔnikəl]	a. 圆锥形的
21.	tinctorial	[tiŋkˈtɔiriəl]	a. 着色的、色泽的、染色的
22.	pickle	[ˈpikl]	n. 盐水、酸洗液
			vt. 腌渍、酸洗
23.	rouge	[ruːʒ]	n. 胭脂、口红、红铁粉、铁丹
24.	chelate	[ˈkiːleit]	a. 螯合的、螯形的;
			n. 螯合物; v. 与……结合成螯合物
25.	umber	[ʌmbə]	n. 棕土、红棕土、赭色
26.	ochre	[ˈoukə]	n. 赭色、赭石、黄锡色
27.	sienna	[siˈenə]	n. 浓黄色、赭色

Phrases

1. refractive index 折光指数
2. covering power 遮盖力
3. tinting strength 着色力
4. oil-based 油基的
5. through-dry 流通干燥
6. cold rubber 冷聚合橡胶(一般指丁苯橡胶)

6.23 SYNTHETIC DYES

The practice of using dyes is an ancient art. Substantial evidence exists that plant dyestuffs were known long before man began to keep written history. Before this century, practically all dyes were obtained from natural plant or animal sources. Dyeing was a complicated and secret art passed from one generation to the next. Dyes were extracted from plants mainly by macerating the roots, leaves, or berries in water. The extract was often boiled and then strained before use. In some cases, it was necessary to make the extraction mixture acidic or basic before the dye could be liberated from the plant tissues. Applying the dyes to cloth was also a complicated process. Mordants were used to fix the dye to the cloth, or even to modify its color.

Madder is one of the oldest known dyes. Alexander the Great was reputed to have used the dye to trick the Persians into overconfidence during a critical battle. Using madder, a root bearing brilliant red dye, he simulated blood-stains on the tunics of his soldiers. The Persians, seeing the apparently incapacitated Greek army, became overconfident and much to their surprise were overwhelmingly defeated. Through modern chemical analysis, we now know the structure of the dye found in madder root. It is called alizarin and is very similar in structure to another ancient dye, henna, which has been responsible for a long line of synthetic redheads.

Indigo is another plant dyestuff with a long history. This dye has been known in Asia for more than 4 000 years. By the ancient process for producing indigo, the leaves of the indigo plant are cut and allowed to ferment in water. During the fermentation, indican is extracted into the solution, and the attached glucose molecule is split off to produce indoxyl. The fermented mixture is transferred to large open vats, in which the liquid is beaten with bamboo sticks. During this process, the indoxyl is air-oxidized to indigo. Indigo, a strong blue dye, is insoluble in water, and it precipitates. Today, indigo is made synthetically, and its principal use is in dyeing denim to produce "blue jeans" material.

Many plants yield dyestuffs that will dye wool or silk, but there are few of these that dye cotton well. Most do not dye synthetic fibers like polyester or rayon. In addition, the natural dyes, with a few exceptions, do not cover a wide range of colors, nor do they yield "brilliant" colors.

Even though some people prefer the softness of the "homespun" colors, from natural dyes, the synthetic dyes, which give rise to deep, brilliant colors, are much in demand today. Also, synthetic dyes that will dye the popular synthetic fibers can now be manufactured. Thus today we have available an almost infinite variety of colors, as well as dyes to dye any type of fabric.

Before 1856, all dyes in use came from natural sources. However, an accidental discovery by W. H. Perkin, an English chemist, started the development of a huge synthetic dye industry, mostly in English and Germany. Perkin, then only aged 18, was trying to synthesize quinine. Structural organic chemistry was not very well developed at that time, and the chief guide to the structure of a compound was its molecular formula. Perkin thought, judging from the formulas, that it might be possible to synthesize quinine by the oxidation of allyltoluidine. He made allyltoluidine and oxidized it with potassium dichromate. The reaction was unsuccessful, because allyltoluidine bore no structural relation to quinine. He obtained no quinine, but he did recover a reddish brown precipitate with properties that interested him. He decided to try the reaction with a simple base, aniline. On treating aniline sulfate with potassium dichromate, he obtained a black precipitate, which could be extracted with ethanol to give a beautiful purple solution. This purple solution subsequently proved to be a good dye for fabrics. After receiving favorable comments from dyers, Perkin resigned his post at the Royal College and went on to found the British coal tar dye industry. He became a very successful industrialist and retired at age 36(!) to devote full time to research. The dye he synthesized became known as mauve. The structure of mauve was not proved until much later. From the structure it is clear that the aniline Perkin used was not pure and that it contained the o-, m-, and p-toluidines also.

Mauve was the first synthetic dye, but soon (1859) the triphenylmethyl dyes pararosaniline, malachite green, and crystal violet were discovered in France. These dyes were produced by treating mixtures of aniline or of the toluidines or of both, with nitrobenzene, an oxidizing agent, and in a second step, with concentrated hydrochloric acid. The triphenylmethyl dyes were soon joined by synthetic alizarin (Liberman, 1868), synthetic indigo (Baeyer, 1878), and the azo dyes (Griess, 1862). The azo dyes, also manufactured from aromatic amines, revolutionized the dye industry.

The azo dyes are one of the most common types of dye still in use today. They are used as dyes for clothing, as food dyes, and as pigments in paints. In addition, they are used in printing inks and in certain color printing processes. Azo dyes have the basic structure $Ar-N=N-Ar'$. The unit containing the nitrogen-nitrogen bond is called an azo group, a strong chromophore that imparts a brilliant color to these compounds.

Producing an azo dye involves treating an aromatic amine with nitrous acid to give a diazonium ion intermediate. This process is called diazotization. The diazonium ion is an electron-deficient (electrophilic) intermediate, and an aromatic compound, suitably rich in electrons (nucleophilic) will add to it. The most common species are aromatic amines and phenols. Both these types of compounds are usually more nucleophilic at a ring carbon than at either nitrogen or oxygen. The addition of the amine or the phenol to the diazonium ion is called the diazonium coupling reaction.

Azo dyes are both the largest and the most important group of synthetic dyes, in the formation of the azo linkage, many combinations of $ArNH_2$ and $Ar'NH_2$ (or $Ar'-OH$) are possible. These combinations give rise to dyes with a broad range of colors, encompassing yellows, oranges, reds, browns, and blues.

The azo dyes, the trimethyl dyes, and mauve are all synthesized from anilines (aniline, o-, m-, and p-toluidine) and aromatic substances (benzene, naphthalene, anthracene). All these substances can be found in coal tar, a crude material that is obtained by distilling coal. Perkin's discovery led to a multimillion-dollar industry based on coal tar, a material that was once widely regarded as a foul-smelling nuisance. Today these same materials can be recovered from crude oil or petroleum as by-products in the refining of gasoline. Although we no longer use coal tar, many of the dyes are still widely used.

Vocabulary

1. macerate   ['mæsəreit]   vt. 浸解
2. mordant   ['mɔːdənt]   n. 媒染剂
3. madder   ['mædə]   n. 茜草染料
4. alizarin   [ə'lizərin]   n. 茜草红、茜素
5. henna   ['henə]   n. 散沫花染料
6. indigo   ['indigəu]   n. 靛蓝
7. indican   ['indikæn]   n. 脲蓝母
8. indoxyl   [in'dɔksil]   n. 吲哚酚
9. denim   ['denim]   n. 斜纹布
10. quinine   [kwi'niːn]   n. 奎宁
11. toluidine   [tɔ'luːdiːn]   n. 甲苯胺
12. aniline   [ænilin]   n. 苯胺
13. mauve   [məuv]   n. 苯胺紫
14. pararosaniline [pærərəu'zænilin]   n. 副玫瑰红
15. malachite   ['mæləkait]   n. 孔雀石绿
16. azo   [æzəu]   n. 偶氮(类)
17. chromophore   ['krəuməfɔː]   n. 发色团
18. diazotization [daiæzətai'zeiʃən]   n. 重氮化(反应)

Phrases

1. diazonium ion   偶氮离子(盐)
2. homespun   家里纺的

3. electron-deficient (electrophilic)　亲电的
4. nucleophilic　亲核的

6.24　ADHESION

Theories of Adhesion

No one knows why adhesives join objects together although there are many theories, each with its supporters and detractors. It is at least agreed that adhesion is an interfacial phenomenon, and that wetting of substrates is essential to adhesion. Substantial loads may be carried across properly designed adhesive bonded joints, certain substrate materials require particular adhesives, and many substrates can produce stronger joints with special surface treatment procedures.

(1) Electrical Theory. The electrical theory presumes that the adhesive and substrates are like the plates of a capacitor that become charged due to the contact of two substances. The theory fails to predict the strong joints that result when a layer of water is frozen to join two blocks of ice, or when an epoxy adhesive is used to join two previously cured blocks of cast epoxy.

(2) Diffusion Theory. The diffusion theory calls for the penetration of the substrates by the adhesive prior to the solidification of the adhesive. The diffusion of a solvent into many porous plastics is readily visualized; however, diffusion of an adhesive into metal, glass, or glazed ceramic is difficult to accept.

(3) Adsorption Theory. The adsorption theory depends upon the concept of molecular forces, such as van der Waals forces, acting across the space between molecules in a material. The adhesive, being a liquid at some stage of the bonding process, supposedly displaces gases so as to fill the voids and make intimate contact with the substrates so that such forces can take effect.

(4) Rheological Theory. The rheological theory suggests that the re-

moval of weak boundary layers in plastics leaves the mechanical properties of the joint to be determined by the mechanical properties of the materials making up the joint and by local stresses[1]. In the absence of weak boundary layers, joint failure must be cohesive within the bondline or the substrates.

Each of these theories has contributed to the beginnings of a science of adhesion and has stimulated the gathering of a mass of experimental data. The works of Zisman, from a surface chemistry and wetting viewpoint, and Bikerman, for a theory of weak boundary layers, give insights into the origins of the science of adhesion. Schonhorn gives a detailed exposition of the various theories and more recent articles may be found in the general references of the bibliography.

The Question of the Ideal Adhesive

There is no ideal adhesive and it is doubtful that there will ever be such a product because of conflicting requirements. Many bonding requirements are for permanent structural adhesives, but many temporary or demountable adhesives, such as masking tapes, must be easily removed. For more than two decades, there have been hundreds of technical papers in which the terms "structural and nonstructural adhesives" have been used. Yet no one has any precise idea of the difference.

Low viscosity is required when the adhesive must penetrate small cracks or be used in very thin bondlines, but heavy bodied or thixotropic adhesives are required to fill gaps and resist sag in vertical applications. The terms thin and thick carry no precise definition. In capacitor adhesives $1.0\ \mu m$ is a thin bondline and 0.1 mm is a standard and 1.0 mm is a thick bondline, but in bonding highway dividers, 1.0 mm is a thin bondline and 10 mm is a thick one.

Many adhesives must have fast grab or tack, but must be easily repositionable for a controlled time prior to the tacky stage[2]. Adhesives must readily wet the bonding area, but not bleed out and contaminate areas not to be bonded. Heat-curing adhesives should cure at low temperatures, but have unlimited shelf stability at temperatures normally attained in transportation and storage. They should remain flexible at cryogenic temperatures, but not creep at elevated temperatures. Low, normal, and high temperature ranges vary. Adhesives should be easy to apply and require no special application or fixturing equipment. They should also be able to bond through oil or other contamination to all solid materials, and be available at low cost.

Special Properties

The adhesive formulator may build-in many properties by controlling the molecular weight of resins or the addition of various constituents.

With epoxy adhesives, even the user may combine epoxy resins of various molecular weights with dozens of amines, polyamines, anhydrides, boron trifluoride complexes or other curing agents. The resulting compositions may have viscosities varying from those suitable for fine needle hypodermic syringe use to highly viscous compounds applicable by a spatula. Potlife, useful application life, can vary from almost instant cure to several days at room temperature. Heat resistance from 70 to 200°C may be obtained by sacrificing flexibility and low-temperature impact strength. Pigments and thixotropic agents ground in epoxy resins are available so that color, gloss, and opacity may be controlled along with sag and gap filling properties. Proprietary formulations contain other fillers, such as quartz or metal powders to match the expansion properties of substrates, or to contribute special thermal or electrical properties. In one user-formulated epoxy adhesive, Seger and Sharpe, incorporated a thixotropic agent to prevent flow of adhesive into a porous ceramic and a silane coupling agent to ensure long-term bond durability.

Other adhesive materials are not as amenable to formulation changes as the epoxies. With some materials, such as cyanoacrylates, urethanes, phenolics, and silicones, slight changes in pH or moisture content of additives may cause shelf stability problems or even spontaneous gelling. Fillers containing active metal ions will cause similar problems with anaerobics. The expert formulator may take advantage of these limitations by including such destabilizing compounds in materials called primers or accelerators. These are often applied to substrates by the user prior to adhesive application. Upon later contact with the adhesive, accelerated curing takes place. An entirely new type of acrylic adhesive has recently evolved using this technique. The two-part system is applied, one part to each substrate; and within minutes after assembling the substrates, the adhesive cures to form a high strength joint.

In only one area, very high temperature adhesives for aerospace, have really significant sums of money been spent for research and development of sophisticated new adhesive systems[3]. The resulting adhesives include polyimides, aramid-imides, polybenzimidazoles, silicones, and a number of organic-inorganic hybrids. In this case, the necessary funds came from government agencies, as commercial use of large quantities of such adhesives is many years in the future.

Vocabulary

1. adhesive   [əd'hi:siv]   n. 胶粘剂
2. detractor   [de'træktə]   n. 减损物、毁损者
3. capacitor   [kə'pæsitə]   n. 电容器
4. visualize   ['vizjuəlaiz]   v. 形象化
5. bondline   [bɔndlain]   n. 接缝
6. void   [vɔid]   a. 空的、无人占用的、空闲的、缺乏的；
                                    n. 空间、空隙
7. intimate   ['intimeit]   a. 密切的、内部的、个人的
8. cohesive   [kou'hi:siv]   a. 粘合的、内聚的
9. bibliography [ˌbibli'ɔgrəfi]   n. 文献目录、目录学
10. demountable [di:'mauntəbl]   a. 可拆卸的

11. masking	['mɑːskiŋ]	n. 遮蔽、掩蔽、伪装
12. precise	[priˈsais]	a. 精确的、明确的、恰好的
13. thixotropic	[ˌθiksəˈtrɔpik]	a. 触变性的、摇溶(现象)的
14. sag	[sæg]	v. 下垂、倾斜、松垂
15. tacky	[ˈtæki]	a. 有些粘性的
16. repositionable	[riˈpɔzitineibl]	a. 保存的、复位的
17. cryogenic	[ˌkraiəˈdʒenik]	a. 低温学的、低温实验法的
18. hypodermic	[haipəˈdəmik]	a. 皮下注射用的
19. spatula	[ˈspætjulə]	n. 抹刀
20. opacity	[əˈpæsiti]	n. 不透明性
21. proprietary	[prəˈpraiətəri]	a. 所有的、私人拥有的
22. amenable	[əˈmiːnəbl]	a. 应服从的

Phrases

1. cast epoxy 浇铸体环氧
2. masking tape 遮蔽胶带

Notes

1. The rheological theory suggests ... by local stresses. 句中"to be determined"带有两个"by + n."构成的被动形式宾语。全句可译为：弱边界层理论假定塑料中弱边界层的转移导致连接处的机械性能由构成连接的材料的机械性能和局部应力来确定。

2. Many adhesives must have ... to the tacky stage. 全句可译为：许多胶粘剂必须具有快速变粘或牢固的特性，但在到达粘滞阶段之前，在所控时间内必须具有灵活可移位性。

3. In only one area, very high temperature adhesive ... new adhesive systems. 句中由于使用了"only"，"have"放在了"really significant of money"前面，此为倒装。

6.25 COATINGS

Introduction

Resistant or high-performance coatings or linings are specialty products used to give long-term protection under difficult corrosive conditions to industrial structures such as chemical plants, paper plants, atomic power plants, food plants, tanks, tank cars, barges, the interior of ships transporting chemicals or strongly corrosive products, etc.. This contrasts with paint, which is used for general appearance and shorter term protection against milder atmospheric conditions and industrial coatings, i.e., coatings that are applied to manufactured products, such as refrigerators, washing machines, bicycles, automobiles, etc., during the manufacturing process[1].

A high-performance coating or lining is one that goes beyond paint in adhesion, toughness, resistance to continuing exposure to industrial chemicals or food products, resistance to water or sea water, and to weather and high humidity. It is designed for difficult exposures and to prevent serious breakdown of an industrial structure even though there may be abrasion damage or holidays (gaps) and imperfections in the coating. It must be inert and noncontaminating to materials with which it is in contact; must be dense and have a minimum of absorption of contacting materials; have a high resistance to the transfer of chemicals through the coating such as various anions and cations; must be able to expand and contract with the surface over which it is applied; must maintain generally good appearance even though subject to severe weather or chemical conditions[2]. It must be and do these things for a sufficient period of time so as to be economically feasible and justify its price and applications costs.

The following definitions will be used in this article.

A resistant coating is a film of material applied to the exterior of structural steel, tank surfaces, conveyor lines, piping, process equipment or other surfaces which is subject to weathering, condensation, fumes, dusts, splash or spray, but is not necessarily subject to immersion in any liquid or chemical. The coating must prevent corrosion or disintegration of the structure by the environment.

A resistant lining is a film of material applied to the interior of pipe, tanks, containers or process equipment and is subject to direct contact and immersion in liquids, chemicals, or food products. As such, it must not only prevent disintegration of the structure by the contained product, but must also prevent contamination of it. In the case of a lining, preventing product contamination may be its most important function.

The function of a high-performance coating or lining is to separate two highly reactive materials, i.e., to prevent strongly corrosive industrial fumes or actual liquids, solids, or gases from contact with the reactive structure or underlying surface[3]. The concept that a coating is a very thin film separating two highly reactive materials brings out the vital importance of the coating and its need to be completely continuous in order to fulfill its function. Any imperfection in the coating becomes a focal point for corrosion and breakdown of the structure or a focal point for the contamination of the contained liquid. The relatively thin, continuous film concept takes on even greater importance when it is understood that these protective coatings are applied to very large areas of structural steel, tank surfaces and similar areas[4]. Many thousands of square meters may be involved in a single coating use.

Basic Methods to Protect the Surface

There are two basic methods by which coatings or linings protect the surface. The first is based on the principle of impermeability. The coating must have excellent adhesion and be inert to chemicals and impervious not only to air, oxygen, water, and carbon dioxide, but also to the passage of ions and to the passage of electrons or electricity. Such a coating prevents corrosion of steel by interrupting or providing a block to the normal processes necessary for the corrosion (Figure 6.11).

The second method uses anodically active or inhibitive pigments in the primer or in the coating to regulate corrosion. Corrosion is prevented not necessarily by the nature of the binder film, but by the use of pig-

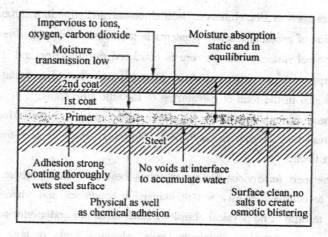

Figure 6.11 Impervious coating

ments which, when subject to moisture or humidity, ionize sufficiently to react with (Figure 6.12) or cathodically protect the steel or metal surface

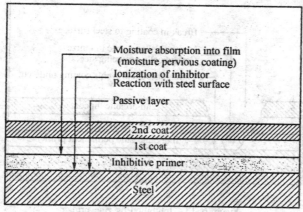

Figure 6.12 Inhibitive coating

(Figure 6.13) to maintain it in a passive state. This takes advantage of the water absorbed by ionizing the active pigment and forming a passive layer on the steel surface. The active pigments are metal salts of various chromates such as zinc chromate, lead chromate, and strontium chromate. The action of anodically active pigments is to cathodically protect the underlying steel surface. The pigment is powdered metallic zinc and it may be incorporated into either organic or inorganic vehicles. Zinc metal may be used also in the form of galvanizing or metallic zinc spray.

These methods in practice have both been proven to be effective. In many areas they overlap in usefulness. However, there is a distinction between them.

The inert impervious system performs best as a lining where it is subject to continual moist, wet or immersion conditions and where it is subject to little or no physical abrasion. Such areas are underwater sewage structures, water tanks, petroleum tanks, chemical tanks or tank cars, food tanks, wine or beer storage tanks, etc.

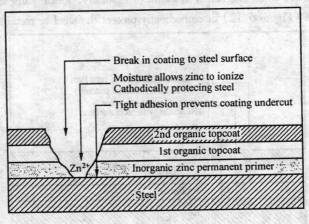

Figure 6.13 Inhibition by zinc primer

Inhibitive coatings perform best in areas where the coating is subject to weathering, atmospheric conditions, high humidities or chemical fumes. Such uses are generally for the exterior maintenance of structures or tanks where the coating may be subject to physical abrasion and coating damage. Under these conditions the inhibitive pigments and particularly the cathodic protection provided by zinc aid materially in the prevention of corrosion.

Vocabulary

1. performance   [pəˈfɔːməns]   n. (机械的)性能、特性
2. lining   [ˈlainiŋ]   n. 涂层、镀覆、覆盖
3. barge   [baːdʒ]   n. 驳船
4. abrasion   [əˈbreiʒən]   n. 磨损
5. condensation   [kɔndenˈseiʃən] n. 浓缩
6. disintegrate   [disintiˈgreiʃən] n. 瓦解
7. focal   [ˈfoukəl]   a. 焦点的
8. impermeability [ˌimpəːmjəˈbiliti] n. 不可渗透性
9. impervious   [imˈpəːvjəs]   a. 不可渗透的、不受干扰的
10. passage   [ˈpæsidʒ]   n. 交换、通过、过渡
11. primer   [ˈpraimə]   n. 底层涂料
12. static   [ˈstætik]   a. 静态的、静电的、固定的
13. osmotic   [ɔzˈmɔtik]   a. 渗透的
14. blister   [ˈblistə]   v. 起泡
15. anodically   [æˈnɔdikəli]   ad. 阳极地
16. cathodically [kæˈθɔdikəli]   ad. 阴极地
17. passive   [ˈpæsiv]   a. 钝性的、消极的
18. strontium   [ˈstrɔnʃiəm]   n. 锶
19. sewage   [ˈsjuːidʒ]   n. 污水、污物

Notes

1. This contrasts with paint ... during the manufacturing process. 全句可译为：这一点与作为一般表面涂层和在较温和介质环境中提供短期保护作用的涂料以及工业涂层不同。这一涂层是在生产过程中用于生产产品的物质。比如，生产电冰箱、洗衣机、自行车、汽车等。句中"contrasts with"带"paint"和"industrial coatings"两个宾语。

2. It must be ... or chemical conditions. 课文中出现了"subject to"和"be subject

to"两种用法,前者有"有待于……;须经……的"的意思;而后者是"以……为条件,受……的支配,易遭(经)受"的意思。
3. The function of a high-performance coating... or underlying surface. 句中,"is"后接两个"to do sth"的形式,表示目的;第二个"to"后带的是"prevent ... from"结构。
4. The relatively thin, continuous film concept ... surfaces and similar areas. 句中主句谓语为"takes on",为"具有、呈现出"之意。

7 化学化工文献介绍

随着互联网的出现和计算机的普及,通过 Internet 获取信息已成为主要的信息获取方式。目前的化学化工文献主要有印刷版和 Internet 上的电子版两类。

7.1 印刷版期刊

与化学相关的期刊可分为综合性化学期刊、化学工程和应用化学期刊和化学化工各专业期刊三大类。

7.1.1 综合性化学期刊

综合性化学期刊刊载的文献内容较广,涉及化学的各个方面。各国化学会的会志都属于此类。比较重要的综合性化学期刊有:

Journal of American Chemical Society(《美国化学会志》,简称 JACS 1879~):主要刊载化学领域各个方面的原始研究论文、简讯和书评。

Journal of Chemical Society(《英国化学会志》,简称 JCS, 1841~):1999 年起由原来的六辑改为四辑出版:Dalton Transactions(道

尔顿汇刊,无机化学);Perkin Transactions 1(Perkin 汇刊 1,有机化学);Perkin Transactions 2(Perkin 汇刊 2,物理有机化学);Chemical Communications(化学通讯)。

Bulletin de la Société Chimeque de France(《法国化学会通报》,1858～):刊登会员的研究论文、重要综述、书评和国内外其它学会的会议报告等。

Chemische Berichte(《化学学报》,1868～)原名 Berichte der Deutschen Chemischen Gesellschaft(《德国化学会志》),由德国化学家学会编辑。

Pure & Applied Chemistry(《纯粹与应用化学》,1960～):国际纯粹与应用化学联合会(IUPAC)的机关刊物,收载该会及其分支机构各种会议上提出的报告、论文和特邀讲演,也包括该会所属命名、符号及标准分析程序等委员会所做的重要建议。

Acta Chemica Scandinavica(《斯堪的纳维亚化学学报》,1947～):由丹麦、芬兰、挪威、瑞典四国化学会联合编辑出版,1974年起分为二辑:A 辑:物理化学与无机化学;B 辑:有机化学与生物化学。

Известия Академии Наук СССР:Серия Химическая(《俄罗斯(苏联)科学院通报:化学类》,1936～)和 Журнал общеи химии(《普通化学杂志》,有英文全译本,刊名为:Russian Journal of General Chemistry)。这两种综合性化学杂志由前苏联和俄罗斯科学院出版。

Bulletin of the Chemical Society of Japan(《日本化学会通报》,1926～)和《日本化学会志》(1972～),由日本化学会出版,前者用英文,后者用日文出版。

《化学学报》(1933～)原名《中国化学会志》,是中国化学会主办的化学科学综合性的学术性期刊。

《化学通报》(1934～)原名《化学》,是中国化学会编辑的另一种综合性学术期刊。

《高等学校化学学报》(1964~),原名《高等学校自然科学学报:化学化工版》,现在由中华人民共和国教育部主办,该学报编辑部编辑出版,月刊。

7.1.2 化学工程和应用化学期刊

化学工程和应用化学期刊的主要种类有:

A.I.Ch.E.Journal(《美国化学工程师学会志》,1955~)和 Chemical Engineering Progress(《化学工程进展》,1908~):由美国化学工程师协会(A.I.Ch.E)编辑出版。前者是协会的机关刊物,主要论述化工方面重要的基础研究、开发性研究的论文和述评;后者是刊载化工领域最新发展的月刊。

Industrial & Engineering Chemistry(《工业与工程化学》,简称 I.E.C.,1909~1970),由美国化学会(ACS)出版,主要刊载化学工业和化学工程的最新研究和发展趋势的原始论文与述评。Chemical & Engineering News(《化学与工程新闻》,1923~)是 ACS 出版的综合报导美国与世界化学和化工各方面情报消息的周刊,内容包括美国化工团体动态、各国化工生产与管理技术、新颖化学产品和仪器设备等;ACS 出版的另一种期刊是 Journal of Chemical & Engineering Data(《化学与工程资料杂志》,1956~)主要介绍化学化工领域重要的技术数据。

Transaction of the Institution of Chemical Engineering(《化学工程师协会汇刊》,1923~):由英国化学工程师协会出版,专门刊载化工基本原理的最新研究和实验成果。1987 年改名为 Chemical Engineering Research & Design(《化学工程研究与设计》)。

Angewandte Chemie(《应用化学》,1888~)由德国化学家学会出版。主要刊载理论与应用化学的新发展、化学分析与试验研究及化学工艺等方面的原始论文。1962 年增出英文版:Angewandte Chemie (International edition in English)。

Journal of Chemical Technology & Biotechnology(《化学工艺学与生物技术杂志》,1951~):由英国化学工业学会编辑出版。原名

是 Journal of Applied Chemistry，主要刊载化学工艺与生物技术领域中的研究论文和技术成就。

《化学工学》(《化学工程学》,1936~)：由日本化学工学协会出版，主要介绍化学工程和化工机械设备。1975年分出《化学工学论文集》单独出版；日本化学工学协会出版的英文刊物为 Journal of Chemical Engineering of Japan，正文后附有《化学工学论文集》相应各期刊载文章的英文摘要。

Журнал прикладной химии(《应用化学杂志》,1928~)和 Химическая прамышленность(《化学工业》,1924~)是最常用的应用化学和化工方面的期刊。前者由前苏联科学院出版,后者由前苏联化学工业部出版。

中国化工学会编辑的《化工学报》(1951~)主要刊登代表我国化工技术基础研究和应用研究方面创造性成果的学术论文。该刊另出英文版,选登中文版的部分内容。《应用化学》(1983~)主要刊载化学学科各领域中在应用基础研究方面的论文,由中国科学院长春应用化学研究所编辑。

7.1.3 化学化工各专业期刊

(1) 无机化学、有机化学方面的期刊

Inorganic Chemistry (《无机化学》,1962~)(美)

Journal of Inorganic & Nuclear Chemistry (《无机与核化学杂志》,1955~)

Журнал неорганической химии(《无机化学杂志》,1956~)(前苏联、俄罗斯),英译版：Russian Journal of Inorganic Chemistry

《无机化学学报》,1985~,中国化学会主办。

Journal of Coordination Chemistry (《配位化学杂志》,1921~)(英)

Journal of Organic Chemistry(《有机化学杂志》,1936~)(美)

Жрунал органической химии(《有机化学杂志》,1965~)(前苏联、俄罗斯),英译版：Journal of Organic Chemistry of the USSR

Journal of Organometallic Chemistry (《有机金属化学杂志》,1963~)(瑞士)

Tetrahedron(《四面体》,1957~)(英)

Tetrahedron Letters(《四面体快报》,1959~)

Journal of Heterocyclic Chemistry (《杂环化学杂志》,1964~)(美)

Synthesis (《合成》,1969~)(德国)

Synthesis Communication (《合成通讯》,1971~)(美)

《有机化学》,1957~(中)

(2)分析化学方面的期刊

Analytical Chemistry (《分析化学》,1929~)(美)

Analytical Letters(《分析通讯》,1967~)(美)

Analytica Chimica Acta(《分析化学学报》,1947~)(荷兰)

The Analyst (《化验师》,1877~)(英)

Talanta (《塔兰塔》,1958~)(英)

Fresenius' Zeitschrift Für Analytische Chemie(《弗来生牛斯分析化学杂志》,1862~)(德国)

Журнал аналитической химии(《分析化学杂志》,1946~),(前苏联),英译版:Journal of Analytical Chemistry

《分析化学》,1952~(日)

ふんせき (《分析》,1975~)(日)

《分析化学》,1972~(中)

Заводская лаборатория(《工厂实验室》,1939~)(前苏联)

Journal of Chromatography (《色层学杂志》,1958~)(荷兰)

Journal of Chromatographic Science(《色谱科学杂志》,1963~)(美)

CRC Critical Review in Analytical Chemistry (《化学橡胶公司分析化学评论》,1970~)(美)

(3)物理化学方面的期刊

Journal of Physical Chemistry (《物理化学杂志》,1896~)(美)

Annual Review of Physical Chemistry (《物理化学年鉴》,1950~)(美)

Zeitschrift für Physikalische Chemie, Neue Folge(《物理化学杂志》,1954~)(德国)

Журнал физической химии(《物理化学杂志》,1930~)(前苏联),英译版:Russian Journal of Physical Chemistry

The Journal of Chemical Physics(《化学物理杂志》,1933~)(美)

Journal of Catalysis (《催化杂志》,1962~)(美)

《触媒》(《催化》,1946~)(日)

Journal of Electrochemical Society (《电化学学会志》,1902~)(美)

《物理化学学报》,1985~(中)

《催化学报》,1980~(中)

(4) 生物化学方面的期刊

The Journal of Biological Chemistry (《生物化学杂志》,1905~)(美)

The Biochemical Journal (《生物化学学报》,1906~)(英)

Biochemistry (《生物化学》,1962~)(美)

European Journal of Biochemistry (《欧洲生物化学杂志》,1967~)(德国)

Biochimica et Biophysica Acta (《生物化学与生物物理学报》,1947~)(荷兰)

Journal of Biochemistry (《生物化学杂志》,1922~)(日)

Биохимия(《生物化学》,1936~)(前苏联)

《生物化学与生物物理学报》,1958~(中)

(5) 高分子化学化工方面的期刊

Journal of Polymer Science(《高分子学报》,1946~)(美)

Journal of Macromolecular Science (《大分子科学杂志》,1967~)(美)

Polymer(《聚合物》,1960~)(英)

Macromolecular Chemistry and Physics (《大分子化学与物理学》,1945~)(瑞士)

《高分子論文集》,1944~(日)

Polymer Engineering & Science (《聚合物工程与科学》,1961~)(美)

Polymer Preprints(《聚合物预印本》,1960~)(美)

Journal of Applied Polymer Science (《应用聚合物科学杂志》,1956~)(美)

Modern Plastics (《现代塑料》,1925~)(美)

Kunststoff(《塑料》,1911~)(德国)

ズラスチックス(《塑料》,1950~)(日)

Plastics Engineering (《塑料工程》,1945~)(美)

Rubber Chemistry & Technology (《橡胶化学与工艺》,1928~)(美)

《高分子学报》(原名《高分子通讯》),1957~(中)

(6) 石油化学化工方面的期刊

Нефтехимия(《石油化学》,1961~)(前苏联)

《石油と石油化学》(《石油与石油化学》,1957~)(日)

Preprints: Division of Petroleum Chemistry of ACS(《美国化学会石油化学分会会议录预印本》,1956~)(美)

Hydrocarbon Processing (《烃加工》,1922~)(美)

Erdöl & Kohle, Erdgas, Petrochemie (《石油与煤炭、天然气、石油化学》,1948~)(德国)

《石油学会志》,1958~(日)

《石油化工》,1970~(中)

(7) 硅酸盐科学与工艺方面的期刊

World Cement (《世界水泥》,1970~)(英)

Цемент(《水泥》,1901~)(前苏联)

Cement & Concrete Research (《水泥与混凝土研究》,1971~)(英)

Zement-Kalk-Gips(《水泥-石灰-石膏》,1911~)(德国)

Journal of the American Ceramic Society (《美国陶瓷学会会志》, 1918~)(美)

American Ceramic Society Bulletin (《美国陶瓷学会通报》, 1922~)(美)

Transactions & Journal of the British Ceramic Society (《英国陶瓷学会汇刊》, 1901~)(英)

《窑業協會誌》(《窑业协会杂志》, 1891~)(日)

Glass Technology (《玻璃工艺学》, 1960~)(英)

Physics & Chemistry of Glass (《玻璃物理学与化学》, 1960~)(英)

Refractories Journal (《耐火材料杂志》, 1925~)(英)

《硅酸盐学报》, 1957~(中)

(8) 环境化学方面的期刊

Environment Science & Technology (《环境科学与技术》, 1967~)(美)

International Journal of Environment Analytical Chemistry (《国际环境分析化学杂志》, 1971~)(英)

Chemosphere (《化学圈》, 1972~)(英)

Journal Water Pollution Control Federation (《污水处理联合会会刊》, 1928~)(美)

Water & Waste Treatment (《水与废水处理》, 1948~)(英)

CRC Critical Reviews in Environment Control (《化学橡胶公司环境控制评论》, 1970~)(美)

Water Research (《水研究》, 1967~)(英)

Water Pollution Control (《水污染控制》, 1927~)(英)

《环境化学》, 1982~(中)

《环境科学学报》, 1981~(中)

利用检索工具检索到课题所需论文的出处以后,就要进一步查找刊载论文的期刊。可以利用全国性的、地区性的和专业系统的期刊联合目录和订购目录,以及某一情报单位、图书馆的馆藏目录和订购目录查找。国内常用的有《全国西文期刊联合目录》、《全国中文期刊联合目录》、《全国俄文期刊联合目录》、《全国日文期刊联合目录》以及各年的外文现期期刊征订目录等。

在查找期刊时,经常会遇到诸如从期刊的缩写刊名查全称,从非拉丁文期刊的音译刊名查找原文刊名及英译刊名等问题,要解决这些问题,需要使用查找期刊的工具书。常用的期刊工具书主要是:《乌利希国际期刊指南》(Ulrich's International Periodicals Directory)、《外国报刊目录》和《美国化学文摘资料来源索引》(The Chemical Abstracts Service Source Index, 简称 CASSI)。

7.2 印刷版文摘刊和索引刊

7.2.1 《化学文摘》

美国《化学文摘》(Chemical Abstracts,简称 CA)专门报导世界各国有关化学、化工方面的文献。摘录的内容有期刊文章、专利说明书、专题论文、技术报告和新书通报等。

CA 创刊于 1907 年。自 1972 年起 CA 为周刊,一年两卷,每卷 26 期,内容分为 80 类,单数期次刊载 1~34 类,包括生物化学(1~20)和有机化学(21~34)两大部分。双数期次刊载 35~80 类,包括大分子化学(35~46)、应用化学和化学工程(47~64)以及物理化学、无机化学和分析化学(65~80)三大部分。从 1997 年第 126 卷开始,每期都刊载 1~80 类的内容。

每期 CA 由文摘和索引两个部分组成。

1. 文摘(Abstracts)

文摘款目按类编排,首页有当期文摘分类目次表:

Biochemistry Sections

① Pharmacology
② Mammalian Hormones
③ Biochemical Genetics
④ Toxicology
⑤ Agrochemical Bioregulators
⑥ General Biochemistry

⑦ Enzymes
⑧ Radiation Biochemistry
⑨ Biochemical Methods
⑩ Microbial Biochemistry
⑪ Plant Biochemistry
⑫ Nonmammalian Biochemistry
⑬ Mammalian Biochemistry
⑭ Mammalian Pathological Biochemistry
⑮ Immunochemistry
⑯ Fermentation and Bioindustrial Chemistry
⑰ Food and Feed Chemistry
⑱ Animal Nutrition
⑲ Fertilizers, Soils, and Plant Nutrition
⑳ History, Education, and Documentation

Organic Chemistry Sections

㉑ General Organic Chemistry
㉒ Physical Organic Chemistry
㉓ Aliphatic Compounds
㉔ Alicyclic Compounds
㉕ Benzene, its Derivatives, and Condensed Benzenoid Compounds
㉖ Biomolecules and Their Synthetic Analogs
㉗ Heterocyclic Compounds (One Hetero Atom)
㉘ Heterocyclic Compounds (More Than One Hetero Atom)
㉙ Organometallic and Organometalloidal Compounds
㉚ Terpenes and Terpenoids
㉛ Alkaloids
㉜ Steroids
㉝ Carbohydrates
㉞ Amino, Acids, Peptides, and Proteins

Macromolecular Chemistry Sections

㉟ Chemistry of Synthetic High Polymers
㊱ Physical Properties of Synthetic High Polymers
㊲ Plastics Manufacture and Processing
㊳ Plastics Fabrication and Uses
㊴ Synthetic Elastomers and Natural Rubber
㊵ Textiles and Fibers
㊶ Dyes, Organic Pigments, Fluorescent Brighteners, and Photographic Sensitizers
㊷ Coatings, Inks, and Related Products
㊸ Cellulose, Lignin, Paper, and Other Wood Products
㊹ Industrial Carbohydrates
㊺ Industrial Organic Chemicals, Leather, Fats, and Waxes
㊻ Surface-Active Agents and Detergents

Applied Chemistry and Chemical Engineering Sections

㊼ Apparatus and Plant Equipment
㊽ Unit Operations and Processes
㊾ Industrial Inorganic Chemicals
㊿ Propellants and Explosives
(51) Fossil Fuels, Derivatives, and Related Products
(52) Electrochemical, Radiational, and Thermal Energy Technology
(53) Mineralogical and Geological Chemistry
(54) Extractive Metallurgy
(55) Ferrous Metals and Alloys
(56) Nonferrous Metals and Alloys
(57) Ceramics
(58) Cement, Concrete, and Related Building Materials
(59) Air Pollution and Industrial Hygiene
(60) Waste Treatment and Disposal
(61) Water
(62) Essential Oils and Cosmetics

(63) Pharmaceuticals

(64) Pharmaceutical Analysis

Physical, Inorganic, and Analytical Chemistry Sections

(65) General Physical Chemistry

(66) Surface Chemistry and Colloids

(67) Catalysis, Reaction Kinetics, and Inorganic Reaction Mechanisms

(68) Phase Equilibriums, Chemical Equilibriums, and Solutions

(69) Thermodynamics, Thermochemistry, and Thermal Properties

(70) Nuclear Phenomena

(71) Nuclear Technology

(72) Electrochemistry

(73) Optical, Electron, and Mass Spectroscopy and Other Related Properties

(74) Radiation Chemistry, Photochemistry, and Photographic and Other Reprographic Processes

(75) Crystallography and Liquid Crystals

(76) Electric Phenomena

(77) Magnetic Phenomena

(78) Inorganic Chemicals and Reactions

(79) Inorganic Analytical Chemistry

(80) Organic Analytical Chemistry

每一类文摘的编排顺序是:第一部分为期刊、技术报告和专题论文等,第二部分为新书通报,第三部分为专利。每部分之间用横线隔开。每卷第一期开头都详细说明各种文献文摘的标头,这里仅举例作简略介绍。

【例 1】 期刊文摘(Journal-Article Abstracts)

见 1978 年,Volume 89(第 89 卷),Number 6(第 6 期),Section 47(第 47 类),Page 93(第 93 页)。

89:① 45522a ② Plant tests point the way to better filter performance ③ Bonem, J. M. ④ (Plast. Technol. Div. Exxon Chem. Co., Baytown. Tex.). ⑤ Chem Eng. (N. Y.) ⑥ 1978, 85(4), ⑦ 107~109(Eng). ⑧ Equations are presented for detg., with plant test data, how to increase the capacity or decreasing the operating cost of a drum filter.

① 文摘号码;② 文章题目;③ 著者姓名;④ 著者所属机关单位和所在地;⑤ 文章所在期刊;⑥ 期刊年份和卷号(1978 年 85 卷第 4 期);⑦ 起讫页数和语种;⑧ 文摘。

文摘内容一般都较简单,并使用很多缩写字[例: detg. = determining, Chem. = Chemical, Eng. = engineering 等]。每期均有缩写词表。非英文文献、题目和文摘均为英译文。著者、期刊名称、单位和所在地采用音译。要知道期刊名称的英文译名可查阅"化学文摘资料来源索引"(Chemical Abstracts Service Source Index,简称 CASSI)。

【例2】 新书通报文摘(New Book Announcement Abstracts)

见 1978 年,Volume 89, Number 9, Page 103.

89: ① 66082x ② Advance Materials in Catalysis. ③ Burton, James J.; Garten, Robert L.; Editor ④ (Academic: New York, N. Y.). ⑤ 1977. ⑥ pp.329

① 文摘号码;② 书名;③ 著者或编者;④ 出版单位及出版地点;⑤ 出版年份;⑥ 全书页数

【例3】 专利文摘(Patent Abstracts)

见 1978 年, Volume 88, Number 24, Page 194

88: ① 173369x ② Sulfur Removal from Hot Gases Derived from Coal Gasification. ③ Sims, Anker Victor ④ (Thermo-Mist Co.) ⑤ Ger. Offen. ⑥ 2 632 064 ⑦ (Cl. C10K 1/20) ⑧ 19 Jan, 1978 ⑨ Appl. ⑩ 16 Jul. 1976 ⑪ 16pp. ⑫ Hot fuel gas to be cooled prior to desulfurization is contacted in a heat exchanger with fluidized Al_2O_3. The heat released to the Al_2O_3 used to heat the desulfurized gas for use in a power plant. Two heat exchangers are required in the operation. The desulfurization is done by a com. process, such as the amine process.

① 文摘号码;② 专利题目;③ 发明人姓名;④ 专利权受让人或企业单位;⑤ 专利的国别;⑥ 专利号码;⑦ 专利分类号(国际专利分类号,若是美国专利,先列出美国专利分类号,后列国际专利分类号);⑧ 专利颁发日期和发表日期;⑨ 专利申请号(有时无专利申请号,仍列 Appl.项,如本例);⑩ 专利申请日期;⑪ 说明书页数;⑫ 文摘。

2. 索引(Index)

CA 索引有期索引、卷索引和累积索引三种。

(1) 期索引(Issue Index)

从 1981 年 94 卷起,CA 每期的文摘后面有三种索引。

ⅰ 关键词索引(Keyword Index):关键词短语是从文献题目或摘要中选出来的,一篇文摘往往选用几条关键词短语来表示,使用本索引时,应按文题适当多选几个词作关键词进行检索。索引中关键词短语按英文字母顺序排列。利用该索引可从关键词检索文摘号码。

【例4】 检索题为 Jet fuel manufacture 一文的文摘号码,查 1974 年,Volume 81, Number 10,CA 刊后附录Ⅰ。

Keyword Index

① Fuel jet manuf hydroalkylation   ② 52136r

① Jet manuf hydroalkylation   ② 52136r

① Hydroalkylation arom hydrocarbon   ② 52136r

其中:① 关键词短语;② 文摘号码。

ⅱ 著者索引(Author Index):CA 刊后附录三是著者索引,利用该索引可从著者姓名检索文摘号码。该索引中姓名的排列是姓在前,名在后,以英文字母顺序排列,著者索引很简单,仅① 著者姓名和② 文摘号码二项(如:① Coulson CJ ② 117606f)。

ⅲ 专利索引(Patent Index):在 1980 年 93 卷以前的期索引中,有专利号索引(Numberical Patent Index)和专利对照(Patent Concordance)。从 1981 年 94 卷起,将这二个索引合并为专利索引。专利

化学化工文献介绍　353

索引的用途是：从已知的专利号查找专利的文摘号（如阅读文摘内容）、该专利在其它国家申请的等同专利和与 CA 首次摘录的专利在技术内容上有一定的改进并与原专利有联系的相关专利（Related Patent）。专利索引的内容包括文摘专利、等同专利和相关专利（用 Relalted 表示）。CA 用两个字母作为国家代号，如 CN 代表中国，JP 为日本、US 为美国，EP 欧洲专利局，WO 世界知识产权组织等。专利索引按国家代号的字母顺序排列，每个国家下面以专利流水号排列，格式如下：

【例 5】　专利索引（Patent Index）
① **JP（Japan）**
② 54/012643 A2　③ (56/0073446B4)
　　　　　　　　　④ [79 12643]　⑤ 92：**180658g**
　54/012643B4　⑥ See US 3986977A
　54/012653 A2　　[79 12653],90：11514 h
　　　　　　　　　⑦ AT 337140 B(Related)
　　　　　　　　　⑧ FR 2270206 A1(B1)
　　　　　　　　　JP 54/072654 A2(Related)
　　　　　　　　　　[79 72654],92：11515i
　　　　　　　　　JP 54/072655 A2(Related)
　　　　　　　　　　[79 72655],92：180658g
⑨ **NL 75/05204 A(Nonpriority)**
　　　　　　　　　US 4173268 A(Related)92：27346w
⑩　US 4173972 A(Continuation-in-part；Related)，
　　　　92：108691y
⑪WO 79/002741 A1（Designated states：
　　　BR,SE；Designated Regional states：
　　　EP(DE,GB)；Elected states：
　　　BR；Elected Regional states：
　　　EP(DE),Related）

说明：① CA 采用的专利国代号，JP 为日本国；② 日本昭和 54 年的公开特许公报(A2)；③ 日本特许公报(B4)公告号，如公开号和公告号相同，括号中只写 B4；④ CA 采用公元年份的专利号。昭和年代加 25 即公元；⑤ 黑体文摘号示为 CA 中首次出现；⑥ 与美国专利为相同专利，以美国专利为主，在本期中能查到；⑦ 此奥地利专利与日本专利有相关内容；⑧ 此法国专利与日本专利有相同关系；⑨ 此荷兰专利也为相同专利。括号中注明非优先权国家。黑体意义与⑤相同，⑩ 此美国专利与日本专利有接续相关内容；⑪ 此世界知识产权组织(WO)专利与日本专利有相关内容。注明 WO 中登记国有 BE(巴西)及 SE(瑞典)及欧洲专利组织中的 DE(德国)及 GB(英国)。

(2) 卷索引(Volume Index)

CA 的卷索引每卷单独出版，是检索当卷各期中全部文摘的索引，目前主要有 5 种：General Subject Index (缩写 GS，普通主题索引)，Chemical Substance Index (缩写 CS，化学物质索引)，Author Index (缩写 A，著者索引)，Formula Index(缩写 F，分子式索引)，Patent Index (缩写 P，专利索引)。专利索引在卷索引和期索引中的编排方式相同。

ⅰ 主题索引(Subject Index)：CA 从 1972 年 76 卷起将主题索引分为普通主题索引和化学物质索引两种。

a. 普通主题索引(General Subject Index)：本索引中的标题包括概念性主题和化学物质类(成分未定或不确定的物质，某一类物质)名称，不用某一个具体化学物质名称。索引标题按英文字母顺序排列。

【例6】

见 1974 年，Volume 81，General Suject Index (A - Z) Page 680 GS
① Fuels,　　② rocket,　　③ jet
　③ Manuf. of, from arom hydrocarbon by hydroalkylation,
　　④ P 52136r
① 索引标题；② 副标题，由著录标题中再分出来的标题；③

索引说明语,对索引标题内容略加补充。读索引条文,其词序需根据语法加以调整;④ 文摘号码。

b. 化学物质索引(Chemical Substance Index):整个索引按化学物质的英文名称的字序排列。

【例7】 见 1973 年, Volume 79, Chemical Substance Index (A—L) Page 386 CS.

① Benzene, methyl—③ [108 - 88 - 3]. ② preparation
　④ alkenes removal from, in petroleum reformates by
　　hydrogenation catalysts for, ⑤ P　83335u

① 索引标题;② 副标题;③ 化学物质登记号:列入该索引中的一切结构明确的化学物质都经化学文摘服务社(CAS)登记编号;④ 索引说明语;⑤ 文摘号码。

普通主题索引和化学物质索引是使用最多的索引,使用时往往联用到索引指南(Index Guide)和索引指南补编(Index Guide Supplement)。例如:一种化合物往往有几种名称,可以从索引指南中查得 CA 所选用的索引名称,如 Propylene 一词在化学物质索引中的标题中找不到,索引指南将指出"propylene see propene"。索引指南除了交叉参考外,还包括标题注和结构式图解等。

ⅱ 著者索引(Author Index)

【例8】 见 1976 年, Volume 85, Author Index (M—Z) Page 2237 A。

① Smith, Alan

② —;Larner, R.W.,

③ Conversion of cephalosporin compounds　④ P61406p

其中:① 著者姓名:姓在前,名在后,二者之间用逗号分开;② 长破折号代表首著者姓名,其后分为合著者姓名;③ 文献标题;④ 文摘号码(号码前 B - Book, P - Patent, R - Review.)

ⅲ 分子式索引(Formula Index):本索引是按分子式元素符号的字母序编排。在一个分子式的下面列出一个或几个化合物的名称,在每一名称后面注明登记号及文摘号,可以根据文摘号直接查阅文摘内容,也可根据索引中所提供的化合物的名称查阅有关主题索引或化学物质索引,所以本索引实际上是对化学物质索引或主题索引的补充。

【例9】 见1975年,Volume 83, Formula Index (A—C_{15})
① $C_8H_8O_2$
　　② Benzeneacetaldehyde, ③ 4-hydroxy – ④ [7339 – 87 – 9],
　　　⑤ 55614x
　　② Benzoic acid
　　⑥ —, 3 – methyl[99 – 04 – 7], 98575f

① 分子式。无机化合物的分子式中的元素符号是按字母顺序排列的,如 H_4N、H_3O_4S。有机化合物的分子式中的元素符号是将碳C排在第一位,氢H排在第二位,其它元素符号按字母顺序排列;② 母体化合物的名称;③ 取代基;④ CAS 登记号;⑤ 文摘号;⑥ 横线代表索引标题为母体

(3) 累积索引(Collective Index)

目前每10卷(5年)单独出版一次累积索引,已出至第13次累积索引(116~125卷),它包括的索引种类与卷索引相同,它是卷索引的累积本,其优点是检索效率高,省时间。但有些内容并未收录,如欲求全,最好查每卷的索引。

3. CA 使用方法例

CA 有多种索引,从哪一种着手,得看具体情况。如需要了解某一课题,知道某一作者在这方面曾有著作或经常发表这方面的文章,就查著者索引。知道化合物的分子式,先查分子式索引。较多使用的是主题卷索引,通常从近期查到早期。有累积索引的,可使用累积索引。新近出版的CA,未有卷索引,使用关键词主题索引。

例如:要了解一篇 1974 年发表的有关重油加氢精制的文章。重油英文是 heavy oil,加氢是 hydrogenation,精制是 purification,将这三个词分别作主题到主题卷索引中去查,但未能查到,考虑到它的内容属石油炼制,就到 petroleum refining 标题下检索。

① 查 1974 年 Vol.81, General Subject Index 1177GS 页
Petroleum refining
 Hydrogenation-refining
  of heavy oils, R65786c

② 查 1974 年 CA 81 卷载有该文摘号码(65786c)的期次(12期)Vol.81(12)1974,139 页

65786c Hydrodemetallization of heavy oil. Ueda, Shigeru; Yoshida, Yuji (Gov, Ind. Dev. Lab., Hokkaido, Japan) Kagaku Kogyo 1974,25(3), 379~83(Japan). A review with 37 refs. is presented on improved methods for heavy oil desulfurization by removing metals(e.g., V, Ni) from the system. The metals and S present in the asphaltenes, and model structures for these compds. are also discussed.

③ 查 Chemical Abstracts Service Source Index 934 页, Kagaku Kogyo 英文名称是 Chemical Industry

④ 查《英汉化学化工略语词典》附录 2,知该刊物的日文原名为"化学工业"

⑤ 查"化学工业"1974 年 25 卷第 3 期 379 页,见原文全文。

CA 目前已经成为化学化工学科一套完整的文献检索体系,图 7.1 给出了这一文献体系的概貌。图中 A、B、C…表示检索点,由检索点确定检索途径。图中的阿拉伯数字的含义是:1—普通主题索引标题;2—索引化合物名称;3—化学物质登记号;4—化学物质分子式;5—文摘号;6—文摘款目中有关文献出处的著录资料;7—刊登原始文献的出版物的馆藏情况资料(指美国图书馆馆藏情

况)。

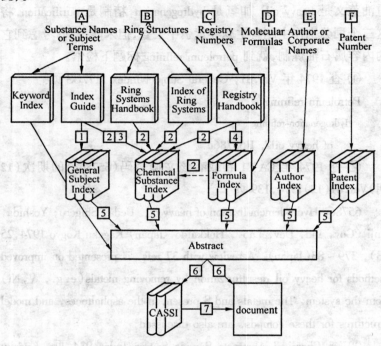

图 7.1 CA 检索体系示意图

7.2.2 《科学文摘》

英国《科学文摘》(Science Abstracts,简称 SA)创刊于 1898 年,由英国电气工程师学会(The Institution of Electrical Engineers,简称 IEE)和英国物理学会(The Physical Society of London)合作编辑出版。本刊主要报导基础理论专著、期刊论文、会议文献、科学报告、学位论文等。1903 年第 6 卷起改为 A、B 辑,A 辑为《物理文摘》(Physics Abstracts,简称 PA),B 辑为《电气与电子学文摘》(Electrical and Electronics Abstracts,简称 EEA)。1966 年起,英国电气工程师学会(IEE)与美国电气与电子学工程师学会(Institution of Electrical

& Electronics Engineers)、英国电子学与无线电工程师学会(Institution of Electronics & Radio Engineers)和国际自动控制联合会(International Federation of Automatic Control)等单位联合编辑出版 C 辑：《控制文摘》。1969 年起，C 辑改名为《计算机与控制文摘》(Computer and Control Abstracts,简称 CCA 月刊)。这三辑与化学工作者的关系都十分密切。例如，利用 A 辑可检索到物理化学、高分子化学、现代分析化学、材料科学等方面的文献，利用 C 辑可检索到计算机在化学化工领域中应用方面的文献。其索引系统包括著者索引、主题指南、主题索引、图书索引、会议索引、参考文献索引、专利索引、报告号码索引、团体著者索引等。

7.2.3 《工程索引》

美国《工程索引》(The Engineering Index,简称 Ei)创刊于 1884 年，是工程技术领域中的综合性检索工具，由美国工程索引公司(The Engineering Index in Corporation. USA)编辑出版。

《工程索引》是整个工程学科的多类型原始文献检索刊。目前 Ei 收录了 50 多个国家、15 种文字，3 500 种科技期刊和世界范围的会议录、论文集、学术专题报告、科技图书、年鉴和标准等出版物,每年报导量约 10 万篇。文献涉及 175 个学科,主要包括：机械、土木工程、环境工程、电工电子、结构学、材料科学、固体物理和超导、生物工程、能源、化工、光学、大气和水污染防治、危险废物处理、运输和安全等等。

Ei 报导的内容与化学化工、生物、环境相关的主题有化学工程、食品工程、环境工程、生物工程等,而且它重点收录的电子、电气和控制工程等方面的文献也涉及在化学化工的应用。

Ei 的印刷版出版形式主要有：月刊、年刊和累积索引等。

一般文摘刊物常按类目排列文摘,而 Ei 是按主题字母顺序排列文摘,划分类目,即文摘条目不是按学科类目编排,而是直接排在各相应主题词和副题词下面。这是 Ei 在编制形式上与其它科技文摘最大的不同点。因此,根据《工程标题词表》(Subject Headings for Engineering, 简称 SHE)确定检索课题的主、副标题后,即可直接查阅 1992 年之前的文摘。从 1993 年起,主题词采用该公司出版的《Ei Thesaurus》(工程索引叙词表),目前使用它的第三版,约有 2 400 个主题词,但没有副题词。

Ei 的索引系统包括期索引和年度索引。

期索引有作者索引(Author Index)和主题索引(Subject Index)。主题索引是 1987 年开始设立的,它的索引标题除了《工程标题词表》中的主、副标题词外,还有非规范化的自由词,从而提高了主题途径检索的灵活性,并可利用其检索跨类的文献。从 1993 年起,主题索引按主题词(叙词和自由词)的字母顺序排列。

年度索引在年刊的最后几个分册中,其中有:

① 主题索引(Subject Index);② 著者索引(Author Index);③ 著者单位索引(Author Affiliation Index)(1988 年起取消); ④ 号码对照索引(Number Translation Index)(1987 年起取消);⑤ 工程出版物索引(Publications Indexed for Engineering, 简称 PIE)。

7.2.4 《科学引文索引》

美国《科学引文索引》(Science Citation Index, 简称 SCI)是美国科学情报研究所(Institute for Scientific Information, ISI)出版的一部世界著名的期刊文献检索工具。

SCI 收录全世界出版的数、理、化、农、林、医、生命科学、天文、地理、环境、材料、工程技术等自然科学各学科的核心期刊约 3 500 多种。ISI 通过严格的选刊标准和评估程序挑选刊源,每年略有增减,以做到能够全面覆盖全世界最重要和最有影响力的研究成果,即报道这些成果的文献能够大量地被其它文献引用。SCI 与其他检索工具通过主题或分类途径检索文献的常规做法不平,设置了"引文索引"(Citation Index)。即通过先期的文献被当前文献的引用,来说明文献之间的相关性及先期文献对当前文献的影响力。SCI 不仅作为一部文献检索工具使用,而且成为科研评价的一种依据。科研机构被 SCI 收录的论文总量,反映整个机构的科研,尤其是基础研究的水平;论文被 SCI 收录的数量及被引用次数,反映论文作者的研究能力与学术水平。SCI 中影响因子(Impact Factor)是指在某一年的引文统计数据中,某种期刊前两年发表论文的被引数量与该刊在同一时期内发表论文数量之比。

依据来源期刊数量和类型,SCI 可分为 SCI 和 SCI – E。SCI 是指来源期刊为 3 600 种左右的 SCI 印刷版和 SCI 光盘版。SCI – E 的全称是 SCI – Expanded,又称 SCI Search,是 SCI 的扩展库,有来源期刊 5 900 种左右,可通过国际联机和国际互联网进行检索。印刷版 SCI 分双月刊、年度累积本和多年度累积本三种形式,内容包括:引文索引,专利引文索引,来源索引,机构索引和轮排主题索引。

7.2.5 世界专利索引

世界专利索引(World Patent Index,简称 WPI)是由英国德温特出版公司(Derwent Publication Ltd.)编辑出版的。德温特是英国一家专门经营专利信息的出版公司。该公司于 1951 年开始经营出版业务。它收集报导 27 个国家和两个国际性专利组织的专刊文献。并以《世界专利索引周报》(WPI Gazette Weekly)、《专利文摘周报》(Alerting Abstracts Bulletins)以及各类分册的形式出版。

德温特的一系列出版物已成为查找世界各国专利文献的最权威、最系统的检索工具。其主要出版物可如下表示:

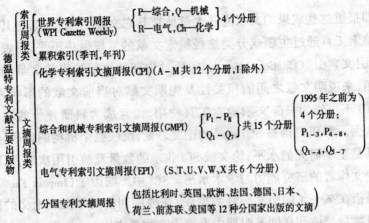

德温特除收录基本专利外(指相同专利中德温特最早收载的一篇),还报道与该基本专利内容相同的、在其它国家重复申请或内容有改进和增加、并在同一国家再申请的专利。这对检索者选用适合自己阅读的语种,以及了解某项专利的国际垄断范围,评价某项专利的实用价值是很有帮助的。

对于各国专利的查找,可通过分类途径检索实现。其具体步骤如图7.2所示。

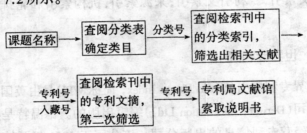

图 7.2 分类途径检索专利文献的步骤

其中确定分类类目是从检索课题内容出发检索专利文献的必经步骤。目前,我国和世界上绝大多数实行专利制度的国家都是依据和采用《国际专利分类法》确定分类类目的。

《国际专利分类法》由"世界知识产权组织"(World Intellectual

Property Organization,简称 WIPO)出版。其《国际专利分类表》由八个部(Section)组成,每部单独成一分册,共有八个分册;另外,第九分册是"使用指南"(Guide)。八个部的内容及标记符号是:A 部——人类生活必需(农、轻、医);B 部——作业、运输;C 部——化学、冶金;D 部——纺织、造纸;E 部——固定建筑物(建筑、采矿);F 部——机械工程(发动机和泵、一般工程、照明与加热、武器和爆破);G 部——物理(仪器、核子学);H 部——电学。

每个部都由很多不同等级的细分类目组成,它们的类级结构及标记符号如表 7.1 所示。

《世界专利索引》已形成完整配套的检索体系。本索引以出版物所包括的技术领域区分,分为四大部类:综合(P - General);机械(Q - Mechanical);电气(R - Electrical);化学(Ch - Chemical)。

表 7.1 《国际专利分类表》类级结构

类目等级	等级英文名称	标 记 符 号
部	Section	A—H(每一字母为一部)
分 部	Subsection	无标记符号
类	Class	部级标记符号加阿拉伯数字,例如:C01
小 类	Subclass	类级标记符号加英文辅音字母,例如:C01F
主 组	Main Group	小类级标记符号加阿拉伯数字,然后加一斜线,再加两个零,例如:C01C12/00
分 组	Subgroup	主组级标记符号中的两个零改为阿拉伯数字,例如:C01C1/02

以上四大部类各出版一个分册。在这 4 个分册中,Ch 分册是查找化学化工方面专利文献的主要分册。它与 CPI 配套。由于这

4个分册是根据德温特分类体系划分的,所以化学化工有关的类目在其它3个分册中也不少。

按文献加工深度或检索功能区分,可把德温特专利文献出版物分为三大系列:

(1)题录周报(WPI Gazette)

题录周报分为P、Q、R、Ch四个部类分别出版,即每一部类为一分册。

每分册均包括四个索引:① 专利权人索引(Patentee Index);② 国际专利分类索引(IPC Index);③ 入藏登记号索引(Accession Number Index);④ 专利号索引(Patent Number Index)。

(2)文摘刊系列

文摘刊系列现通称为"Alerting Abstracts Bulletin"。当与题录周刊配合检索时,必须注意卷、期号的一致性。这一系列出版物较多,其刊名如表7.2所示。

(3)累积索引系列

累积索引系列一般按索引种类(如:国际专利分类索引)单独编制,有年度索引和多年累积索引。

以前述的四个部类为基础建立了"德温特分类体系"(Derwent Classification)。其中化工部类(Ch)的分类体系称为"CPI分类表"。

CPI分类表是德温特分类表中Ch部类的类表,它是CPI分类途径检索专利的基本工具。Ch部类首先将化学化工技术分为12个大类:

  A Polymers, plastics

  B Pharmaceuticals

  C Agricultural chemicals

  D Food, detergents, water treatment and biotechnology

  E General chemicals

表7.2 文摘刊系列的刊名及分册名表

总刊名：Alerting Abstracts Bulletin	对应的题录周报分册
GENERAL & MECHANICAL PATENTS INDEX 共15个分册： 　P1：Agriculture, Food, Tobacco 　P2：Personal, Domestic 　P3：Health, Amusement 　P4：Separating, Mixing 　P5：Shaping Metal 　P6：Shaping Non-metal 　P7：Pressing, Printing 　P8：Optics, Photography, General	P
Q1：Vehicles in General 　Q2：Special Vehicles 　Q3：Conveying, Packing Storing 　Q4：Buildings, Construction 　Q5：Engines, Pumps 　Q6：Engineering Elements 　Q7：Lighting, Heating	Q
ELECTRICAL PATENTS INDEX (CLASSIFIED) 共6个分册： 　S：Instrumentation, Measuring & Testing 　T：Computing & Control 　U：Semiconductors & Electronic Circuitry 　V：Electronic Components 　W：Communications 　X：Electric Power Engineering **ELECTRICAL PATENTS INDEX (Country Bulletin)** 　S－T：Instrumentation Computing 　U－V：Electronic Components Circuitry 　W－X：Communications Electric Power	R
CHEMICAL PATENTS INDEX (CLASSIFIED) 　A－M 共12个分册（I 除外）	Ch

F   Textiles and paper-making
G   Printing, coating, photographic
H   Petroleum
J   Chemical engineering
K   Nucleonics, explosives and protection
L   Refractories, ceramics, cement and electro(in)organics
M   Metallurgy

每一大类下,根据各类所属技术的实际需要进行细分:有的是二级,有的是三级。表7.3 给出了类目的中、英文对照。

表7.3 德温特分类表 CH 部类类目表

分类号	类目英文名称及国际专利分类号	类目中译名
A	POLYMERS AND PLASTICS	聚合物、塑料分册
A1	Addition and Natural Polymers	加聚物和天然聚合物
A2	Condensation Polymers	缩聚物
A3	Processing: General Additives and Applications	加工:一般添加剂及应用
A4	Monomers and Condensants	单体和缩合物
A6	Additives and Compounding Agents	添加剂和配合剂
A8	Applications – Part One	应用——第一部分
A81	Adhesives and binders	胶粘剂
A82	Coatings, impregnations, polishes (excl. textile finishing)	涂复、浸渗、上光材料(纺织品除外)
A83	Clothing, footwear	衣服、鞋
A84	Household and office fittings	家用和办公室设备

续表 7.3 德温特分类表 CH 部类类目表

分类号	类目英文名称及国际专利分类号	类目中译名
A85	Electrical applications	电气应用
A86	Fancy goods, games, sports, toys	精制工艺品、文娱用品、体育用品，玩具
A87	Textile auxiliaries	纺织用辅助设备
A88	Mechanical engineering and tools	机械工程和工具
A89	Photographic, laboratory equipment, optical	照相、实验室设备、光学应用
A9	Applications—Part Two	应用第二部分
A91	Ion-exchange resins, polyelectrolytes	离子交换树脂、聚合电解质
A92	Packaging and containers	包装和容器
A93	Roads, building construction, flooring	道路、建筑结构、地板
A94	Semi-finished materials	半成品材料
A95	Transport	运输应用
A96	Medical, Dental, Veterinary	内科、牙科、兽医方面的应用
A97	Miscellaneous goods not specified elsewhere	未具体规定应用范围的杂类
B	PHARMACEUTICALS	药物
C	AGRICULTURAL CHEMICALS	农业化学
D	FOOD, DETERGENTS, WATER TREATMENT AND BIOTECHNOLOGY	食品，洗涤剂，水处理和生物技术
D1	Food and Fermentation	食品和发酵
D2	Disinfectants and Detergents	消毒剂、去污剂
E	GENERAL CHEMICALS	一般化学品
E1	General Organic	一般有机物
E2	Dyestuffs	染料
E3	General inorganic	一般无机物

续表 7.3　德温特分类表 CH 部类类目表

分类号	类目英文名称及国际专利分类号	类目中译名
F	TEXTILES AND PAPER-MAKING	纺织、造纸、纤维
F1	Threads and fibres-natural or artificial; spinning(D01)①	天然、人造的丝和纤维及其纺织
F6	Chemical-type treatment of textiles (D06blmpq)	织物化学处理
F7	Other textile applications (D06cfghjlm)	其它织物应用
F8	Flexible sheet materials	软薄材料——由纤维夹层的聚合物组成
G	PRINTING, COATING, PHOTOGRAPHIC	印刷、涂层、照相
G1	Inorganic pigments and non-fibrous fillers(C09c)	无机颜料和不含纤维的体质颜料
H	PETROLEUM	石油、燃料
H1	Obtaining crude oil and natural gas (C10g, E21b)	原油和天然气的勘探和钻采
H2	Unit operations (C10g)	单元操作
H3	Transportation and storage	运输和储存
H4	Petroleum processing (C10g)	石油加工
H5	Refinery engineering	精制工程
H6	Gaseous and liquid fuels (C1011)	气液燃料
H7	Lubricants and lubrication (C10m)	润滑剂和润滑技术
H8	Petroleum products, other than fuels and lubricants(C10m)	除燃料、润滑剂以外的其它石油产品
H9	Fuel products, not of petroleum origin	非石油来源的燃料产品
J	CHEMICAL ENGINEERING	化学工程

续表7.3 德温特分类表 CH 部类类目表

分类号	类目英文名称及国际专利分类号	类目中译名
J1	Separation (B01d, B03, B04, B07b)	分离
J2	Mixing, crushing, spraying (B01f B05bc)	混合、破碎、喷雾
J3	Electrochemical processes and electrophoresis (B01b)	电化学工艺和电泳
J4	Chemical/physical processes/apparatus (B01j1)	化学/物理过程和设备
J5	Boiling and boiling apparatus (B01b)	蒸煮和蒸煮设备
J6	Storing or distributing gases or liquids (F17)	气、液储存和分布
J7	Refrigeration; ice; gas liquefaction/solidification (F25)	冷冻；结冰；气体液化和固化
J8	Heat transfer and drying (F26, F28)	热传递和干燥
J9	Furnaces, kilns, ovens, retorts (F27)	熔炉、烘箱、窑、蒸馏釜
K	NUCLEONICS, EXPLOSIVES AND PROTECTION	原子能、爆炸物防护
L	REFRACTORIES, CERAMICS, CEMENT AND ELECTRO(IN) ORGANICS	耐火材料、陶瓷、水泥
L1	Glass (C03)	玻璃
L2	Refractories, ceramics, cement (C04)	耐火材料、陶瓷、水泥
L3	Electro-(in)organic, chemical features	材料的(无)有机电化学性能
M	METALLURGY	冶金

① 括号内为相应的国际专利分类号。

总之，WPI 中化学方面的专利检索可查阅《CPI 文摘快报，分类版》(CHEMICAL PATENTS INDEX, Alerting Abstracts Bulletin, CLASSIFIED);及《WPI 题录周报,Ch 分册》(WORLD PATENTS IN-

DEX，WPI Gazette，A – M Chemical)。

7.2.6 其它重要文献检索刊

① 日本《科学技术文献速报》：由日本科学技术信息中心编辑出版，是综合性和文摘性的检索工具，创刊于1958年。目前共出版12个分册，其中和化学化工关系最密切的有两个分册：化学·化学工业编（国内编）和化学·化学工业编（国外编）。

化学·化学工业编（国内编），代号 J，也称 J 分册。1964年创刊，半月刊。文摘的内容包括化学和化工各方面。有期索引和年度索引。它只收录日本国内化学化工文献，其面广，速度快。

化学·化学工业编（国外编），代号 C，也称 C 分册。1958年创刊，旬刊。它专门收录世界各国的重要期刊中最新发表的化学化工文献。

② 俄文《文摘杂志：化学》(Реферативный Журнал：Химия)：由前苏联和俄罗斯出版。俄文《文摘杂志》创刊于1953年。该杂志是多学科的庞大的文摘检索工具。它分为综合本，分册本和单卷本。综合本按学科划分，化学是综合本19，简称 РЖХ，它创刊于1953年，半月刊，它由19个分册组成（内容分成16大类），РЖХ是综合的化学化工文摘，内容较详细，有较完备的月度、年度索引。

③《化学题录》(Chemical Titles)：由美国化学会编辑出版，1960年创刊，双周刊。本刊是美国《化学文摘》出版前能迅速获取最新文献线索的检索刊。

④《化合物索引》(Index Chemicus)：原名《最新化学文摘和化合物索引》(Current Abstracts of Chemistry & Index Chemicus)，周刊，由美国科学情报研究所出版，1960年创刊，本刊选自有机、药物和医药化学方面世界上最重要的100多种学术期刊，所选者均是上述各个方面重要的学术论文。

⑤《中国化学化工文摘》：原名《中国化工文摘》，月刊，由中国化学化工信息中心编辑出版，是检索国内化学化工文献的最完善的检索刊。

7.3 Internet 上的电子期刊和图书

随着网络的发展,在国内外网络上已建立了化学化工电子期刊和图书的相关网站以及可检索数据库和虚拟图书馆。

7.3.1 美国《化学文摘》收录的电子期刊

目前,美国《化学文摘》中收录了 31 种与化学化工有关的网络版电子期刊(没有印刷版)。读者可根据各期刊的 URL 进行免费访问和阅读。其名称和对应 URL 分别为:

Alzheimer's Disease Review (http://www.coa.uky.edu/ADReview/);

Asahi Garasu Zaidan Josei Kenkyu Seika Hokoku(http://www.af-info.or.jp/JPN/subsidy/report.html);

Biochemistry On-Line(http://biochem.arach-net.com/essays/);

Chemical Educator(http://journals.springer-ny.com/chedr/);

Complexity International(http://www.csu.edu.au/ci/);

Earth Interactions(http://EarthInteractions.org/);

EJB Electronic Journal of Biotechnology(http://www.ejb.org/);

Electronic Journal of Geotechnical Engineering(http://geotech.civ-en.okstate.edu/ejge/); Electronic Journal of Theoretical Chemistry (http://ejtc.wiley.co.uk/);

Experimental Biology Online (http://link.springer.de/link/service/journals/00898/index.htm);

Frontiers in Bioscience(http://www.bioscience.org/);

HYLE-International Journal for Philosophy of Chemistry (http://www.uni-karlsruhe.de/philosophie/hyle.html);

Internet Journal of Chemistry(http://www.ijc.com/);

Internet Journal of Science-biological chemistry(http://www.netscijournal.com/);

Journal of Contemporary Neurology (http://mitpress.mit.edu/e-journals/CONE/);

Journal of Corrosion Science and Engineering(http://www.cp.umist.ac.uk/JCSE/);

Journal of Molecular Modeling(http://link.springer.de/link/service/journals/00894/index.htm);

Journal of Quantitative Trait Loci(http://probe.nalusda.gov:8000/otherdocs/jqtl/index.html);

Molecular Vision (http://www.molvis.org/molvis/); Molecules (MDPI)(http://www.mdpi.org/molecules/);

Molecules (springer) (http://link.springer.de/link/service/journals/00783);

Molecules Online (http://link.springer.de/link/service/journals/00783/);

MRS Internet Journal of Nitride Semiconductor Research (http://nsr.mij.mrs.org/);

Network Science(NetSci)(http://www.netsci.org/);

Neuroscience-Net(http://www.neuroscience.com/);

Nonlinear Science Today(http://www.springer-ny.com/nst/home.html);

Optics Express(http://epubs.osa.org/opticsexpress/);

Technical Tips Online(http://www.biomednet.com/db/tto);

Weeds World(http://nasc.nott.ac.uk:8300/home.html);

World Wide Web Journal of Biology(http://epress.com/w3jbio/);

Zphys-e.A:Hadrons and Nuclei(http://link.springer.de/link/service/journals/10025)。

7.3.2 Elsevier Science 电子期刊全文数据库

Elsevier Science 公司出版的期刊已在我国清华大学和上海交通大学建立镜像,教育网用户可以通过校园网阅览 1998 年以来 Elsevier 公司 1 100 余种电子期刊全文数据库服务,即 ScienceDirect

OnSite(SDOS)服务。网上浏览期刊全文(PDF 格式)需要使用 Acrobat Reader 软件。访问 SDOS 数据库采用校园网范围的 IP 地址控制使用权限,不需要账号和口令。

Elsevier Science 的 1 100 种全文电子期刊中有 220 种属 Chemistry and Chemical Engineering。

以"电化学噪声"(ECN)检索为例,在"Quick Serach"栏中输入检索词"ECN",在检索范围中选择"Abstract"和"just this journal"(Corrosion Science),按"Submit",开始检索,点击检索结果中的"Bibliographic page",可得到文章的文摘,点击"Article Full Text PDF"可以得到 PDF 格式的文章全文。

7.3.3 万方数据库数字化期刊子系统

北京万方数据股份有限公司的万方数据库数字化期刊子系统可以浏览 2 000 多种科技期刊的全文。

以检索化学期刊为例,首先在"数字化电子期刊库"中的"基础科学"栏选取"化学"选项,进入后选择期刊名就可进入期刊主页。读者可以阅读或下载一篇或多篇论文,文章为 PDF 格式。此外,万方数据库还提供了简单查询功能。点击"查询"按钮,进入检索界面,检索范围有刊名、标题、作者、作者单位、关键词及文摘,逻辑关系有"与"和"或"。点击"PDF 文件"可以得到论文原文。

7.3.4 中国期刊全文数据库

中国期刊全文数据库是目前世界最大的连续动态更新的中国期刊全文数据库,收录国内 6 600 种核心与专业特色中英文期刊中自 1994 年至今发表的全文,积累全文文献 500 多万篇,题录 1 500万余条,分九大专辑,126 个专题文献数据库,提供《中国期刊网专题全文数据库》、《中国学术期刊(光盘版)》和《中国学术期刊专题全文数据库光盘》三种产品形式。

覆盖理工 A(数理科学)、理工 B(化学化工能源与材料)、理工 C(工业技术)、农业、医药卫生、文史哲、经济政治与法律、教育与社会科学、电子技术与信息科学等学科。

读者可通过 CAJ 全文浏览器阅读。

1. 基本检索功能

(1) 检索范围

在题录、题录摘要、专题全文三个层次中选择检索,同时检索若干年内若干个专题数据库。

(2) 初级检索

导航检索、篇名检索、作者检索、关键词检索、机构检索、中文摘要检索、中文刊名检索、年检索、期检索及全文检索。

(3) 二次检索

对初级检索任何方式的检验结果,可以在此结果范围内用新的检索词进行逐次逼近检索。

2. 高级检索

在浏览器基本检索界面中,提供多个检索词检索项目的逻辑组合(与,或)检索。

3. 输出功能

(1) 显示题录

以文本方式显示检索结果的题录。在同一界面显示多篇文章的题录,包括文章的篇名、刊名、年、期等内容。

(2) 显示题录摘要

以文本方式显示文章的题录摘要。除题录外还包括作者、关键词、机构、中文摘要、引文等详细信息,无摘要的文章显示文章首页前 500 个字。

(3) 网上浏览全文

检索到结果后,通过 CajViewer 全文浏览器直接在网上浏览文章的全文原版内容。

(4) 下载全文

将检索到的文章的全文原版文件下载到本地计算机中,然后,脱机浏览文章的原版内容。

(5) 机上摘录功能

浏览文章原版时,可用鼠标直接从屏幕上抓取文章的内容,以文本方式自动临时存入剪贴板中,然后使用各种文字处理软件,如 Word、Wps、方正等,进行编辑后保存到磁盘中。

(6) 排序输出

根据方式不同,排序输出可分为按相关度输出和更新日期输出两种。

① 按相关度输出:根据检索词与检索主题的密切程度(根据词频、词位等因素综合评价)进行排序,可大大提高检索结果的准确性,既保证了查全率,又消除了检索垃圾的影响。

② 更新日期输出:最近更新的记录排列在最前面,这样可以快速检索到最近更新的文章。

4. 辅助功能

(1) 检索结果保存为文件

将题录摘要以超文本的形式保存到磁盘中。

(2) 打印结果

将题录摘要打印输出。

7.3.5 化学化工虚拟图书馆

通过 Internet 查询和检索化学化工图书馆的馆藏图书目录,可方便快捷获取图书信息资料。Internet 上的化学化工图书馆主要有美国国家医学图书馆、中国国家科学数字图书馆和化工"虚拟图书馆"。

美国国家医学图书馆(National Library of Medicine, NLM, http://www.nlm.nih.gov)的主页中设有分类目录。关于化学的有 ChemIDplus 数据库(含有 350 000 个化学品记录)、Hazardous Substances Databank Structures 数据库(含有以 2D 结构表示的 4 500 种物质)、NCI-3D 数据库(含有以 2D 和 3D 结构表示的 126 554 种物质)等。

中国国家科学数字图书馆中心门户网站(http://159.226.100.51/)由中国科学院文献情报中心承担建设,是国家科学数字图书馆建设之中的一个重要项目。国家科学数字图书馆中心门户网站设有化学学科信息门户、数理学科信息门户、生命科学学科信息门户及资源环境学科信息门户等。

化工"虚拟图书馆"(http://www.che.ufl.edu/www-CHE/outline.html)由美国佛罗里达大学建立,主要为用户提供化工、生物、环境、给排水、能源等方面的技术资料,同时还提供有关标准、专利以及化学制品的价格、制造商和相关服务信息,用户还可免费订阅"化学品交易信息"。该网站连接了许多著名化工站点,通过它可进一步搜寻有关化工信息。

7.4 Internet 上的化学数据库

化学数据库是化学工作者获取化学信息的重要工具。化学数据库包括化学文献库、化学结构数据库、化学物质的物性数据库及化学常用数据数据库等。

7.4.1 剑桥化学专业型数据库

1. 剑桥结构数据库(CSD)

剑桥结构数据库(Cambridge Structural Database, CSD)是剑桥晶体结构数据中心(Cambridge Crystallographic Data Centre, CCDC, http://www.ccdc.cam.ac.uk)建立的有机小分子和金属结构的数据库。它是全球最大的晶体结构的数字化数据库,库中数据均经过了 X 射线或中子衍射实验的分析。CSD 是同类科学数据库中应用最广泛和最负盛名的一个。

CSD 提供的结构数据是收费服务,其提供的联系电子信箱为:deposit@ccdc.cam.ac.uk

2. 化合物基本性质数据库(CS ChemFinder)

CS ChemFinder(http://chemfinder.cambridgesoft.com/)是 CambridgeSoft 公司在 Internet 上推出的化学数据库检索服务。ChemFinder WebServer 含有化合物的基本数据库。可通过化合物名称、相对分子质量、CAS 登录号和化学结构等进行查询。可以得到该化合物的基本性质,包括分子结构、相对分子质量、熔点、沸点、密度、水中溶解度等。

输入化合物英文名称进行 CS ChemFinder 检索,即可得到该化合物的基本物性数据。使用"Tools"中的"View Chemdraw Struct",还可以看到该物质的平面(2D)结构图;使用"View Chem3D Model",可以看到三维空间(3D)结构图。

此外,还可以得到生产厂家、包装说明及购买方法等方面的信息。

3. Brookhaven 蛋白质数据库

Brookhaven(http://www.rcsb.org)蛋白质数据库 PDB(protein data bank)是生物大分子的三维结构数据库。它包括了原子坐标、文献出处、一级序列及二级结构信息、晶体结构因子及 2D - NMR 实验数据等。所涉及的大分子包括蛋白质、DNA、RNA、病毒和碳水化合物等。

该蛋白质数据库还有一个专门的通信讨论组,讨论与该数据库有关的问题,电子信箱为:listserv@pdb.pdb.bnl.gov

4. Chemistry WebBook 物性数据库

Chemistry WebBook(http://webbook.nist.gov/chemistry)是美国国家标准与技术研究院(NIST)开发的基于 Web 的物性数据库,被认为是 Internet 上著名的免费化学数据库之一。

Chemistry WebBook 提供了多种检索途径,有分子式、物质名称、CAS 登录号、作者、结构或子结构检索等。如输入苯分子的分子式(C_6H_6),则可以查到其 16 种同分异物体及苯的其它化学名称;选择苯,则得到苯的各项物性数据及相关光谱图。谱图还可以根据需要进行局部放大显示。

5. 化学文摘服务及其数据库

化学文摘服务(Chemical Abstracts Service, CAS, http://www.cas.org)是美国化学会的一个部门。CAS 是世界上最大、最广泛和最全面的化学信息数据库的制作者,主要数据库是 CA 数据库、化学物质(CAS REGISTRY)、化学反应(CASREACT)、商业用化学品、一般系列化学品和专利申请的化合物等大型数据库。CAS 是注册收费资源网点。

6. 中国科学院科学数据库

中国科学院科学数据库建在中科院上海有机化学研究所,数据库地址为:http://202.127.145.69/database/sjk.htm。数据库包括:化学核心期刊论文数据库、化工产品数据库、中药与有效成分数据库、药物与天然产物数据库、化工产品性质价格和厂商数据库、红外数据库、化学配方数据库及化学综合数据库系统。

Internet 上还有许多化学数据库,如有害化学品数据库(Hazardous Chemical Database, http://ull.chemistry.uakron.edu/erd),可查阅 1 300 多种化学物质的毒性、保存及销毁方法,它可根据名称分子式及 CA 登记号等进行检索,即:

美国 Rutgers(http://nbserver.rutgers.edu:80)的核酸数据库;

纳米技术数据库(http://itri.loyola.edu/nanobase);

光谱学数据库(http://www.acdlabs.com);

氨基酸数据库(http://www.chemie.fu-berlin.de/chemistry/bis/amino-acids.html);

试剂库(http://www.lof.com/bp/bpintl.html);

生物分子学、生物基因顺序库(哈佛大学)(http://golgi.harvard.edu/sequences.html);

合成方法数据库(http://euch6f.chem.emory.edu/)等等。

7.4.2 专利型数据库

1. 化学专利大全(CPP)

从 CA 中可以查阅化学专利大全(Chemical Patents Plus, CPP, http://www.cas.org/casdb.html)专利数据库。CA 的专利数据库含有 29 个国家专利局和两个国际联合体的专利文档。查阅 CPP 必须注册,以获得 CAS 用户账号的身份验证与口令,但不需要交注册费和月收费等。使用 CPP 时,所有的搜索都是免费的,专利标题和摘录的显示也是免费的。

2. 美国专利数据库(USPTO Web Patent Databases)

美国专利数据库(http://www.uspto.gov/patft/index.html)是由美国专利和商标局(United State Patent and Trademark Office)提供的。专利库由文献库(Bibliographic Database)、文本全文库(Full-Text Database)组成,此外已经开始提供 300 点/in TIFF(Tagged Image File Format)格式的专利全文扫描图像 Patent Full-page Images。

每个库的检索都有三种检索方式,即快速检索、高级检索及专利号检索。

3. 欧洲专利局(EPO)的 esp@cenet

欧洲专利局(EPO)的 esp@cenet(http://ep.espacenet.com)向 Internet 用户提供免费的专利服务。专利全文扫描图像数据格式为 PDF,可以用 Acrobat Reader 阅读器来浏览。

4. 中国专利文摘数据库

中国专利文摘数据库(http://www.patent.com.cn)始建于 1998 年 5 月,它集专利检索、专利知识、专利法律法规、项目推广、高技术传播、广告服务等功能为一体。用户既能实时了解和中国专利相关的任何信息,又能方便快捷地查询专利的详细题录内容,以及下载专利全文资料。

在网站主页上选择"专利查询",输入主题词后,点击"搜索"按钮,即可显示查到的专利标题。双击所要选择的专利标题,就能得到具体的专利文摘内容。中国专利文摘数据库还提供了"高级查询方式"。如果想得到专利的全文,需要付费。

其他专利数据库还有:

IMB 知识产权信息网:http://www.patents.ibm.com;
加拿大专利数据库:http://patents1.ic.gc.ca/intro-e.html;
世界知识产权组织(WIPO)的 IPDL:http://ipdl.wipo.int;
日本专利数据库:http://www.ipdl.jpo-miti.go.jp/homepg-e.ipdl 等。

7.4.3 通用型数据库

1. 科学引文索引(SCI)

美国科技情报所(ISI:http://www.isinet.com)出版 SCI 的形式

除印刷版期刊和光盘版外，还包括联机数据库，现在还发行了互联网上 Web 版数据库(http://wos.isiglobalnet.com)。通过国际联机和国际互联网进行 SCI 检索，一般需由图书馆代理，用户支付系统费用。

2. 工程索引(Ei)

美国工程信息公司(Ei:http://www.ei.org)出版的 Ei 相关数据库主要有 Ei Comp endex Plus 数据库、Ei Page One 数据库和 Ei Village。

Ei Compendex Plus 数据库：Compendex 是 Computerized Engineering Index 的缩写。该数据库文字出版物即为《工程索引》，其标引方法从 1993 年开始由标题词法改为叙词法，一些大的联机系统，如 DIALOG、DATA-STAR、ESA-IRS、STN International、OCLC 等，都可以检索到该数据库。

Ei Page One 数据库：该库每年收集 32 万条文献的题录，这些文献来自世界范围内 5 400 种期刊、会议论文和技术报告，该数据库只收题录，无文摘，就收集范围而论，它是世界上最大的数据库之一，该数据库无文字出版物，其光盘出版物有两种：一种在 Windows 环境下运行(Ei Page One Windows)，另一种在 DOS 环境下运行(Ei Page One on DIALOG On Disc)，读者可通过刊名、自由词、作者姓名、作者单位来检索文献。所有被 Ei Compendex 和 Ei page one 数据库收录的文献原文在 Ei 可做有偿的服务，将原文用 Fax 或压缩图像方式传给读者。

Ei 工程信息村(Engineering Information Village)：随着 Internet 的普及和扩大，Ei 在 Internet 上建立了 Engineerin Information Village，以便能在网上检索到 Ei 和其他科技信息。

EI 新的简化检索界面分为两块检索模板——基本检索模板(Basic)和高级检索模板(Advanced)，点击界面上方的提示条，即可在两块检索模板之间进行切换。

用户在使用基本检索模板时，可以用单词和词组检索，但不能在检索窗口中使用位置算符(adj,w/n,near)，系统默认使用 NEAR

连接检索词。

使用 All Fields、Author、Author Affiliation 和 Serial Title 字段检索时,可使用 AND、OR 和 NOT 将输入到同一检索窗口的词或词组连接起来。

使用 All Fields、Title Words、Abstracts 和 Publisher 字段检索时,不能使用 AND、OR 和 NOT 连接检索词。检索时系统默认词根运算(如输入检索词 manager,将会检索到含有 management 或 managerial 的文献)。

可以用 AND、OR 和 NOT 对两个检索窗口中的检索式进行逻辑运算。

3. 中国科学引文数据库

中国科学引文数据库是由中国科学院和国家自然科学基金委共同资助,中国科学院文献情报中心承建的。中国科学引文数据库是一个集多种功能为一体的综合性文献数据库,收录了中国出版的数、理、化、天、地、生、农林、医药、卫生和工程技术领域的 1 000 余种中英文期刊。目前已积累自 1989 年以来来源数据 71 万余条,引文数据有 222 万余条。1991~1994 年数据约 13 万条,其中引文数据有 45 万条,提供免费查询服务。

中国科学引文数据库可查询论著被引用情况(包括专著、期刊论文、会议文献、学位论文、专利文献等)、机构发文量、国家重点实验室和部门开放实验室发文量、科技期刊被引情况等,是进行科技文献检索、文献计量研究和科学活动定量分析评价的有力工具。中国科学引文数据库可检索核心期刊的发文情况和科技文献被核心期刊引用的情况;可检索多种文献类型(期刊论文、会议文献、学位论文、专利文献等)的被引用情况;可检索全部作者的发文情况;可检索全部作者所在机构的发文情况;可检索国家重点实验室和部门开放实验室的发文情况;用关键词为检索点,通过引文专题进行文献扩检。

8 科技英语写作及翻译

8.1 科技英语特点

要掌握好科技英语写作和翻译,提高用英语进行科技交流和科研工作的能力,首先必须对科技英语的特点有所了解。虽然科技英语在语言、词汇、语法上和日常英语有不少共同之处(在用词、造句、行文上强调简洁、明确和直截了当),但由于其在科技领域的长期使用,逐渐形成了自身的一些特点。

8.1.1 科技英语词汇特点

为了准确地反映自然界的客观规律,并进行探讨和研究,科技英语的用词要求词义明确专一,尽量避免词义模糊或一词多义的现象。例如:

speculate—consider　exceed—go beyond　respiration—breath　collide—run into one another　circulate—circle　supervise—watch over　mobile—movable　synthetic—man-made　aviation—flying　illuminate—light up　edible—eatable　decompose—go to pieces

了解并掌握科技英语的用词特点,有助于在科技英语写作时用词更为规范,表达更为地道和清楚。

8.1.2 科技英语语法特点

英语科技文章的句子结构较一般文章的句子结构更为复杂,且长句较多。这是因为长句更为周密细腻,包容量大,有利于表达复杂意思和更精确地揭示事物间的内在联系。例如:

Although high density and pure mullite materials have been obtained from small laboratory batches of Al_2O_3-SiO_2 gel, using hot pressing as a consolidation route, little attention has been paid in the literature to evaluate the parameters that control the overall processing of mullite gels.

再如:

The general mechanism for coating degradation and substrate corrosion is characterized by an increase in electrolyte uptake, development of microscopic porosity and an increase in ionic conductance in the coating followed by corrosion (an electrochemical process) at the coating/metal interface.

由此可见,要掌握科技英语的写作,必须具备较好的英语基础和造句能力。支离破碎的句子难以表达缜密的思想,也会妨碍信息的准确交流。

另外,从上面的例子中亦可看出,为比较客观地进行描述和讨论,避免主观武断,科技英语中的被动语态用得尤为广泛。

8.1.3 科技英语文体结构特点

(1) 描述要求具体、准确。

例如:Winds between 15 and 30 mph, when accompanied by snow and temperature between 10°F and 30°F, often create unstable slabs in avalanche-starting zones.

(2) 用词造句力求简洁、明了。

例如:A series of runs made under identical conditions often yielded different results.

(3) 运用图表、公式、符号、缩写词语等来替代和简化文字描述，使论述和说明更为直观和简洁。

(4) 较多地使用各类复合词结构。

例如：eight-cylinder engine, linear-expansion, metal-cutting machine, fine-grained steel, light-tight material, moderator-reflector.

(5) 科技英语写作依内容的不同往往有一些固定的格式和要求，如论文、说明书、实验报告、信函等。掌握一些通用的固定格式，对科技文章的写作有很大的帮助。

8.2 科技英语写作素材

进行学术论文或科研报告的写作时，很重要的一点是运用他人的工作和思想，即运用他人已发表的信息材料来支持自己的观点和判断。在选取素材时有三个基本技巧：引用(quoting)、意译(or 间接引用)(paraphrasing)和参考文献(documenting)。

写作时如果需要直接运用他人信息，逐字拷贝他人信息，即为引用；如果只需要原始信息的内涵意思而不是精确的原词汇，那么首先是试图吸收全部信息，然后，用自己的语言表达，即为意译；在引用和意译时，为使他人确信信息的可靠性，以告知他人引用信息源于何人、何处，须引出材料的出处，即为参考文献。

8.3 科技英语写作题材

8.3.1 技术描述

技术描述是科技英语写作中常用的方法。技术描述一般要求对某物体、装置、设备等的尺寸、形状、结构、材料及制作装配方法等各个细节加以说明。与带有情感和可激发人想像的文字描述不同，技术描述要求客观，简练、精确。

技术描述一般分为：构造描述(Structure Description)、物体和物质描述(Objects and Substance Description)、实验解说描述(Experimental and Explanatory Description)和过程描述(Process Description)。

(1) 物体和物质描述

一般是针对某事物或某物质及特性进行描述。其内容包括物体的物理特性和物质的化学特性。此类描述的步骤是：
① 定义；② 物理或化学特性；③ 主要用途（不一定均有用途）。

例如：Nitric acid

Nitric acid is a colorless, fuming liquid with a boiling point of 80℃. It used to be called "aqua fortis." It has the chemical formula HNO_3. Nitric acid is a powerful oxidizing agent. It attacks most metals, producing fumes of nitrogen dioxide. Its low boiling point indicates that it is highly volatile. The reaction of nitric acid with organic substances produces important compounds such as TNT and celluloid. It is also widely used in the fertilizer industry.

(2) 实验描述

实验描述主要是说明某一实验的对象、器具以及实验的步骤。在实验描述中，常用一般现在时态，也可以用一般过去时，但在同一篇描述中，不能将两个时态混合使用。此外，句子的人称主语应始终保持一致。一般形式可以是祈使、被动、第一人称单数、复数、第三人称复数。

(3) 过程描述

过程描述主要是解释某事物的产生过程、工作原理和从事某事的步骤或机械运行程序等。过程描述是科技英语写作的重要一环。过程描述的结构是：① 引言；② 构成整个过程的各主要步骤及各步骤的解释；③ 结论（或结果）。

引言可以是正式定义，也可以指出某一过程产生的时间、地点、原因或实行者。必要时，引言中也可提出实现某一过程的特定时间、特定条件（如温度、湿度、无尘、通风等）或实现某一过程的必要准备等。

各个步骤的描述和解释一般均按时间顺序列出。在描述中，常用到 first, second, finally 等表示先后的时间接续词。例如：

Sulphur Extraction by the Frasch Process

In some parts of the world sulphur deposits lie too deep to be mined in the ordinary way. However, in about 1900 an American engineer called Heman Frasch developed a process for the extraction of this deep-lying sulphur. The Frasch process depends on the fact that the melting point of sulphur is only a little above the boiling point of water. The process consists of three basic operations. First, large amounts of water are super-heated, in other words, the water is heated under pressure to above its normal boiling point. Secondly, this super-heated water is pumped down the well so that it melts the sulphur. Finally, the molten sulphur is pumped to the surface.

上述关于提取硫磺的描述可分为两部分：第一部分简要介绍深层矿藏和提取方法的历史，第二部分描述主要开采过程。句子与句子之间的衔接结构既能帮助作者展开描述，也利于读者清楚地了解描述的内容。

8.3.2 技术报告

技术报告是科技英语写作中的重要部分。可根据报告内容和目的分为实地调查报告(Field Report)、进度报告(Progress Report)、实验报告(Laboratory Report)、测试报告(Test Report)等。技术报告主要提供进展情况，调查、实验或测试的结果。报告可以用书信形式书写，也可以用报告形式书写。

实地调查报告是根据现场调查或对事物的实际考察，了解具体情况之后所写的报告。其内容可以是关于某产品的试制或使用情况、某种方法的采用情况、某事物的状况以及提出的建议等。

进度报告是报导某项工作或工程的进展情况和进一步的计划。主要叙述工作背景、至报告时完成的情况、此后的任务、可能出现的问题和解决的办法等。

实验报告一般包括在实验室验证某一理论的实验报告书和对某一仪器设备检测结果的报告书。实验报告一般含有下列几个项目:标题、实验目的、理论根据、实验方法、结论和建议、参考资料等。

测试报告类似于实验报告,主要内容是对某一仪器设备或工具的测试结果。测试报告一般包括:目的、测试对象、测试方法、结果、结论等。

8.3.3 摘要

摘要是论文、学术报告、技术文章等的简短叙述。通过摘要,可以很快地了解文章的主要内容,判断文章的性质并决定是否有必要阅读此文。因此,摘要本身也是科技文章中重要的一部分。

根据目的,摘要一般分为说明性摘要(Descriptive Abstract)和资料性摘要(Informational Abstract)。说明性摘要只说明文章的主题,不介绍文章的内容(问题、方法、结果等),故不能替代文章。通常所说的摘要(Abstracts)多指说明性摘要,放于正文之前,篇幅较短,一般不超过 200 个单词。资料性摘要是整篇文章主要内容的简述,除了点明文章的主题外,还介绍文章的目的、研究的问题、采用的方法、主要论点、结论以及建议采取的措施等,包括了文章的主要思想和数据,因而可以作为文章的替代品。通常将此类摘要称为 Summary,其篇幅较 Descriptive Abstracts 稍长,其长度约为文章的 1/10 或更少一点。摘要主要包括三个部分:

① 目的;② 内容;③ 结论或建议。

摘要的开头一般可用下列句型:

The author /writer/paper/purpose/of this article ···

The paper /essay/article describes/explores ···

The author investigates /presents ···

An investigation was designed to ···

结尾的常用句型有:

The results suggest /show that ···

It is suggested /recommended that ···

The paper concludes that ···

摘要中包括的主要内容有：

a. The problem to be solved or the process/ mechanism/phenomena to be described;

b. The scope of your work;

c. The significant findings or results;

d. Any major conclusions;

e. Any major recommendations.

Two Model Abstracts:

Example one

Descriptive Abstracts:

Chemical modification of the alkyl side-chain of cardanol can lead to the formation of polyfunctional compounds. Cardanol was reacted using maleic anhydride as a dienophile under various exprimental conditions with products being obtained in up to 70% yield. IR, NMR and mass spectroscopy techniques have been used for preparing water-soluble binders and modified alkyd resins. The properties of the coatings so formed from these binders such as flexibility, hardness and resistance to water and chemicals were found to be superior to those of coatings based on conventional cardanol media.

Informational Abstracts:

Chemical modification of the alkyl side-chain in the meta position of cardanol allows the formation of products with a potential industrial value.

Polyfunctional compounds were obtained in substantial yields by the reaction of cardanol with maleic anhydride as a dienophile. Spectroscopic studies indicate the formation of a reaction product which can easily be converted into water-soluble binders. The coatings obtained from these binders exhibit better physical and chemical properties than conventional [reference number] coatings based on cardanol media.

The reaction product of cardanol and maleic anhydrides is useful in the modification of conventional alkyd resins by replacing part of the phthalic anhydride. Air-drying products with better film properties are thus obtained.

Example Two

Descriptive Abstracts:

Studies of polyester/melamine films using ATR FT-IR spectroscopy indicate that the melamine distribution across the films is uniform unless the hydroxyl content is low and the cure temperatures are relatively high. With polyester resins having an OH number as low as 6, curing at 250℃ with 30 w/w% melamine gives rise to melamine enrichment at the film-air interface. Enrichment is believed to result from melamine self-condensation near the interface. Although the distribution is a complex function and depends upon the amount of acid catalyst, the hydroxyl number of polyester, the film thickness and the reaction rate difference between the film-air and film-substrate interfaces, the primary factor leading to self-condensation is the loss of amine at the surface. Apparently, these factors play a key role in melamine self-condensation near the film-air interface and may control such properties as the cross-link density and the modification of properties such as flexibility and stain resistance.

Informational Abstracts:

ATR FT-IR spectroscopy has been utilized in this study to monitor the distribution of melamine across polyester/melamine films. Apparently the distribution of the melamine in a low OHV value polyester/melamine coating is such that excess melamine is found near the film-air interface, giving rise to improved stain resistance. This behavior is contrary to conventional polyester/melamine coatings which may have a uniform melamine distribution but poorer stain resistance properties.

Although the distribution of melamine may also depend upon other factors such curing conditions, concentration of acid catalyst and film thickness, the primary source of the non-uniform distribution is the volatility of the amine which is believed to result from a fast self-condensation reaction of melamine at the film-air interface. Although for the same coating formulations, improved stain resistance is usually attributed to the increased crosslink density, in this case stain resistance is largely affected and improved by the film-air-rich melamine content. Although we have recently illustrated how the surface tension of a substrate and other factors may govern the behavior of surfactants in latex films [reference numbers], to our knowledge the present is the first study which shows that by controlling the chemical reactions occuring during film formation, a non-uniform distribution of species across films may be obtained. This, in turn, may affect coating properties.

8.3.4 科技论文

科技论文和报告是很重要的交流方式和手段。撰写科技期刊的论文主要步骤包括：选题(Choosing a Subject)；收集资料(Gathering Information)；整理资料(Organizing Information)；实际撰写(Actual Writing)。写作时，论文格式和其他细节还要遵从不同期刊的具体要求。

(1) 选题(Choosing a Subject)

在选题时首先要明确的一点是你的思想和研究工作取得了理论或实际的进展，并对其他研究人员的工作是有帮助的。下面给出选择一个新而有创造性的主题时的有益的建议：

① Read widely in your major field;

② Have a wide aquaintance and communication with your instructors, colleagues and other professionals in your major field;

③ Use your imagination and be critical of whatever you read or discuss;

④ Pay attention to the impact that information from other fields may have on your field.

(2) 选择期刊(Selecting a Journal)

选择适于文章发表的期刊，要了解：

① What Journals publish material of the kind you have;

② Analyze the Journals you have chosen as possible targets.

选择期刊后，应按其要求写作，并注意以下两点：

① Analyze the style of the Journal you have chosen to see if there are any special references or prejudices in style;

② Analyze the physical format (subheads, footnotes and bibliographical forms, whether or not an abstract is used, types of illustrations, how numbers are written, and what abbreviations are used, preferred by the chosen Journal.

(3) 收集资料(Gathering Information)

在实际写作前,收集资料(包括收集自己工作领域中前人已完成的工作资料)是很重要的一环,应查阅大量文献来获取足够的信息。具体步骤是:

① The card catalogue;

② Reference works $\begin{cases} \text{a Reference books} \\ \text{b Encyclopedias} \\ \text{c The various almanacs,} \\ \quad \text{yearbooks and handbooks} \\ \text{d Atlases} \end{cases}$

③ Periodicals;

④ Computerized information retrieval.

(4) 整理资料(Organizing your Paper)

整理资料可以通过以下两个步骤完成:

① Making an outline;

② Planning a format.

(5) 写作(Writing your Paper)

一般讲,实际写作时要遵从下列步骤:

① Examining the outline and making any necessary changes;

② Writing the draft;

③ Revising your draft;

④ Editing your next-to-last draft.

8.4 科技英语翻译

学习专业英语不仅仅是学习英语专业词汇。在科技英语中,专业词汇仅占 20%,其余 80% 都是我们经常使用的词汇。科技英语的初学者往往碰到这样的难题:借助字典查出了一个句子中所有的专业词汇,所有的词意也都明白,但对整个句子的意思仍然模糊不清。这主要是由于没有掌握科技英语的语法特点和翻译方式。

科技英语主要应用于科技报告和论文的写作中,在表达上具有简洁、准确的特点,这就使科技英语的语法具有一定的特殊性。通过专业文献的阅读翻译来掌握科技英语语法的基本特点,可以使我们在以后的文献阅读和科技英语写作中收到事半功倍的效果,这也是专业英语教学的主要目的之一。

这里并不是对科技英语的语法进行具体系统地讲解,只是提供一些阅读和翻译的小技巧,以尽快适应专业英语的学习。

翻译的过程就是译者理解原文,并把这种理解恰当地传递给读者的过程。要想做到翻译准确而完整,就必须有一个衡量译文质量的尺度,即翻译标准。目前,在科技翻译界,"信"与"顺"是大家公认的一条十分重要的翻译标准。"信"是指准确、忠实,而"顺"则是指通达顺畅。标准的译文必须在含义上与原文贴切,在行文上阅读起来流畅。相比之下"信"对科技翻译尤为重要。因为科技书刊和文章的任务在于准确而系统地论述科技学术问题,本身就要求有高度的准确性。

8.4.1 一般翻译方法

(1) 增词法

由于英、汉两种语言的词汇在涵义、搭配、习惯等方面的差异都很大。因此翻译时,译者必须根据句法上、意义上或修辞上的需要增加一些原文中虽无其词,但内含其意的词,以便能更加忠实通顺地表达原文的思想内容。

例如:Steel and iron products are often coated lest they should rusts.

钢铁制品常常涂上油漆以免生锈。(原文 coated 本身含有"上油漆"之意,而汉语"涂上"没有这种意思,故汉译时需要加上"油漆"两个字意思才明白)

再如:An atom consists of a positively charged uncleus surrounded by electrons in orbit.

原译:原子由被轨道电子围绕的带正电的原子核构成。

改译:原子由带正电的原子核和在轨道中围绕其运动的电子组成。

分析:原译未作补充,造成逻辑上、概念上的失误。根据汉语习惯,只有两个以上的部分才能用"构成"这个词,故改译要好些。

(2) 重复法

重复法实际上也是一种增词法,其修辞意味很强,所增添的词都是上文出现过的词,在译文中之所以重复原文中关键的词,是因为要使译文更加顺畅。

例如:All bodies consists of molecules and these of atoms.

一切物体都是由分子组成的,<u>而分子</u>由原子组成。(英语中忽略了 consist,并用 these 代替 molecules,汉译时重复译出)

再如:Atmospheric pressure decreases with increase in altitude and so does the density of the atmosphere.

大气压力随高度增加<u>而降低</u>,大气密度也随高度增加<u>而降低</u>。

(3) 省略法

英汉两种语言在遣词造句、方式和章法结构方面存在很大差别。比如,有些词(冠词、介词、连词等)在英语中经常出现,但译成汉语时就需酌情加以删减。省略就是按汉语的修辞习惯。在不损害原文思想内容前提下,将英语中的某些词语或成分略去不译。

例如:When the pressure <u>gets</u> low, the boiling point becomes low.

气压低,沸点就低。(省略动词)

再如:The critical temperature is different <u>for</u> different kinds of steel.

不同种类的钢,临界温度各不相同。(省略介词)

再例如:The mechanical energy can be changed back into electrical energy by means of a generator <u>dynamo</u>.

机械能可以利用发电机再转变成电能。(省略意义上重复的词)

8.4.2 常见句型分析

被动句、否定句和强调句是科技英语中的常见句型。

科技英语中叙述推理的文章较多,因而强调指出行为发生者的情况较少。当不需要或不可能指出行为的发生者时,或者需要突出动作接受者时,往往应使用被动句。英译汉时,很多情况下可译成主动句,有时也可译成被动句或无主句。

例如:Several elements and compounds may be extracted directly from sea water.

有些元素和化合物可直接从海水中提取。(译成主动句)

再如:Up to now, sulphur dioxide has been regarded as one of the most serious of these pollutants.

到目前为止,二氧化硫一直被看作是这些污染物中最严重的一种。(译成被动句)

这种方法常用于一些固定的句型,如:

it is believed   有人认为……
it is suggested   有人建议……
it is said   有人说……
it is estimated   据估计……
it is well known   众所周知……
……

在科技英语翻译中,对否定句应特别注意。英语有全部否定、部分否定和双重否定之分,也有名为肯定,实为否定,反之亦然的情况。翻译时须辨别清楚。

① 英语中的 all, both, every, each, many, much, always, often 等与 not 在一起使用时,表示部分否定。

例如,All these metals are not good conductors.

这些金属并不都是良导体。

② 英语中有的句子形式上是肯定的,但却包含带否定意义的词或短语。翻译时应译成否定句。

例如:On freezing water becomes larger in volume instead of smaller.

水在结冰时体积增大而不是缩小。

再如：Pure metal have few useful properties.

纯金属没有多少有用的性能。

类似表示否定的词和短语还有 fail(没有)，but for(如果没有)，rather than(而不是)，free from(没有,免于)，short of(缺少)，far from(远非,一点也不)等。

强调句的强调方法主要有三种：一是用助动词 do 或 did；二是利用句型 it is (was)… that …；三是利用倒装语序。

第一种方式，常译成"确实"、"一定"、"必须"等，以加强语气；第二种方式，常译成"正"、"正是"；第三种方式，翻译时可以利用语气词"就"、"也"等来表示强调，但一般采用顺译法即可。

例如：Hence is the way to think of the actions taking place in the germanium.

这便是研究发生在锗中的那些反应的方法。

8.4.3 长句译法

在英语科技文献中，从科技内容的严密性、准确性和逻辑性出发，常常会使用一些长句子。在翻译英语长句时，就需要尽可能加以拆散，并按照汉语习惯，重新加以组织，力求在"明确"的基础上，做到简明顺畅。

简单的方法往往很有效，在科技英语的长句翻译中也是这样。对于比较复杂的句子，尤其是长的单句，"缩句"往往是行之有效的方法。例如：

Suitable polyols including aliphatic diols such as ethylene glycol, 1,4-butane diol, diethylene glycol, hexamethylene glycol and the like; aliphatic triols such as trimethylol methane, trimethylol propane, 1,2,6-hexane triol and the like are essential for the purpose of extending the chain length of the urethane prepolymer.

尽管该句子很长，但缩句后却很简单。

Polyols are essential for purpose.

主句译出以后，句子中其他各部分也清楚了，分别为 polyols

和 purpose 的定语。全句翻译为：

适当的多元醇，包括脂肪族二元醇，如：乙二醇，1.4-丁二醇，一缩二乙二醇，1,6-己二醇等，以及脂肪族三元醇，如：三羟甲基甲烷，三羟甲基丙烷，1,2,6-己三醇等，主要是用来扩展聚氨酯预聚体的链长。

在较复杂的科技英语句子中，并列关系非常容易混淆，为了确保句子的准确性，避免产生歧义，科技英语的并列关系之间往往会有暗示，比如两个并列关系都使用现代分词、过去分词或不定式为引导，或者在两个并列关系的后者中加入一个看似多余的词或词组，来明确并列成分。比如下列句子：

The adhesive is effective to bond thermoplastic copolyester to flexible substrates including natural and synthetic elastomers and leather and to nonflexible substrates, such as metal, at ambient temperature.

在这个句子中，nonflexible substrates 前面的 to 是多余的，但是这个介词的加入能够明确表明由 and 所引出的这个并列成分为 to 的介词宾语，是与前面 to 的介词宾语并列的，在本句的前半部分有两个 to，第一个 to 是不定式，第二个 to 才是介词，因此，nonflexible substrates 是与 flexible substrates 并列的。这样，该长句可译成：

这种胶粘剂能有效地将热塑性共聚酯与天然或合成弹性体和皮革等韧性衬底以及金属等非韧性衬底在室温下粘在一起。

词语搭配是在科技英语中另一个需要注意的问题。熟练掌握一些固定词组，可以在阅读中节省很多时间，比如：effect(s) of A on B 是一个非常常见的固定搭配。在科技英语中，A 往往是并列项的罗列，或者带有较长的定语。在阅读时跳过 of 以后的内容先找到 on B，可在总体上掌握句子的结构和内容，对句子各部分的成分和结构有更清楚的认识。

非固定的词语搭配则需要对词性和语法有更深刻的认识。比如：

The underground water is very like to have <u>dissolved in it materials</u> that help it dissolve certain rock materials. The water dissolved the limestone that was once here and carried most of it away.

在第一句的划线部分中 materials 的成分要从 dissolve 的词性来考虑。dissolve 是及物动词,后面应接动词宾语,而本句中 dissolve 后接介词 in 和介词宾语 it,并没有动词宾语,因此,这一部分是倒装结构,正常的结构是… dissolved materials that … in it. 这个句子应译成:地下水很可能已经在其中溶解了一些可以帮助它溶解某些特定岩石材料的物质。这样的地下水溶解了曾经在这里的石灰石,并把大部分石灰石带走。

这种倒装形式在科技英语中并不少见,其主要特点是及物动词与介词搭配,同时具有动词宾语和介词宾语,而动词宾语有较长的定语。这时将介词和介词宾语提前而将动词宾语置后,可以避免在阅读中产生歧义。同时,避免句子结构中的大头小尾,使句子结构平衡。比如:

If we denote by f, the fraction of the free radicals produced that is effective in initiating the chain growth, the R_i can be modified as $R_i = 2fk_d[I]$.

最后,再举一个有代表性的例句:

The process for producing laminated films employing a two component nonsolvent adhesive system, comprises the steps of applying a thin film of one polyether having at least two terminal isocyanate groups and a molecular weitht between 2 000 and 5 000 to one of two films to be bonded together, applying a thin film of one long chain compound having at least two reactive terminal amino groups to the other film to be bonded, the application of both thin films being such amount that the molar ratio of isocyanate groups to amino groups is from 1:1 to 5:1 and the two components together being present in an amount of 0.5 to 5.0 $g \cdot m^{-2}$, pressing the coated side of the two coated films together and curing the laminated films.

该句较长,令人望而生畏。实际上,长句不一定结构复杂。首先要对句子缩句:

process comprises steps

仔细分析可以发现本长句在整体结构上比较简单,process 与 comprises 之间为 process 的定语,而 steps 后面部分为 steps 的定语。

由于 process 的定语结构简单,重点看 steps 的定语。注意这里的第一个提示:steps 为复数形式,说明工序不只一道,这就要有并列关系。简单地浏览 steps 的定语部分可以看到这样的结构:applying …, applying …, pressing … and curing …. 各个工序是以现在分词引导的,这是第二个提示。现在分析各工序:

工序 1:apply 有固定的搭配形式 apply A to B(将 A 涂在 B 上)。因此,跳过 applying 之后的部分首先找与之搭配的 to B,在本句中应为 to one of the two films,那么 applying 与 to 之间则为 A。对 A 加以分析可以看出,A 的主体为 a thin film,其余为主体的定语。而工序 1 中的最后一部分——to be bonded together 则为 B 中 films 的定语。

工序 2:按照在工序 1 中所用的分析方法找出 apply A to B 结构中的 A 与 B,再确定各自的修饰结构。还应注意到,工序 2 中 the other film 是与工序 1 中 one of the two films 相对应的。

在工序 2 之后的 the applications of … 一部分并非以现在分词引导,可暂时将这部分跳过不看。

工序 3 与工序 4 的结构非常简单,没有难点。只须注意工序 3 中 the coated sides 和 the two coated films 是与工序 1 和工序 2 对应的。

对四道工序过程了解清楚后,再看跳过部分,就可以很清楚地看出,这一部分与四道工序并非并列关系。根据 applying 与 application 的对应关系,这一部分是对工序 1 与工序 2 的补充说明。

这个长句的内容结构完全清楚后,整个长句可汉译如下:

使用双组分无溶剂胶粘剂体系生产复合膜的工艺包括以下几道工序:首先将一至少含有两个端异氰酸酯基的分子量为 2 000~5 000 的聚醚薄层涂在要粘在一起的两片薄膜的一片上;再将一薄层含有反应性端胺基的长链化合物涂在另一片要粘的薄膜上,这两薄层的涂敷量是异氰酸酯基与胺基的物质的量比例为 1:1~5:1,两种化合物总计为 $0.5 \sim 5.0 \text{ g} \cdot \text{m}^{-2}$,然后将被涂敷薄膜和被涂敷面压在一起,最后将复合膜固化。

这里介绍的只是一些翻译的方法,要想准确地翻译英文科技文献,还需要在实践中锻炼、摸索和体会。

8.5 科技论文范例

Expansion of the Porous Solid Na$_2$Zn$_3$[Fe(CN)$_6$]$_2\cdot$9H$_2$O: Enhanced Ion-Exchange Capacity in Na$_2$Zn$_3$[Re$_6$Se$_8$(CN)$_6$]$_2\cdot$24H$_2$O

Miriam V. Bennett, Matthew P. Shores, Laurance G. Beauyais, and Jeffrey R. Long*

Contribution from the Department of Chemistry. University of California. Berkeley. California 94920-1460 Received February 18, 2000

Abstract: A technique for increasing the porosity in solid structures by replacing octahedral metal ions with hexanuclear cluster cores is extended with the expansion of the ion-exchange material Na$_2$Zn$_3$[Fe(CN)$_6$]$_2\cdot$9H$_2$O. At high relative cluster concentrations, the reaction between [Zn(H$_2$O)$_6$]$^{2+}$ and Na$_4$[Re$_6$Se$_8$(CN)$_6$] produces Na$_2$Zn$_3$[Re$_6$Se$_8$(CN)$_6$]$_2\cdot$24H$_2$O(1), a compound exhibiting a porous three-dimensional framework isotypic with that of Na$_2$Zn$_3$[Fe(CN)$_6$]$_2\cdot$9H$_2$O. Its framework is characterized by hexagonal bipyramidal cages, each enclosing two Na$^+$ ions and 24 water molecules in a volume of 1.340 nm^3—more than triple the volume of the cages in the original structure. The expanded cavities and framework openings in compound 1 are shown to facilitate absorption of larger cationic complexes, specifically [M(H$_2$O)$_6$]$^{n+}$ (M = Mg^{2+}, Cr^{3+}, Mn^{2+}, Ni^{2+}, Zn^{2+}) and [Cr(en)$_3$]$^{3+}$, via exchange for the Na$^+$ ions. Indeed, when the preparation of 1 is attempted using a slight excess of the zinc reactant, [Zn(H$_2$O)$_6$]$^{2+}$ incorporates instead of Na$^+$, leading to direct formation of the ion-exchanged compound [Zn(H$_2$O)$_6$]Zn$_3$[Re$_6$Se$_8$(CN)$_6$]$_2\cdot$18H$_2$O (2). The crystal structure of 2 reveals the [Zn(H$_2$O)$_6$]$^{2+}$ complexes to reside at the exact center of the hexagonal bipyramidal cages. With prolonged exposure to air, compound

1 and its ion-exchanged variants undergo a color change from orange to green, which is attributed to the one-electron oxidation of the $[Re_6Se_8]^{2+}$ cluster cores in the solid framework. In a further parallel with ferrocyanide chemistry, the reaction between $[Zn(H_2O)_6]^{2+}$ and $Na_4[Re_6Se_8(CN)_6]$ at low relative cluster concentrations is found to yield $[Zn(H_2O)]_2[Re_6Se_8(CN)_6] \cdot 13H_2O$ (3), a phase exhibiting a two-dimensional framework structure isotypic with that of $[Zn(H_2O)]_2[Fe(CN)_6] \cdot 0.5H_2O$.

Introduction

Over the past decade, increasing attention has been devoted to investigating solution-based methods of solid synthesis, wherein simple ligand substitution reactions are used to generate extended framework structures.[1] Much of this research has been prompted by the prospect of using molecular design principles for tailoring functional solid materials—particularly porous materials exhibiting zeolite-type behavior. The strategy commonly adopted for introducing porosity involves expanding the framework of a known solid structure by enlarging one or more of its components. Unfortunately, such efforts are often thwarted by interpenetration and architectural frailty, difficulties that arise from the skeletal nature of the framework obtained when the enlargement is in only one or two dimensions.[2] It has been suggested that these difficulties might be reliably avoided by utilizing multinuclear clusters as replacement components that are isotropically enlarged in all three dimensions.[3] While there now exist

---

(1) (a) Hoskins, B. F.; Robson. R. *J. Am. Chem Soc.* 1990, 112, 1546. (b) Zaworotko, M. J. *Chem. Soc. Rev.* 1994, 283. (c) Moore, J. S.; Lee, S. *Chem. Ind.* 1994, 556. (d) Bowes, C, L.; Ozin, G. A. *Adv. Mater.* 1996, 8, 13. (e) Yaghi, O. M.; Li, H. L.; Davis, C. Richardson. D.; Groy, T. L. *Acc. Chem. Res.* 1998, 31, 474.

(2) Batten, S. R.; Robson, R. *Angew. Chem., Int. Ed. Engl.* 1998, 37, 1461 and references therein.

(3) (a) Beauvais, L. G.; Shores, M. P.; Long, J. R. *Chem. Mater.* 1998, 10, 3783. (b) Shores, M. P.; Beauvais, L. G.; Long. J. R. *J. Am. Chem. Soc.* 1999, 121, 775.

numerous examples of porous solids with frameworks featuring cluster components that have been assembled in situ.[4~6] only a few such materials have been formed via ligand substitution reactions employing an intact cluster precursor.[3,7,8] The latter process, however, offers a clear advantage in its potential for controlling product struture. As a demonstrative example, we recently showed that the cubic structure of Prussian blue ($Fe_4[Fe(CN)_6]_3 \cdot 14H_2O$)[9] can be expanded by replacing the $[Fe(CN)_6]^{4-}$ reactants in its aqueous assembly reaction with face-capped octahedral $[Re_6Q_8(CN)_6]^{4-}$ (Q = Se, Te) clusters.[3b] The ensuing cluster-expanded materials $Fe_4[Re_6Te_8(CN)_6]_3 \cdot 27H_2O$ and $Ga_4[Re_6Se_8(CN)_6]_3 \cdot 38H_2O$ were found to be at least as thermally robust as Prussian blue, and to exhibit enhanced inclusion properties commensurate with their increased pore size.[3b] Herein, we extend this approach by applying it in the expansion of a second important metal-cyanide structure: the anionic framework of the porous ion-exchange solid $Na_2Zn_3[Fe(CN)_6]_2 \cdot 9H_2O$.[10]

(4) Selected examples assembled under standard solution condiuions; (a) Dance, I. G.; Garbutt, R. G.; Craig, D. C.; Scudder, M. L. *Inorg. Chem.* 1987, 26, 4057. (b) Yaghi, O. M.; Davis, C. E.; Li, G.; Li, H. *J. Am. Chem. Soc.* 1997. 119. 2861. (c) Li, H.; Davis. C. E.; Groy, T. L.; Kelley, D. G.; Yaghi, O. M. *J. Am. Chem. Soc.* 1998, 120, 2186. (d) Khan, M. I.; Yohannes. E.; Powell, D. *Inorg. Chem.* 1999, 38, 212. (e) Müller, A.; Krickemeyer. E.; Bögge, H.; Schmidtmann, M.; Beugholt, C.; Das. S. K.; Peters, F. *Chem. Eur. J.* 1999, 5, 1496. (f) Li, H.; Eddaoudi, M.; O'Keeffe, M.; Yaghi, O. M. *Nature* 1999, 402, 276. (g) Zhang, Q. F.; Leung, W.-H.; Xin, X.-Q.; Fun, H.-K. *Inorg. Chem.* 2000, 39, 417.

(5) Selected examples assembled under hydrothermal conditions; (a) Parise, J. B.; Ko, Y. *Chem. Mater.* 1994, 6, 718. (b) Tan, K.; Darovsky, A.; Parise, J. B. *J. Am. Chem, Soc.* 1995, 117, 7039. (c) Bowes, C. L.; Huynh, W. U.; Kirkby. S. J.; Malek, A.; Ozin, G. A.; Petrov, S.; Twardowski, M.; Young, D.; Bedard, R. L.; Broach, R. *Chem. Mater.* 1996, 8, 2147. (d) Cahill, C. L.; Parise, J. B. *Chem. Mater.* 1997, 9, 807. (e) Cahill, C. L.; Ko, Y.; Parise, J. B. *Chem. Mater.* 1998, 10, 19. (f) Li, H.; Yaghi, O. M. *J. Am. Chem. Soc.* 1998, 120, 10569. (g) Li, H.; Laine, A.; O'Keeffe, M.; Yaghi, O. M. *Science* 1999, 283, 1145. (h) Li, H.; Eddaoudi, M.; Laine, A.; O'Keeffe, M.; Yaghi, O. M. *J. Am. Chem. Soc.* 1999, 121, 6096. (i) Chui, S. S.-Y.; Lo, S. M.-F.; Charmant, J. P. H.; Orpen, A. G.; Williams, I. D. *Science* 1999, 283, 1148.

(6) Zeolites with framework structures readily described in terms of cluster building units (e.g., sodalite cages) could even be classified as such.

(7) (a) Yaghi, O. M.; Sun, Z.; Richardson, D. A.; Groy, T. L. *J. Am. Chem. Soc.* 1994, 116, 807. (b) Naumcv, N. G.; Virovets, A. V.; Sokolov, M. N.; Artemkina, S. B.; Fedorov, V. E. *Angew. Chem., Int. Ed. Engl.* 1998, 37, 1943. (c) MacLachlan, M. J.; Coombs, N.; Bedard, R. L.; White, S.; Thompson, L. K.; Ozin, G. A. *J. Am. Chem. Soc.* 1999, 121, 12005.

(8) (a) Shores, M. P.; Beauvais, L. G.; Long, J. R. *Inorg. Chem.* 1999, 38, 1648. (b) Beauvais, L. G.; Shores, M. P.; Long, J. R. *J. Am Chem. Soc.* 2000, 122, 2763.

(9) Buser, H. J.; Schwarzenbach, D.; Petter, W.; Ludi, A. *Inorg. Chem.* 1977, 16, 2704.

(10) Garnier, E.; Gravereau, P.; Hardy, A. *Acta Crystallogr.* 1982, B38, 1401.

This compound belongs to a family of isostructural phases with formulas $A_2Zn_3[Fe(CN)_6]_2 \cdot xH_2O$ (A = H, Na, K, Cs), and is readily prepared from aqueous solution as follows.[10,11]

$$3[Zn(H_2O)_6]^{2+} + 2Na_4[Fe(CN)_6] \longrightarrow Na_2Zn_3[Fe(CN)_6]_2 \cdot 9H_2O + 6Na^+ \qquad (1)$$

Its structure consists of octahedral $[Fe(CN)_6]^{4-}$ units linked through tetrahedral Zn^{2+} ions to form a three-dimensional framework characterized by large hexagonal bipyramidal cages (see Figure 1) encapsulating the Na^+ ions and water molecules.[10] The solid can be dehydrated without loss of crystallinity, and the potassium-containing analogue has been shown to display type I adsorption isotherms typical of microporous zeolites.[12] Due to the tetrahedral Zn^{2+} ion coordination, the cavities defined by the $\{Zn_3[Fe(CN)_6]_2\}^{2-}$ framework are significantly more accessible than those present in Prussian blue. Alkali metal cations can easily move through the structure, and consequently, these compounds have been widely studied for their ion-exchange properties.[11b,13] In particular, much research has focused on their ability to selectively absorb radioactive Cs^+ cations from nuclear waste and other contaminated solutions.[14,15]

Experimental Section

Preparation of Compounds. $Na_4[Re_6Se_8(CN)_6]$ was prepared

---

(11) (a) Tananaev, I. V.; Korolkov, A. P. *Izv. Akad. Nauk SSSR. Neorgan. Mater.* 1965, 1.100. (b) Kawamura, S.; Kuraku, H.; Kurotaki, K. *Anal. Chim. Acta* 1970, 49, 317 (c) Bellomo, A. *Talanta* 1970, 17, 1109 (d) Gravereau, P.; Garnier, E.; Hardy, A. *Acta Crystallogr.* 1979, B35, 2843 (e) Gravereau, P.; Garnier, E. *Rev Chim. Min.* 1983, 20, 68 (f) Loos-Nekovic, C.; Fedoroff, M.; Garnier, E.; Gravereau, P. *Talanta* 1984, 31, 1133 and references therein. (g) Loos-Nekovic, C.; Fedoroff, M.; Garnier, E. *Talanta* 1989, 36, 749.

(12) (a) Renaud, A.; Cartraud, P.; Gravereau, P.; Garnier, E. *Thermochim. Acta* 1979, 31, 243. (b) Cartraud, P.; Caintot, A.; Renaud. A. *J. Chem. Soc., Faraday Trans. I.* 1981, 77, 1561.

(13) (a) Vlasselaer, S.; d'Olieslager, W.; d'Hont, M. *J. Inorg. Nucl. Chem.* 1976, 38, 327. (b) Marei, S. A.; Basahel, S. N.; Rahmatallah, A. B. *J. Radioanal. Nucl. Chem.* 1986, 104, 217. (c) Loos-Neskovic, C.; Fedoroff, M. *React. Ploym.* 1988, 7, 373. (d) Loos-Neskovic, C.; Fedoroff, M.; Mecherri, M. O. *Analyst* 1990, 115, 981.

(14) (a) Barton, G. B.; Hepworth, J. L.; McClanahan, E. D. Jr.; Moore, R. L.; van Tuyl, H. H. *Ind. Eng. Chem.* 1958, 50, 212 (b) Nielsen, P.; Dresow, B.; H. C. Z. *Naturforsch.* 1987, B42, 1451. (c) Loos-Neskovic, C.; Fedoroff, M. *Radioact. Waste Manage. Nucl. Puel Cycle* 1989, 11, 347. (d) Loos-Neskovic C.; Fedoroff, M.; Garnier, E.; Jones, D. J. *Adv. Mater. Proc.* 1998, 282, 171 and references therein.

(15) The strong affinity for cesium likely arises from interactions between the soft Cs^+ cations and the π-electron clouds of the framework cyanide ligands, as evident in the crystal structure[11c] of $Cs_2Zn_3[Fe(CN)_6]_2 \cdot 6H_2O$.

as described previously.[3b,16] as was $Cr(en)_3(SCN)_3$.[17] $Na_3[Re_6Se_8(CN)_6]$ was prepared by oxidizing the $[Re_6Se_8(CN)_6]^{4-}$ cluster with iodine in methanol. Water was distilled and deionized with a Milli-Q filtering system. All other reagents were used us purchased. Product identity and purity were verified by comparison of the observed X-ray powder diffraction pattern with a calculated pattern generated from the single-crystal results. The solvent content of each compound was determined by thermogravimetric analysis. Once prepared, all products were stored under dinitrogen to prevent oxidation.

$Na_2Zn_3[Re_6Se_8(CN)_6]_2 \cdot 24H_2O(1)$. Single crystals of this compound were obtained directly by layering reactant solutions in a narrow diameter tube, however, the material was best isolated in pure form as follows. A solution of $[Zn(H_2O)_6](ClO_4)_2$ (0.067 g, 0.25 mmol) in 5 mL of 3.5 M aqueous $NaClO_4$ was added to a solution of $Na_4[Re_6Se_8(CN)_6]$ (2.0 g, 1.0 mmol) in 10 mL of 3.5 M aqueous $NaClO_4$. After the solution was left standing for 30 min, the resulting orange precipitate was collected by centrifugation, washed with methanol, and quickly dried in air to give 0.23 g (63%) of $Na_2Zn_3[Re_6Se_8(CN)_6]_2 \cdot 12H_2O \cdot 4MeOH$. IR(KBr): ν_{CN} 2 149 cm^{-1}. Anal. Calcd for $C_{16}H_{40}N_{12}Na_2O_{12}Re_{12}Se_{16}Zn_3$: C, 4.37; H, 0.92; N, 3.82; Na, 1.05; Zn, 4.46. Found: C, 4.37; H, 1.04; N, 3.95; Na, 0.92; Zn, 4.33.

$[Zn(H_2O)_6]Zn_3[Re_6Se_8(CN)_6]_2 \cdot 18H_2O(2)$. Single crystals of this compound were obtained by layering a 0.075 M aqueous sloution of $[Zn(H_2O)_6](ClO_4)_2$ over a 0.015 M aqueous solution of $Na_4[Re_6Se_8(CN)_6]$ in a narrow diameter tube. However, the material was best prepared in pure form as follows. A solution of $[Zn(H_2O)_6](ClO_4)_2$ (0.060 g, 0.22 mmol) in 2 mL of methanol was added to a solution of $Na_4[Re_6Se_8(CN)_6]$ (0.060 g, 0.030 mmol) in 2 mL of methanol. After the solution was left standing for 12 h, the resulting orange precipitate was collected by centrifugation, washed with methanol, and quickly dried in air to give 0.046 g (68%) of $[Zn(H_2O)_6]Zn_3[Re_6Se_8(CN)_6]_2 \cdot$ 14$H_2O \cdot 2MeOH$. IR(KBr): ν_{CN} 2 150 cm^{-1}. Anal. Calcd for $C_{14}H_{48}N_{12}$

---

(16) Mironov, Y. V.; Cody, J. A.; Albrecht-Schmitt, T. E.; Ibers, I. A. *J. Am. Chem. Soc.* 1997, 119, 493.

(17) Rollinson, C. L.; Bailar, J. C., Jr. *J. Am. Chem. Soc.* 1943, 65, 250.

$O_{22}Re_{12}Se_{16}Zn_4$: C, 3.74; H, 1.08; N, 3.74; Na, 0.00; Zn, 5.82. Found: C, 3.71; H, 1.12; N, 3.75; Na < 0.06; Zn, 5.94.

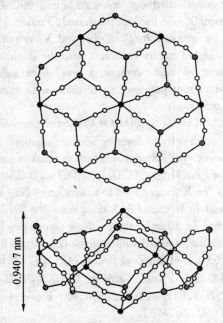

Figure 1. Top and side views of a hexagonal bipyramidal cage unit in the structure of $Na_2Zn_3[Fe(CN)_6]_2 \cdot 9H_2O$.[10] Black, shaded, white, and crosshatched spheres represent Fe, C, N, and Zn, respectively. The cage is situated on a $\bar{3}$ symmetry site. Selected mean interatomic distances (nm) and angles (deg): Fe—C 0.188 1(3), C—N 0.116(1), Zn—N 0.196 85(7), Fe—C—N 17.88(7), Zn—N—C 15.97(6), and N—Zn—N 11(4).

$[Zn(H_2O)]_2[Re_6Se_8(CN)_6] \cdot 13H_2O$ (3). A solution of $[Zn(H_2O)_6](ClO_4)_2$ (0.10 g, 0.37 mmol) in 10 mL of water was added to a solution of $Na_4[Re_6Se_8(CN)_6]$ (0.10 g, 0.050 mmol) in 10 mL of wates. After the solution was left standing for 12 h, large orange-red crystals of product had formed. The crystals were collected by centrifugation, separated from a small amount of orange powder impurity (compound 2) by sonication, washed with water, and dried in air to give 0.076 g (68%) of $[Zn(H_2O)]_2[Re_6Se_8(CN)_6] \cdot 8H_2O$. IR (KBr): ν_{CN} 2 144 cm^{-1}. Anal. Calcd for $C_6H_{20}N_6O_{10}Re_6Se_8Zn_2$: C, 3.25; H, 0.91; N, 3.79. Found: C, 3.29; H, 0.75; N, 3.65.

Table 1. Crystallographic Data[a] and Structure Refinement Parameters for $Na_2Zn_3[Re_6Se_8(CN)_6]_2 \cdot 24H_2O(1)$, $[Zn(H_2O)_6]Zn_3[Re_6Se_8(CN)_6]_2 \cdot 18H_2O(2)$, and $[Zn(H_2O)_2][Re_6Se_8(CN)_6]_2 \cdot 13H_2O(3)$

	1	2	3
formula	$C_{12}H_{48}N_{12}Na_2O_{24}$-$Re_{12}Se_{16}Zn_3$	$C_{12}H_{48}N_{12}O_{24}$-$Re_{12}Se_{16}Zn_4$	$C_6H_{34}N_6O_{15}$-$Re_6Se_8Zn_2$
formula wt	4 484.47	4 503.86	2 310.01
T/K	154	142	159
space group	$R\bar{3}c$	$R\bar{3}c$	$C2/m$
Z	6	6	2
a/nm	1.708 7(1)	1.719 2(1)	1.898 99(2)
b/nm			1.088 06(3)
c/nm	4.964 3(6)	4.936 9(5)	0.857 88(3)
$\beta/(°)$			108.218(2)
V/nm^3	12.552(2)	12.636(2)	1.683 71(8)
μ/mm^{-1}	25.142	25.241	31.584
d_{calc}/(g·cm^3)	3.560	3.551	4.556
R_1, $wR2$ [b]%	4.46, 8.11	2.78, 6.87	3.99, 8.91

[a] Obtained with graphite monochromated Mo $K\alpha$ ($\lambda = 0.071\ 073$ nm) radiation. [b] $R_1 = \sum \|F_o| - |F_c\| / \sum |F_o|$; $wR_2 = \{\sum[w(|F_o|) - |F_c|)^2]/\sum[(w|F_o|^2)]\}^{1/2}$.

X-ray Structure Determinations. Single crystals were coated in Paratone-N oil, attacthed to quartz fibers, transferred to a Siemens SMART diffractometer, and cooled in a dinitrogen stream. Lattice parameters were initially determined from a least-squares refinement of more than 40 carefully centered reflections. The raw intensity data were converted (including corrections for background, and Lorentz and polarization effects) to structure factor amplitudes and their esd's using the SAINT 5. 00 program. An empirical absorption correction was applied to each data set using SADABS. Space group assignments were based on systematic absences, E statistics, and successful refinement of the structures. Structures were solved by direct methods, with the aid of difference Fourier maps, and were refined with successive full-matrix least-squares cycles. The sodium ions in the structure of 1 and the solvate water molecules in both 1 and 2 are disordered over multiple positions, and were modeled with partial occupancies accordingly. In these structures, disordered atoms were refined with isotropic thermal parameters, and all other atoms were refined anisotropically. The oxygen atoms of solvate and bound water molecules are fully ordered in the structure of 3 and were refined isotropically; all other atoms were refined anisotropically. Hydrogen atoms were not included in any of the refinements. Crystallographic parameters are listed in Table 1.

Ion-Exchange Experiments. A sloution of the perchlorate or nitrate salt of the hydrated metal ion or of $[Cr(en)_3](SCN)_3$ in 10 mL of methanol was stirred over a precisely weighed quantity (ca. 100 mg) of solid 1 for 10 h. The solution and subsequent methanol washes (3×20 mL) were separated from the exchanged solid by centrifugation and combined. The integrity of the crystal structure of the exchanged solid was checked by X-ray powder diffraction. The separated solutions were reduced to dryness, and the resulting residue was dissolved in 25.00 mL of an aqueous solution that was 0.55 M in ammonium hydroxide and 0.33 M in ammonium chloride. An Orion model 97 - 12 Na^+ ion-selective electrode was then used to determine the Na^+ ion concentration of this solution by comparison with calibration curves obtained from standards having concentrations in a similar range. For each paramagnetic metal ion, the effective magnetic moment of the exchanged solid was measured at 295 K using a Quantum Design SQUID magnetometer, and was found to be within 1 % of the expected moment (as calculated using previously reported data for the metal complex).[18]

Other physical Measurements. X-ray powder diffraction data were collected using Cu $K\alpha$ ($\lambda = 0.1546$ nm) radiation on a Siemens D5000 diffractometer. Thermogravimetric analyses were performed under a dinitrogen atmosphere at a ramp rate of 1 ℃/min, using a TA Instruments TGA 2950. Infrared spectra were recorded on a Mattson Infinity System FTIR spectrometer. Cyclic voltammetry was performed using a Bioanalytical systems CV - 50 W voltammagraph, 0.1 M KNO_3 supporting electrolyte, and a glassy carbon working electrode. Potentials were determined vs a Ag/AgCl reference electrode. Diffuse reflectance spectra of solids were acquired relative to $BaSO_4$ on a Perkin-Elmer Lambda 9 spectrophotometer equipped with a 60 mm integrating sphere. Absorption spectra were measured with a Hewlett-Packard 8453 spectrophotometer.

Results and Discussion

The synthesis of $Na_2Zn_3[Re_6Se_8(CN)_6]_2 \cdot 24H_2O(1)$, a cluster-

---

(18) (a) Figgis, B.N. *Introduction to Ligand Fields*; Robert F. Krieger Publishing; Malabar, 1986. (b) Figgis, B.N.; Lewis, J.; Mabbs, F.E. *J. Chem. Soc.* 1961, 3138.

expanded analogue of $Na_2Zn_3[Fe(CN)_6]_2 \cdot 9H_2O$, is accomplished by employing $Na_4[Re_6Se_8(CN)_6]^{3b,16,19}$ in place of $Na_4[Fe(CN)_6]$ in reaction 1.

$$3[Zn(H_2O)_6]^{2+} + 2Na_4[Re_6Se_8(CN)_6] \rightarrow$$
$$Na_2Zn_3[Re_6Se_8(CN)_6]_2 \cdot 24H_2O + 6Na^+ \quad (2)$$

However, to avoid impurities, the reaction is best carried out in the presence of a large excess of both the cluster reactant and Na^+ ions. Similar conditions have been recognized as helpful in obtaining the ferrocyanide-based materials in pure form.[11bc] Reactions analogous to reaction 2 have now been used to synthesize a variety of open framework solids, most of these with unprecedented structure types.[3,7b,8] Interestingly, the reactions are highly sensitive to the choice of transition metal ion, cluster chalcogen element ($Q = S$, Se, or Te), and cluster countercation, as well as to the relative reactant concentrations. As will be shown here, this last factor is of particular importance for reaction 2, leading to formation of no less than three different phases.

Single crystals of compound 1 were grown by layering the reactant solutions in a narrow diameter tube. X-ray analysis revealed a structure in which the framework is perfectly isotypic with that of $Na_2Zn_3[Fe(CN)_6]_2 \cdot 9H_2O^{10}$ (Figure 1), adopting the same $R\bar{3}c$ space group, and differing only in the substitution of $[Re_6Se_8]^{2+}$ cluster cores onto the Fe^{2+} ion sites. This substitution results in the expansion of the hexagonal bipyramidal cages of the framework (see Figure 2), which each now contain two Na^+ cations and 24 water molecules. As calculated[3b] based on the estimated van der Waals radii of the framework atoms, the volume enclosed

---

(19) For information on related $[Re_6Q_8(CN)_6]^{4-}$ clusters, see: (a) Mironov, Y. V.; Virovets, A. V.; Fedorov, V. E.; Podberezskaya, N. V.; Shishkin, O. V.; Struchkov, Y. T. *Polyhedron* 1995, 14, 3171. (b) Slougui, A.; Mironov, Y. V.; Perrin, A.; Fedorov, V. E. *Croat. Chem. Acta* 1995, 68, 885 (c) Imoto, H.; Naumov, N. G.; Virovets, A. V.; Saito, T.; Fedorov, V. E. *J. Struct. Chem.* (*Engl. Trans.*) 1998, 39, 720. (d) Podberezskaya, N. V.; Virovets, A. V.; Mironov, Y. V.; Kozeeva, L. P.; Naumov, N. G.; Fedorov, V. E. *J. Struct. Chem.* (*Engl. Trans.*) 1999, 40, 436.

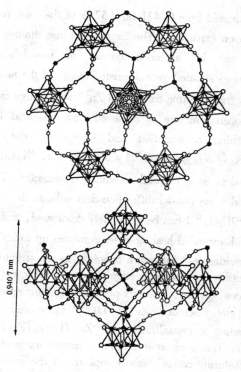

Figure 2. Top and side views of a hexagonal bipyramidal cage unit (with vertices located at the center of each cluster) in the structure of $[Zn(H_2O)_6]Zn_3[Re_6Se_8(CN)_6]_2 \cdot 18H_2O(2)$. Black, large white, shaded, small white, crosshatched, and hatched spheres represent Re, Se, C, N, Zn, and O atoms, respectively. The cage is centered by a $[Zn(H_2O)_6]^{2+}$ complex situated on a $\bar{3}$ symmetry site. The framework in $Na_2Zn_3[Re_6Se_8(CN)_6]_2 \cdot 24H_2O$ (1) is essentially identical, except that its cages each encapsulate two hydrated Na^+ ions rather than a $[Zn(H_2O)_6]^{2+}$ complex. Comparison with Figure 1 shows how this framework is related to that of $Na_2Zn_3[Fe(CN)_6]_2 \cdot 9H_2O$ simply by replacing the Fe^{2+} ions with $[Re_6Se_8]^{2+}$ cluster cores. Mean interatomic distances and angles are listed in Table 2.

by these cages has increased from 0.421 nm^3 (57% of the total volume) to 0.34 nm^3 (64%) upon expansion. The largest openings through which the cavities in 1 can be accessed consist of pseudohexagonal $(Re_6Se_8)_3(CN)_6Zn_3$ rings situated over alternate "faces" of the hexagonal bipyramidal cages. By incorporating tetrahedral Zn^{2+} ions in this fashion, a significantly more open structure is achieved than observed for the frame-works of matching compostition and charge in the phases $Cs_2[M(H_2O)]_3[Re_6Se_8(CN)_6]_2 \cdot xH_2O$ (M = Mn, Fe, Co, Ni, Cd),[3a,7b] where the M^{2+} ions adopt an octahedral coordination geometry. Thermogravimetric analysis and X-ray powder diffraction data indicate that, similar to $K_2Zn_3[Fe(CN)_6]_2 \cdot 9H_2O$,[12] 1 can be completely desolvated, and retains its crystal structure with extended heating at temperatures up to 250 ℃.

The expanded cavities and framework openings in the structure of 1 suggest its potential utility in absorbing large cationic complexes via exchange for the Na^+ ions. In fact, when its preparation is attempted using a slight excess of the zinc reactant. $[Zn(H_2O)_6]^{2+}$ incorporates instead of the Na^+ ions, leading to crystallization of $[Zn(H_2O)_6]Zn_3[Re_6Se_8(CN)_6]_2 \cdot 18H_2O$ (2). This compound exhibits a framework isostructural with that of 1, but featuring cationic zinc complexes at the center of its hexagonal bipyramidal cages (Figure 2), wherein each is surrounded by 18 solvate water molecules. Direct exchange experiments with preformed samples of solid 1 were conducted for a range of $[M(H_2O)_6]^{n+}$ species dissolved in methanol.[20] The extent of the exchange was assessed by measuring the amount of Na^+ released into solution using an ion-selective electrode. The results obtained are summarized in Table 3, and demonstrate the successful uptake of Mg^{2+}, Cr^{3+}, Mn^{2+}, Ni^{2+}, and Zn^{2+}, with no perturbation of the framework structure (as probed by powder X-ray diffraction). For paramagnetic ions, the extent of the exchange was also calculated from the measured effective magnetic moment of the product, and found to be within 1% of the potentiometrically determined

---

(20) Exchanges were studied in methanol, because experiments in water were typically complicated by simultaneous formation of an impurity phase.

value. This exchange behavior is reminiscent of the water-softening properties of zeolites; however, as with traditional zeolites, certain metal ions were found to modify the framework of the material. X-ray powder diffraction data indicate that exchanges attempted with Fe^{2+}, Co^{2+}, and Cu^{2+} all resulted in the loss of framework strcture and formation of an amorphous phase. A satisfactory explanation for why these ions should destroy the framework structure is not immediately forthcoming, particularly in view of the range of stability and lability parameters characterizing those metal ion complexes that simply exchange with no loss of crystallinity. Clearly there is sufficient space for larger cationic complexes to fit within the cavities of the $\{Zn_3[Re_6Se_8(CN)_6]_2\}^{2-}$ framework, and indeed, analogous exchange experiments indicate that 75% of the Na^+ ions in compound 1 can be exchanged for $[Cr(en)_3]^{3+}$.[21] The possibility of building still larger complexes within the cavities by first exchanging in $[M(H_2O)_6]^{n+}$ and subsequently adding the free ligand is currently under investigation.

Table 2. Selected Mean Interatomic Distances (nm) and Angles (deg) from the Structures of $Na_2Zn_3[Re_6Se_8(CN)_6]_2 \cdot 24H_2O$ (1), $[Zn(H_2O)_6]Zn_3[Re_6Se_8(CN)_6]_2 \cdot 18H_2O$ (2), and $[Zn(H_2O)]_2[Re_6Se_8(CN)_6]_2 \cdot 13H_2O$ (3)

	1	2	3
Re—Re	0.262 7(2)	0.263 4(1)	0.263 8(6)
Re—Se	0.251 7(8)	0.252 5(8)	0.252 2(9)
Re—C	0.209 5(7)	0.208 9(7)	0.210 9(1)
C—N	0.116(2)	0.116 7(5)	0.114(1)
Zn—N	0.194 6(5)	0.196 2(3)	0.198 3(2)
Zn—O		0.212 6(1)	0.2
Se—Re—C	9.2(2)	9.2(2)	9.2(3)
Re—C—N	17.75(7)	17.7(2)	17.39(1)
Zn—N—C	16.1(1)	16.1(2)	16.52
N—Zn—N	11(6)	11(4)	11.57(8)
O—Zn—O		9(1)	
N—Zn—O			10.2(4)

---

(21) The absorbed $[Cr(en)_3]^{3+}$ is also evident in the infrared spectrum of the exchanged solid, which exhibits no change in its framework structure (as again probed by X-ray powder diffraction).

Table 3. Exchange of $[M(H_2O)_6]^{n+}$ into Solid 1

	Na^+ released[a]	% exchange	$\mu_{eff}(\mu_B)$[b]
Mg^{2+}	0.85	43	
Cr^{3+}	1.83	92	4.89[c]
Mn^{2+}	1.17	59	3.45
Ni^{2+}	1.50	75	2.46
Zn^{2+}	1.68	84	

[a] Number of equivalents per formula unit of 1. [b] At 295 K. [c] Moment and electronic absorption spectrum (Figure 3) indicate oxidation of 45% of the cluster cores to $[Re_6Se_8]^{3+}$ ($S = 1/2$)

Compound 1 and its ion-exchanged variants all undergo a color change from orange to green upon prolonged exposure to air. This change is not observed when the samples are stored under dinitrogen, and we attribute it to the one-electron oxidation of the $[Re_6Se_8]^{2+}$ cluster cores in the solid framework by dioxygen. An analogous oxidation is apparent in the cyclic voltammagram of $Na_4[Re_6Se_8(CN)_6]$ in aqueous solution, which reveals a $[Re_6Se_8(CN)_6]^{3-/4-}$ couple centered at $E_{1/2} = 0.62$ V vs Ag/AgCl.[22] Furthermore the diffuse reflectance spectra of the green solids exhibit bands matching those centered at 545 and 599 nm in the absorption spectrum of a green aqueous solution of $[Re_6Se_8(CN)_6]^{3-}$ (Figure 3). Typically, emergence of the green color in the solids is gradual, only becoming noticeable after approximately 1 day. However, the exchange of $[Cr(H_2O)_6]^{3+}$ into compound 1 induces a more rapid change, with the green color already apparent immediately following the (10 h) exchange reaction. At this stage, the exchanged product displays an

---

(22) Note that this potential is slightly (~ 0.25 V) more positive than that observed in acetonitrile solution: Yoshimura, T.; Ishizaka, S.; Sasaki, Y.; Kim, H.-B.; Kitamura, N.; Naumov, N. G.; Sokolov, M. N.; Fedorov, V. E. *Chem. Lett*. 1999. 1121.

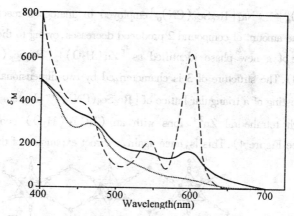

Figure 3. Electronic absorption spectra of the solid compound 1 before (dotted line) and after (solid line) exchange of $[Cr(H_2O)_6]^{3+}$, and of $Na_3[Re_6Se_8(CN)_6]$ in aqueous solution (dashed line). The units along the vertical axis correspond to the solution spectrum, and are arbitrary for the two solid spectra.

electronic absorption spectrum substantially different from that of the original compound 1 (see Figure 3) and an effective magnetic moment indicating oxidation of 45% of the framework clusters. We propose that the increased rate is due to the greater acidity of $[Cr(H_2O)_6]^{3+}$ ($pK_a \approx 4$), which provides protons for the acid-promoted oxidation of the clusters by absorbed dioxygen. Consistent with our claim, exchange of $[Cr(en)_3]^{3+}$, a much less likely source of protons, results in a solid product that oxidizes only very slowly. Significantly, no loss of crystallinity is apparent in the X-ray powder diffraction patterns of these oxidized materials.

When reaction 1 is carried out using an excess of $[Zn(H_2O)_6]^{2+}$, the zinc-enriched compound $[Zn(H_2O)]_2[Fe(CN)_6]\cdot 0.5H_2O$[11bc,23] is obtained instead of $Na_2Zn_3[Fe(CN)_6]_2\cdot 9H_2O$. Similarly, as the molar ratio of

---

(23) (a) Kourim, V.; Rais, J.; Million, B, *J. Inorg. Nucl. Chem.* 1946, 26, 1111; (b) Siebert, H.; Nuber, B.; Jentsch, W. *Z. Anorg. Allg. Chem.* 1981, 474, 96.

$[Zn(H_2O)_6]^{2+}$: $Na_4[Re_6Se_8(CN)_6]$ employed in analogous reactions increases, the amount of compound 2 produced decreases, owing to the gradual emergence of a new phase identified as $[Zn(H_2O)]_2[Re_6Se_8(CN)_6] \cdot 13H_2O$ (3). The structure of 3 is characterized by two-dimensional sheets, each consisting of a triangular lattice of $[Re_6Se_8(CN)_6]^{4-}$ clusters connected through tetrahedral Zn^{2+} ions with an $(NC)_3(H_2O)$ coordination sphere (see Figure 4). This is, once again, a direct expansion of the frame-

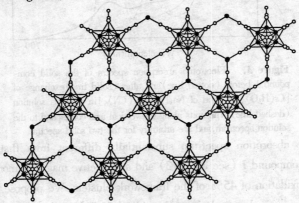

Figure 4. A portion of a two-dimensional sheet in the structure $[Zn(H_2O)]_2[Re_6Se_8(CN)_6] \cdot 13H_2O$ (3). Atom representations are the same as in Figure 2. Each cluster resides on a site of 2/m crystallographic symmetry, with the 2-fold rotational axes running horizontally within the plane of the drawing. Mean interatomic distances and angles are listed in Table 2.

work in the iron-containing analogue,[23b] with $[Re_6Se_8]^{2+}$ cluster cores directly substituted onto the Fe^{2+} ion positions. More simply, the structure can be viewed as an expansion of the CdI_2 structure, with $[Re_6Se_8(CN)_6]^{4-}$ clusters and $[Zn(H_2O)]^{2+}$ moieties residing on the Cd^{2+} and I^- sites, respectively. Interestingly, compound 3 is converted to compound 2 with sustained exposure to methanol; an analogous transformation has been reported for the ferrocyanide-based solids.[24]

In sum, the reaction chemistry between Zn^{2+} and $[Re_6Se_8(CN)_6]^{4-}$ has been found to closely parallel that between Zn^{2+} and $[Fe(CN)_6]^{4-}$, permitting synthesis of $Na_2Zn_3[Re_6Se_8(CN)_6]_2 \cdot 24H_2O$, an expanded form of $Na_2Zn_3[Fe(CN)_6]_2 \cdot 9H_2O$ wiht an enhanced ion-exchange capacity. Future work will focus on assembling reactive cationic complexes in the cavities of this new material with the intention of rendering known homogeneous catalysts heterogeneous.

Acknowledgment. This research was funded by the University of California, Berkely and the University of California Energy Institute. We thank Prof. J. Arnold for use of the thermogravimetric analysis instrument, Prof. A. M. Stacy for access to the X-ray powder diffractometer and SQUID magnetometer, and Prot, T. D. Tilley for use of the infrareo spectrometer.

Supporting Information Available: Tables of crystal data, structure solution and refinement parameters, atomic coordinates, bond lengths and angles, and anisotropic thermal parameters for compounds 1, 2, and 3 (PDF). An X-ray crystallographic file (CIF). This material is available free of charge via the Inhternet at http://pubs.acs.org.

---

(24) Braterman. P. S; Arrhenius. G; Hui. S; Paplawsky, W. *Origin Life Evol. Biosphere* 1995, 25, 531.

Appendixes

附录 I 化学元素表

English name and phonetic symbol	Symbol	Atomic number	Chinese name
Actinium [æk'tiniəm]	Ac	89	锕
Aluminum [ə'lju:minəm], Aluminium [ˌælju'minjəm]	Al	13	铝
Americium [ˌæmə'risiəm]	Am	95	镅
Antimony ['æntiməni] (*Stibium*)①	Sb	51	锑
Argon ['ɑ:gɔn]	Ar	18	氩
Arsenic ['ɑ:snik]	As	33	砷
Astatine ['æstəti:n]	At	85	砹
Barium ['bɛəriəm]	Ba	56	钡
Berkelium ['bə:kliəm]	Bk	97	锫
Beryllium [be'riljəm]	Be	4	铍
Bismuth ['bizməθ]	Bi	83	铋
Bohrium	Bh	107	铍②
Boron ['bɔ:rɔn]	B	5	硼
Bromine ['broumi:n]	Br	35	溴
Cadmium ['kædmiəm]	Cd	48	镉
Calcium ['kælsiəm]	Ca	20	钙

① 表中圆括号内斜体字为元素的拉丁文名称(只有一个例外,74号元素为德文名)。
② 101~109号元素的名称由全国科学技术名词审定委员会公布,参见:化学通报,1998.(11),8。

Californium [ˌkæki'fɔːniəm]	Cf	98	锎
Carbon ['kɑːbən]	C	6	碳
Cerium ['siəriəm]	Ce	58	铈
Cesium, Caesium ['siːziəm]	Cs	55	铯
Chlorine ['klɔːriːn]	Cl	17	氯
Chromium ['kroumiəm]	Cr	24	铬
Cobalt [kə'bɔːlt; 'koubɔːlt]	Co	27	钴
Copper ['kɔpə] (*Cuprum*)	Cu	29	铜
Curium ['kjuəriəm]	Cm	96	锔
Dubnium	Db	105	𨧀
Dysprosium [dis'prousiəm]	Dy	66	镝
Einsteinium [ains'tainiəm]	Es	99	锿
Erbium ['əːbiəm]	Er	68	铒
Europium [juə'roupiəm]	Eu	63	铕
Fermium ['fəːmiəm]	Fm	100	镄
Fluorine ['flu(ː)əriːn]	F	9	氟
Francium ['frænsiəm]	Fr	87	钫
Gadolinium [gædə'liniəm]	Gd	64	钆
Gallium ['gæliəm]	Ga	31	镓
Germanium [dʒəː'meiniəm]	Ge	32	锗
Gold [gould] (*Aurum*)	Au	79	金
Hafnium ['hæfniəm]	Hf	72	铪
Hassium	Hs	108	𨭆
Helium ['hiːljəm]	He	2	氦
Holmium ['hɔlmiəm]	Ho	67	钬
Hydrogen ['haidridʒən]	H	1	氢
Indium ['indiəm]	In	49	铟
Iodine ['aiədiːn]	I	53	碘
Iridium [ai'ridiəm]	Ir	77	铱

Iron ['aiən] (*Ferrum*)	Fe	26	铁
Krypton ['kriptɔn]	Kr	36	氪
Lanthanum ['lænθənəm]	La	57	镧
Lawrencium [lɔ:'rensiəm]	Lr	103	铹
Lead [led] (*Plumbum*)	Pb	82	铅
Lithium ['liθiəm]	Li	3	锂
Lutetium [lju:'ti:ʃiəm]	Lu	71	镥
Magnesium [mæg'ni:zjəm]	Mg	12	镁
Manganese [ˌmæŋɡə'ni:z]	Mn	25	锰
Meitnerium	Mt	109	䥑
Mendelevium [ˌmendə'li:viəm]	Md	101	钔
Mercury ['mə:kjuri] (*Hydrargrum*)	Hg	80	汞
Molybdenum [mɔ'libdinəm]	Mo	42	钼
Neodymium [ˌni(:)ə'dimiəm]	Nd	60	钕
Neon ['ni:ən; 'ni:ɔn]	Ne	10	氖
Neptunium [nep'tju:njəm]	Np	93	镎
Nickel ['nikl]	Ni	28	镍
Niobium [nai'oubiəm]	Nb	41	铌
Nobelium [nou'beliəm]	No	102	锘
Nitrogen ['naitridʒən]	N	7	氮
Osmium ['ɔzmiəm]	Os	76	锇
Oxygen ['ɔksidʒən]	O	8	氧
Palladium [pə'leidiəm]	Pd	46	钯
Phosphorus ['fɔsfərəs]	P	15	磷
Platinum ['plætinəm]	Pt	78	铂
Plutonium [plu:'tounjəm]	Pu	94	钚
Polonium [pə'louniəm]	Po	84	钋
Potassium [pə'tæsjəm] (*Kalium*)	K	19	钾
Praseodymium [ˌpreiziou'dimiəm]	Pr	59	镨

Element	Symbol	Number	中文
Promethium [prə'mi:θiəm]	Pm	61	钷
Protactinium [ˌproutæk'tiniəm]	Pa	91	镤
Radium ['reidjəm]	Ra	88	镭
Radon ['reidɔn]	Rn	86	氡
Rhenium ['ri:niəm]	Re	75	铼
Rhodium ['roudjəm]	Rh	45	铑
Rubidium [ru(:)'bidiəm]	Rb	37	铷
Ruthenium [ru(:)'θi:niəm]	Ru	44	钌
Rutherfordium [rʌðə'fɔ:diəm]	Rf	104	𬬻
Samarium [sə'mɛəriəm]	Sm	62	钐
Scandium ['skændiəm]	Sc	21	钪
Seaborgium	Sg	106	𬭳
Selenium [si'li:njəm]	Se	34	硒
Silicon ['silikən]	Si	14	硅
Silver ['silvə] (*Argentum*)	Ag	47	银
Sodium ['soudjəm] (*Natrium*)	Na	11	钠
Strontium ['strɔnʃiəm]	Sr	38	锶
Sulfur, Sulphur ['sʌlfə]	S	16	硫
Tantalum ['tæntələm]	Ta	73	钽
Technetium [tek'ni:ʃiəm]	Tc	43	锝
Tellurium [te'ljuəriəm]	Te	52	碲
Terbium ['tə:biəm]	Tb	65	铽
Thallium ['θæliəm]	Tl	81	铊
Thorium ['θɔ:riəm]	Th	90	钍
Thulium ['θju:liəm]	Tm	69	铥
Tin [tin] (*Stannum*)	Sn	50	锡
Titanium [tai'teinjəm]	Ti	22	钛
Tungsten ['tʌŋstən] (*Wolfram*)	W	74	钨
Uranium [juə'reinjəm]	U	92	铀

Vanadium [və'neidjəm]	V	23	钒
Xenon ['zenɔn]	Xe	54	氙
Ytterbium [i'tə:bjəm]	Yb	70	镱
Yttrium ['itriəm]	Y	39	钇
Zinc [ziŋk]	Zn	30	锌
Zirconium [zə:'kouniəm]	Zr	40	锆

附录Ⅱ　常用数学符号及数学式写法

+	plus; positive
−	minus; negative
±	plus or minus
×(*)	multiplied by; times
÷	divided by
=	is equal to; equals
≡	is identically equal to
≈(≅, ≐, ≙)	is approximately equal to; approximately equals
( )	round brackets; parentheses
[ ]	square(angular) brackets
{ }	braces
~	difference
⇔	is equivalent to
⇒	implies
→	maps into
$\log_n x$	log x to the base n
$\log_{10} x$	log x to the base 10 (即 common logarithm)
$\log_e x$ 或 $\ln x$	log x to the base e (即 natural logarithm 或 Napierian logarithm)
x^n	the n th power of x; x to the power n
$x^{\frac{1}{n}}$ 或 $\sqrt[n]{x}$	the n th root of x; x to the power one over n
$x \to a$	x approaches the limit a
$|x|$	the absolute value of x

$\bar{x}$	the mean value of x; x bar
b'	b prime
b''	b double prime; b second prime
b_m	b sub m
$f(x), F(x)$	function f of x
$y = f(x)$	y is a function of x
$\frac{dy}{dx}$ 或 $D_x y$	the differential coefficient of y with respect to x; the first derivative of y with respect to x
$\frac{d^2 y}{dx^2}$	the second derivative of y with respect to x
$\int_a^b$	integral between limits a and b
∞	infinity
$\vec{F}$	vector F
$x + y$	x plus y
$(a + b)$	bracket a plus b bracket closed
$a = b$	a equals b; a is equal to b; a is b
$a \neq b$	a is not equal to b; a is not b
$a \approx b$	a is approximately equal to b
$a > b$	a is greater than b
$a \gg b$	a is much (far) greater than b
$a \geq b$	a is greater than or equal to b
$a \ngtr b$	a is not greater than b
$a < b$	a is less than b
$a \perp b$	a is perpendicular to b
$a /\!/ b$	a is parallel to b
$a \propto b$	a varies directly as b
$1:2$	the ratio of one to two
x^2	x square; x squared; the square of x; the second power of x; x to the second power
y^3	y cube; y cubed; the cube of y; y to the third power; y to the third
y^{-10}	y to the minus tenth (power)

$\sqrt{x}$	the square root of x
$\sqrt[3]{a}$	the cube(cubic) root of a
$\sqrt[5]{a}$	the fifth root of a
$\dfrac{x^3}{y} = z^2$	x raised to the third power divided by y equals z squared
2%	two percent
5‰	five per mille
3/4 km	three quarters of a kilometer
20°	twenty degrees
100℃	one(a) hundred degrees Centigrade
32°F	thirty-two degrees of Fahrenheit

附录Ⅲ SI 单位

SI 基本单位(SI Base Units)

Physical Quantity	Unit	Abbreviation
length	meter	m
mass	kilogram	kg
time	second	s
electric current	ampere	A
temperature	kelvin	K
luminous intensity	candela	cd
amount of substance	mole	mol
plane angle	radian	rad
solid angle	steradian	sr

SI 前缀 (SI Prefixes)

Multiple	Prefix	Abbreviation	Submultiple	Prefix	Abbreviation
10^{12}	tera	T	10^{-1}	deci	d
10^{9}	giga	G	10^{-2}	centi	c
10^{6}	mega	M	10^{-3}	milli	m
10^{3}	kilo	k	10^{-6}	micro	μ
10^{2}	hecto	h	10^{-9}	nano	n
10^{1}	deka	da	10^{-12}	pico	p
			10^{-15}	femto	f
			10^{-18}	atto	a

SI 衍生单位 (Derived SI Units)
直接衍生单位 (Directly Derived Unit)

Physical Quantity	Unit	Abbreviation
area	square meter	m^2
volumn	cubic meter	m^3
velocity	meter per second	$m \cdot s^{-1}$
acceleration	meter per second squared	$m \cdot s^{-2}$
density	kilogram per cubic meter	$kg \cdot m^{-3}$
molar mass	kilogram per mole	$kg \cdot mol^{-1}$
molar volumn	cubic meter per mole	$m^3 \cdot mol^{-1}$
molar concentration	mole per cubic meter	$mol \cdot m^{-3}$

特殊衍生单位名称 (Special Names of Derived Unit)

Physical Quantity	Unit	Abbreviation	In Terms of SI Units
frequency	hertz	Hz	s^{-1}
force	newton	N	$kg \cdot m \cdot s^{-2}$
pressure	pascal	Pa	$N \cdot m^{-2}$
energy	joule	J	$kg \cdot m^2 s^{-2}$
power	watt	W	$J \cdot s^{-1}$
electric charge	coulomb	C	$A \cdot s$
electric potential difference	volt	V	$J \cdot A^{-1} \cdot s^{-1}$
electric resistance	ohm	Ω	$V \cdot A^{-1}$

不支持或废止的 SI 单位 (Units to be Discouraged or Abandoned)

Physical Quantity	unit	Abbreviation	Definition in SI Units
length	angstrom	Å	1×10^{-10} m
force	dyne	dyn	1×10^{-5} N
energy	erg	erg	1×10^{-7} J
energy	calorie	cal	4.184 J
pressure	atmosphere	atm	101 325 Pa
pressure	millimeter of mercury	mmHg	133.322 Pa
pressure	torr	torr	133.322 Pa

附录 Ⅳ 实验室常用化学仪器名称

Typical Chemistry Laboratory Equipment

Beaker 烧杯; Cylinder, graduated cylinder 量筒; Burette, buret 滴定管; Pipet, pipette 移液管; Graduated pipet 刻度吸量管

Funnel 漏斗; Buchner funnel 布氏漏斗; Sintered glass funnel 烧结玻璃漏斗, 熔砂漏斗; Dropping funnel 滴液漏斗; Separatory funnel 分

液漏斗

  Flask 烧瓶；Erlenmeyer flask, conical flask 锥（形）烧瓶；Boiling flask 长颈烧瓶；Distilling flask 蒸馏烧瓶；Flat bottom flask 平底烧瓶；Round-bottom(ed) flask, round flask 圆底烧瓶；Two-necked flask 双颈烧瓶；Filter(ing) flask, suction flask 吸滤瓶；Volumetric flask, measuring flask (容)量瓶

  Condenser 冷凝器, 冷凝管；Reflux condenser 回流冷凝器；Ball condenser 球形冷凝器；Coil condenser 蛇管冷凝器

  Column 柱；Distilling column 蒸馏柱；Fractional column, fractionating column 分馏柱；Ion exchange column 离子交换柱；Chromatography column 层析柱

  Soxhlet extractor 索氏提取器

  Head 接头；Still(ing) head, distilling head 蒸馏头；Claisen head, Claisen adapter 克莱森蒸馏头

  Adapter 接受管, 尾接管；Vacuum adapter 真空尾管

  Tube 管(子)；Test tube 试管；Centrifuge tube 离心(试)管；Drying tube 干燥管；Capillary (tube) 毛细管；Melting point tube 熔点管；Color comparison tube 比色管；Glass tube(pipe) 玻璃管

  Bottle 瓶；Dropping bottle 滴瓶；Washing bottle 洗瓶；Weighing bottle 称量瓶；Gas-washing bottle, gas wash bottle, gas bottle 洗气瓶

  Thermometer 温度计

  Drying tower 干燥塔；Desiccator, dryer, exsiccator 干燥器

  Evaporating dish 蒸发皿；Watch glass 表面皿

  Glass rod 玻璃棒；Dropper 滴管

  Mortar 研钵；Pestle (研)杵

  Stopper 塞(子)；Cork (stopper)(软)木塞；(Ground) glass stopper (磨口)玻璃塞；Rubber stopper 胶塞, 橡皮塞；Stopcock 活塞,(活)栓

  Crucible 坩埚；Crucible cover 坩埚盖；Crucible tongs 坩埚钳

  Asbestos gauze 石棉网；Asbestos center gauze 石棉心铁丝网

  Rubber tubing, rubber tube 胶管, 橡皮管

Iron stand 铁架(台);Support stand 支架,(铁)架台
Clamp (铁)夹;Clamp holder 持夹器,十字夹(头)
Barometer,air pressure gauge 气压计;Manometer U形管测压计
Densimeter,density gauge 比重计
Centrifuge,centrifuger 离心机
Balance 天平,秤;Counter balance 托盘天平;Platform balance,platform scale 台秤;Analytical balance 分析天平;Electronic balance 电子天平;Balance weight,weight(s) 砝码
Heater 加热器;Alcohol burner 酒精灯;Bunsen burner 本生灯,煤气灯;Electric hot plate 电热板;Electric mantle 电热套;Heating mantle 加热套;Muffle furnace 马弗炉;Tube furnace 管式炉
Thermostat oven 恒温(加热)箱;Vacuum drying oven 真空干燥(烘)箱
Rotary evaporator 旋转蒸发器
Stirrer 搅拌器;Magnetic stirrer (电)磁搅拌器;Magnetic stirbar 磁性搅拌棒;Mechanical stirrer 机械搅拌器;Stirring paddle 搅拌桨;Stirring rod 搅拌棒
Variac (自偶)调压变压器;Variable transformer 可调变压器
compressed-gas cylinder 压缩气体钢瓶
Water aspirator 水泵;Aspirator pump 吸(抽)气泵;Vacuum pump 真空泵
Filter paper 滤纸;Test paper 试纸;pH test paper pH 试纸
syringe 注射器
spatula 刮铲(勺);spoon 药匙(勺)
pinch clamp 弹簧夹,节流夹;screw clamp 螺旋夹
standard-taper ground glassware 标准锥形磨口玻璃仪器
bench 实验台,实验桌
fume hood,fuming cupboard 通风厨

附录 V 词汇索引

A

abortive	2.5	alkaline	1.2
abrasion	6.25	alkanolamide	6.20
acetaldehyde	3.4	alkene	5.2
acetic	1.2	alkyl	2.5
acetylate	6.21	alkyne	5.2
acoustic	4.1	allude	2.6
acrylate	6.3	almond	6.21
actinide	1.3	alumina	3.4
adhesive	6.1	aluminum	1.3
adiabatic	3.2	ambient	3.5
adjacent	3.2	amenable	6.16
administer	6.18	amine	6.4
admixture	6.20	ammonia	1.1
advent	3.3	amorphous	6.2
aeration	6.8	amperometric	4.1
aerosol	6.19	amplitude	4.3
aesthetic	6.15	anaerobic	3.5
aggregate	1.8	analogy	6.1
aggregation	6.2	anatase	6.22
agitation	3.4	anesthetic	5.1
aldehyde	5.2	aniline	6.23
algae	6.12	anodically	6.25
alicyclic	2.6	anthracene	5.2
alignment	4.2	antiperspirant	6.21
aliphatic	5.2	antipolar	6.10
alizarin	6.23	antiquity	1.2
alkali	1.2	appertain	3.3
		aqua	6.21

aquadag	6.7	benzoic	2.6
aqueous	1.2	bibliography	4.1
aquifer	6.15	billiard	2.3
arc	4.2	binary	1.2
arene	2.6	binder	6.10
argentocyanide	6.7	biomass	6.16
argon	1.3	biota	6.17
aromatics	2.6	bisphenol	6.4
arsenic	1.5	blade	1.5
ascertain	3.3	bleach	1.5
assay	4.4	bleeder	6.5
assess	6.14	blimp	1.5
assimilative	6.13	blister	6.25
astatine	1.4	bombard	4.2
asterisk	6.2	bondline	6.24
asymmetric	3.5	bone marrow	6.18
attrition	6.19	borate	5.1
autoclave	3.4	borax	6.21
axes	2.2	brevity	5.2
axially	3.4	brewery	3.5
axis	3.1	brisk	6.9
azo	6.23	bromine	1.5
		buckle	6.22

B

		buff	6.22
backbone	6.1	buffer	3.2
barbecue	1.5	butadiene	6.1
barge	6.25		

C

base	1.2		
bauxite	6.6	calandria	3.4
beaker	1.10	calcine	6.22
beet	3.5	calibration	4.2

calory	1.7	cinder	6.14
capacitor	6.24	circumvent	6.3
carbanion	2.5	clarifier	6.16
carbonate	1.2	clay	6.5
carbonium	2.5	cleavage	2.6
carbonyl	3.4	cling	1.4
casual	1.5	clinker	6.14
catalyst	6.4	cluster	6.19
cathodically	6.25	coagulation	6.15
cation	2.5	coalesce	6.21
caustic	1.2	coalescence	3.4
cavity	4.2	cobalt	1.5
cellophane	6.1	cocoamide DEA	6.20
cellulose	3.5	coefficient	2.4
centrifuge	3.6	cohesive	6.2
cerotate	6.21	coil	6.3
cetyl	6.21	collagen	2.7
charcoal	1.5	collide	2.3
charge	1.3	collision	2.6
chelate	6.22	combustion	2.6
chlorhydroxide	6.21	commence	2.6
chloride	1.2	comminutor	6.16
chlorination	3.4	compensate	2.3
chlorine	1.4	compressible	3.1
chloroform	2.6	conceptual	6.1
cholera	6.17	condensation	6.1
chromatography	4.1	conduit	6.4
chromium	1.5	configuration	2.5
chromophore	6.23	confinement	6.19
cider	1.2	conformity	2.3

conical	6.22	cubical	6.2
conjugate	1.2	cyclotron	1.4
consecutively	3.3	cylindrical	6.10
consistency	6.20		
consternation	1.9	**D**	
contamination	6.20	dampen	3.1
contour	6.5	damping	6.3
convective	3.2	dash	4.3
convertor	6.9	deactivate	4.2
copiously	6.20	decalin	2.6
corneum	6.21	declining	4.4
corollary	3.1	decompose	6.4
corrosion	6.8	decoration	6.5
corrugate	3.2	defect	6.19
cosmetic	5.2	deficiency	6.9
coulometry	4.1	dehydration	5.2
countercurrent	3.2	dehydrogenation	2.6
counterpart	1.2	deionize	6.20
covalent	1.2	deleterious	6.8
crack	2.6	delocalize	2.5
crevice	6.7	delude	6.9
criteria	6.13	demineralization	6.15
crock	6.22	demolition	6.14
crockery	6.14	demountable	6.24
crosslink	6.3	denim	6.23
crossover	6.5	deodorant	6.21
crumble	1.5	deplete	6.7
cryogenic	6.24	depletion	6.17
crystal	1.6	desalination	3.5
crystalline	1.2	descent	6.12
		desorb	3.5

detergent	6.17
deteriorate	6.4
detractor	6.24
deviation	1.3
devise	2.4
diagnostic	2.6
diagonally	1.3
diaphragm	6.5
diazotization	6.23
dictate	6.15
dicyanoargentate	6.7
diesel	6.10
digitize	4.6
diluent	6.4
dilute	2.6
dilution	6.15
dioxin	2.8
discard	2.4
discern	2.6
discrepancy	1.3
discrete	4.2
discretion	1.4
disinfection	6.15
disintegrate	6.25
dispose	1.1
dissipation	3.1
dissociate	1.2
dissociation	2.6
dissolution	6.1
dissolving	1.6
distillate	3.3
distillation	1.1
disturbance	6.17
donate	1.2
donor	1.2
drier	3.2
duality	4.2
ductility	1.5
duration	6.2
dyestuff	1.1
dynamic	1.6

E

eddy	3.2
effluent	6.13
effusion	2.3
eka	1.3
elastomer	6.3
elastometry	6.1
electrodeposition	6.6
electroform	6.7
electrolysis	1.7
electromagnetic	3.2
electronegative	5.1
electronegativity	1.3
electrophilic	2.6
electroplate	6.7
electroplating	1.7
electropositive	5.1
electrostatic	1.8
electrovalent	1.8

elementary	2.4	evacuate	4.2
elucidate	3.5	excitonic	6.19
emante	6.22	expenditure	2.8
embed	6.19	explicit	1.2
emission	1.3	explosive	2.7
emolliency	6.21	exponent	2.4
enantiomer	2.5	extension	6.3
encase	6.10	extract	1.1
encompass	6.14	extrapolation	2.6
endeavor	1.1	extruder	6.5
endothermic	1.10	exudate	6.20
ensemble	6.19	exude	6.20
entangle	6.3	**F**	
entanglement	6.2	fallout	6.18
enthalpy	1.10	FD & C (food, drug and cosmetic)	6.20
entity	6.1		
entropy	1.10	feint	3.3
envision	2.3	fermentation	1.2
enzyme	2.7	fermenter	3.4
epichlorohydrin	6.4	ferroconcrete	3.2
epidermal	6.20	fetus = foetus	6.17
epoxide	6.4	filament	3.1
equalization	6.13	filtering	4.4
equalize	6.16	filtration	3.1
equilibria	2.5	flake	6.10
equimolecular	3.2	flammable	6.14
erosion	6.12	flexibility	6.2
ethylene	3.3	floc	6.12
ethylene glycol	6.4	flocculating	6.12
eutectic	3.6	flotation	6.16

fluctuation	6.2
fluid	3.1
fluorescence	4.2
fluorine	2.2
focal	6.25
foreshot	3.3
foresighted	1.3
formability	6.4
formaldehyde	5.2
fragment	2.1
fragrance	6.20
francium	1.4
friction	6.12
fungal	6.20

G

gallium	1.3
galvanic	6.8
galvanical	6.10
galvanize	6.8
gasket	3.2
geometry	2.5
germanium	1.3
globule	6.21
glucose	3.5
glue	1.8
glycine	6.21
glycol	6.20
gonad	6.18
gradient	3.1
grain	6.19

gramophone	6.7
granular	3.2
granule	6.4
graphical	3.2
graphite	1.5
grating	4.4
gravitational	6.12
grease	6.7
greenish	1.1
grid	3.2
grist	6.20
grit	6.16
gritty	6.20
groom	6.20
gross	6.8
grudgingly	6.1

H

halide	5.2
halogen	1.9
handful	1.5
hash	6.20
hazardous	2.8
helical	6.5
helium	1.3
henna	6.23
hereditary	6.18
heterogeneous	1.4
hexagonal	6.10
hitherto	6.6
homogeneous	3.4

homologous	5.2
hoop	6.5
humidify	3.2
hybrid	2.5
hydrated	1.2
hydrogen	1.1
hydronium	1.2
hydroxide	1.2
hydroxychloride	6.21
hyperboloid	3.2
hyperfiltration	3.5
hyphenate	4.1
hypodermic	6.24
hypothesis	2.5
hypothetical	6.18
hysteresis	6.3

I

ignitability	6.14
illuminating	2.3
ilmenite	6.22
imperative	6.19
impermeability	6.25
impervious	6.25
impinge	6.22
implication	1.1
impoundment	6.17
impregnate	6.4
impulse	2.3
incineration	6.16
incommensurate	2.5
increment	6.18
indestructible	2.1
indican	6.23
indigo	6.23
indiscriminate	6.14
indoxyl	6.23
inert	1.5
inertial	3.1
infusible	6.1
ingredient	1.1
initiation	2.7
innermost	4.2
innumerable	2.3
inoculate	3.4
insecticide	5.2
instrumentation	4.1
insuperable	6.9
intact	6.2
interface	1.1
interfacial	3.2
interferometry	4.4
interlink	6.1
intermediate	2.4
intermingle	2.1
interplay	1.10
intersperse	6.1
intimate	2.5
intrusion	6.15
intuition	2.4
inviscid	3.1

invoke	2.5	leukemia	6.18
iodic	1.8	levigate	6.22
iodine	1.5	ligand	2.5
ionic	1.2	lime	1.4
irrespective	3.3	lining	6.25
irreversible	3.1	linoleum	6.2
irrigate	6.15	lipid	2.7
isolate	6.1	liquor	3.3
isomerization	3.4	litharge	6.22
isometric	5.2	lithium	6.9
isoprene	6.3	lithography	6.19
isotope	1.3	litmus	1.2
iterative	3.2	logarithmic	3.2
		lotion	6.20

J

joule	2.8
jute	6.5

lubricant	6.5
luminescent	6.22
lump	1.5
luster	1.5
lye	1.5
lymph	6.18

K

kinetic	1.7
krypton	3.3

M

macerate	6.23
macroscopic	3.2
madder	6.23
magnesium	1.2
magnet	1.4
magnitude	4.2
mainframe	3.3
malachite	6.23
maleic (anhydride)	3.4

L

lactose	3.5
lag	4.4
laminal	3.1
lanthanide	1.3
lattice	2.8
laundry	3.2
lauramide DEA	6.20
lawrencium	1.4
leach	6.6

malfunction	2.7
malleability	1.5
mandrel	6.5
mantle	1.2
masking	6.24
mat	6.5
matrices	6.4
matrix	3.2
matrix	6.4
mauve	6.23
mechanics	2.3
mechanism	2.4
medicament	6.20
mercury	1.4
merge	2.2
mesoscopic	6.19
metabolism	2.7
metalloid	1.3
metathesis	1.9
meteorologic	6.17
methane	1.4
microbial	6.16
mileage	6.14
mirage	6.9
miscellaneous	1.9
mischievous	6.1
miscible	6.21
modular	3.2
modulus	4.6
momentum	1.1
monochromator	4.4
mordant	6.23
mould	6.2
mount	6.3
mow	2.8
mutation	6.18
myriad	6.19

N

nail	2.1
naphthalene	2.6
negative	1.2
neutralization	1.9
neutron	2.1
newtonian	3.1
nitrate	1.9
nitrator	3.4
nitric	1.1
nitrile	4.4
nitrogen	1.1
nitroglycerine	3.4
nomenclature	4.5
nonlinear	2.2
nuclei	1.3
nucleophile	2.6
nullify	1.3
numeral	5.1
numerical	2.1

O

obscure	1.1

occlusive	6.21	parentheses	5.1
ochre	6.22	passage	6.25
odorless	1.1	passivation	6.10
olefine	3.4	passive	6.25
opacity	6.22	pathogenesis	6.16
orientation	6.5	pathogenic	6.16
oscillation	3.1	peculiarity	6.1
osmosis	3.6	pellet	6.5
osmotic	6.25	perforate	3.4
oven	6.5	performance	6.25
overlap	1.1	perfume	1.1
overtone	4.4	periodic	1.3
overwhelm	6.12	periodicity	1.3
oxidation	1.9	permeable	3.5
oxo	5.1	peroxide	2.6
oxyacid	1.9	perpendicularly	3.2
oxyhydroxide	6.10	pertinent	4.2
ozone	2.6	perturb	4.3
ozonolysis	2.6	pharmaceutical	1.1

P

		pharmacology	1.1
palette	6.10	phenol	6.4
palladium	2.6	phenolics	6.4
panacea	1.3	phenyl	2.5
panel	6.5	phosphite	5.1
parabolic	3.1	phosphorescence	4.1
paradoxical	3.4	phosphoric	2.5
paraffin	5.2	photoacoustic	4.1
parallel	3.1	photosynthesis	1.7
pararosaniline	6.23	phthalic(anhydride)	3.4
paratyphoid	6.17	phthalocyanine	6.22

pickle	6.22	precise	6.24
piston	2.3	precision	4.4
planar	2.5	precursor	6.19
plaque	6.10	predecessor	4.4
plasma	4.1	prefix	5.1
plasterer	5.1	prerequisite	6.1
plasticiser	6.4	primer	6.25
plateau	6.2	procedure	1.1
platen	6.5	profile	3.1
platinum	1.5	proline	4.4
plexiglas	6.2	promethium	1.4
plumbing	6.14	propagation	2.7
poke	1.5	propensity	6.20
polarizable	4.4	property	1.1
polarography	4.5	proprietary	6.24
polyatomic	4.3	propylene	3.4
polycarbonate	6.5	proton	2.1
polyimide	6.4	protoplasm	6.18
polymerization	6.1	prototype	6.5
polystyrene	6.1	pseudo	3.1
polysulphide	6.1	purge	3.3
porosity	6.10	putrescible	6.14
portable	6.9	pyroelectric	4.4
pose	6.14	pyrolysis	4.6
positive	1.2	**Q**	
postulate	1.2	quantum	4.3
potable	6.17	quiescent	6.16
potential	1.7	quinine	6.23
precede	4.5	**R**	
precipitate	1.9	racemization	2.5

rad	6.18	rescaling	4.6
radial	3.1	reservation	2.6
radiant	1.7	residual	6.16
radical	2.5	resonance	2.5
radii	1.3	retard	3.2
radome	6.4	retort	6.22
rake	2.8	retract	6.3
ramification	1.8	reverberatory	6.22
randomness	1.10	rheo	3.1
rayon	6.1	rheodestruction	3.1
raze	6.14	rheologic	3.1
reactant	1.6	rheostat	3.1
reagent	2.6	rigidity	6.4
rearrange	2.1	roller	6.5
recalcitrate	6.16	rouge	6.22
recipe	5.1	rutherfordium	1.4
reciprocal	2.6	rutile	6.22
reclaim	6.15		
rectangular	6.10	**S**	
rectify	6.14	sacrificial	6.8
recur	1.3	sag	6.24
redundant	4.4	salicylic	5.2
refine	6.6	sandstone	6.5
refinery	3.3	sawdust	6.5
regime	6.19	scandium	1.3
remedial	3.3	scatter	4.4
repertoire	3.4	scavenging	6.12
replenish	6.15	scent	6.20
repositionable	6.24	scrap	6.6
requirement	6.3	screening	6.13
		sealant	6.1

sebaceous glands	6.20	spatula	6.24
sebum	6.20	specimen	6.2
sedimentation	6.16	spectrometer	4.3
segment	6.2	spectrometry	4.1
segregate	6.3	spectroscopy	4.1
selenium	2.6	spectrum	4.2
sequestering (agent)	6.20	sperm	6.18
sewage	6.25	spermaceti	6.21
shear	3.1	spin	2.2
sheen	1.5	spleen	6.18
shell	2.2	spongy	6.7
shingle	6.14	spontaneity	1.10
shistosomiasis	6.17	spray	6.8
shuttle	6.5	squeeze	6.5
sienna	6.22	standstill	1.6
signify	6.2	static	2.5
silvery	1.1	stearate	6.20
sinter	6.10	stellite	1.5
skimmer	6.16	steric	6.2
slough(off)	6.20	sterility	3.5
sludge	6.13	stimulus	4.4
slug	3.4	stir	3.3
sluggishly	6.2	stoichiometric	4.6
slurry	3.4	stoichiometry	3.3
slush	1.10	strain	6.3
sodium laureth sulfate	6.20	strand	2.7
sole	6.3	strata	6.15
solubility	6.1	streamline	3.1
somatic	6.18	stress	6.3
spatial	3.4	stretchable	6.3

strontium	6.25	tertiary	5.2
styrene	3.6	terylene	6.1
submerge	1.7	tetrahedral	2.5
submicron	6.12	therapy	6.18
sucrose	1.4	thermodynamic	1.10
suffix	5.1	thermometer	1.5
sulfate	1.9	thermoplastic	6.1
sulfide	1.4	thermoset	6.3
sulphonate	3.4	thermostat	6.2
sundry	6.12	thiosulfate	5.1
superimpose	3.2	thixotropic	3.1
superposition	4.3	thrust	1.7
sustain	2.6	tier	6.22
symmetrical	2.3	tin	2.1
synthesize	2.6	tinctorial	6.22
syrup	3.5	titration	4.5

T

tabulate	1.3	toiletry	6.20
tacky	6.24	tolerance	6.5
tailored	6.19	toluene	3.4
tannery	6.13	toluidine	6.23
taper	6.13	torch	6.9
tarnish	1.5	toxic	1.1
tart	1.2	transducer	4.6
TEA (triethanolamine)	6.20	transformation	1.7
technetium	1.4	transistor	1.1
template	6.19	translucence	6.4
tensile	6.3	trap	6.5
termination	2.7	trickling	6.16
terrestrial	6.9	trimming	6.14
		troposphere	6.12

tubular	3.4	vigor	4.5
tungsten	1.5	vinegar	1.2
turbulent	3.1	vinyl	2.5
typhoid	6.17	viscoelasticity	6.2
		viscosity	3.1

U

		viscous	6.1
ubiquitous	4.4	visualize	6.24
ultramarine	6.22	vitrified	6.2
umber	6.22	viz	2.6
unambiguously	3.3	void	6.24
underlying	1.1	voltammetric	4.1
unguentum	6.21	volumetric	3.3
unify	1.1	vulcanize	6.1
unsurmountable	6.2		
upgrade	6.16		

W

warrant	2.8
watt	3.2
whey	3.5
winning	6.6

V

vacant	1.3
valence	1.8
validity	2.5
velocity	1.7
ventilation	3.2
versatile	1.5
versatility	6.4
via	2.6
viable	6.9
vibrate	2.8
vibrational	6.3

X

xylene	3.4

Y

yoburt	3.5

Z

zigzag	1.3
zirconyl	6.21

参 考 文 献

1 杨嘉谟,包传平,余卫华等.化学化工专业英语.武汉:武汉大学出版社,1997
2 Ralph H Petrucci, William S Harwood. General Chemistry—Principles & Modern Applications. Sixth Edition. New Jersey: Prentice-Hall Inc., Englewood Cliffs, 1993.
3 Hein M. Foundations of College Chemistry. Fourth Edition. Dickenson, Encino, 1977
4 James E Brady, John R Holum. Fundamentals of Chemistry. Third Edition. Wiley & Sons, Inc., 1988
5 Crow D R. Principles and Applications of Electrochemistry. Third Edition. London: Chapman and Hall, 1988
6 David C Gusche, Daniel J Pasto. Fundamental of Organic Chemistry. New Jersey: Prentice-Hall Inc., 1975.
7 Coulson J M, Richardlson J F, Backhurst J R, Harker J H. Chemical Engineering (Vol. one). Third Edition. Oxford: Pergamon Press, 1977
8 Denbigh K G, Turner J C R. Chemical Reactor Theory—An Introduction. Third Edition. Cambridge: Cambridge University Press, 1984
9 Robert W Field. Chemical Engineering—Introductory Aspects. Houndmills: Macmilian Education Ltd., 1988
10 Charles K Mann, Thomas J Vickers, Wilson M Gulick. Instrumental Analysis. New York: Harper & Row, 1974
11 Hobart H Willard, Lynne L Merritt Jr, John A Dean, Frank A Settle Jr. Instrumental Methods of Analysis. Seventh Edition. Belmont, California: Wadsworth Publishing Company, 1988
12 余向春,许家琪,邹荫生.化学化工文献检索与利用.大连:

大连理工大学出版社,1991
13 冯子良,李秀英,郑品思,冯白云,任其荣.科技文献检索.北京:中国科学技术出版社,1988
14 赖茂生,徐克敏.科技文献检索.北京:北京大学出版社,1985
15 陈光祚.科技文献检索.武汉:武汉大学出版社,1985
16 周季特,庄德君.研究生实用英语系列教程——写译.哈尔滨:哈尔滨工业大学出版社,1994
17 张正举.科技英语教程.长沙:湖南大学出版社,1996.
18 Madhusudhan V, Murthy B G K. Polyfunctional Compounds from Cardanol, Progress in Organic Coatings—An International Review Journal. 1992,20 (1):63
19 Takao Hirayama, Marek W. Urban. Distribution of Melamine in Melamine/ Polyester Coatings; FT-IR Spectroscopic studies, Progress in Organic Coatings—An International Review Journal 1992,20 (1):81
20 王金玉.电化学专业英语.哈尔滨:哈尔滨工业大学校内教材,1992
21 胡立江.专业英语(精细化工类).哈尔滨:哈尔滨工业大学教材科,1997
22 Sir Geoffrey Allen, John C. Boevington. Comprehensive Polymer Science. Pergamon Press,1989
23 马志毅,苏玉民.环境保护——环境工程专业英语.北京:中国环境科学出版社,1995
24 Gowariker V R, Viswanathan N V, Jayadev Sreedhar. Polymer Science. New Delhi: Wiley Eastern Limited, 1986
25 James E Huheey. Inorganic Chemistry, Principles of Structure and Reactivity. Third Edition. Harper and Row Publishers Inc., Harper International SI Edition,1983
26 William L Jolly. The Synthesis and Characterization of Inorganic

Compounds. N J: Prentice-Hall, Inc., Englewood Cliffs, 1970

27 Daniel J Pasto, Carl R Johnson, Marvin J Miller. Experiments and Techniques in Organic Chemistry. N J: Prentice Hall, Englewood Cliffs, 1992

28 科学出版社名词室编.英汉化学化工词汇.第4版.北京：科学出版社,2000

29 魏运洋,南晓平.精细化工专业英语.南京:南京理工大学出版社,1993

30 Zbigniew D Jastrzebski. The Nature and Properties of Engineering Materials. Third Edition. John Wiley & Sons, 1987

31 Miriam V Bennett, Matthew P Shores, Laurance G Beauvais, Jeffrey R Long. J Am. Chem. Soc. 2000, 122(28): 6664

32 李欣,齐晶瑶,韩喜江.基础化学信息学.哈尔滨:哈尔滨工业大学出版社,2003